Agricultural Innovation and Technology

Agricultural Innovation and Technology

Edited by **Jamie Hanks**

SYRAWOOD
PUBLISHING HOUSE
New York

Published by Syrawood Publishing House,
750 Third Avenue, 9th Floor,
New York, NY 10017, USA
www.syrawoodpublishinghouse.com

Agricultural Innovation and Technology
Edited by Jamie Hanks

International Standard Book Number: 978-1-68286-230-8 (Hardback)

Printed in China.

Contents

Preface

Over the recent decade, advancements and applications have progressed exponentially. This has led to the increased interest in this field and projects are being conducted to enhance knowledge. The main objective of this book is to present some of the critical challenges and provide insights into possible solutions. This book will answer the varied questions that arise in the field and also provide an increased scope for furthering studies.

The constant technological advances in the recent times have accelerated the growth of agriculture as a scientific discipline. Researches are being conducted worldwide to devise new methods to improve crop production and enhance quality. This book attempts to assist those with a goal of delving into such advancements with the help of topics like postharvest technology, organic agriculture, food security, genetic engineering, agronomy, etc. It will prove beneficial to students, academicians, scientists and everyone else associated with the field of agriculture.

I hope that this book, with its visionary approach, will be a valuable addition and will promote interest among readers. Each of the authors has provided their extraordinary competence in their specific fields by providing different perspectives as they come from diverse nations and regions. I thank them for their contributions.

Editor

EFFECT OF GAMMA RADIATION ON THE TITRABLE ACIDITY AND VITAMIN C CONTENT OF CITRUS FRUITS

Iftekhar Ahmad[1*], Mizanur Rahman[1], Md. Mahfuzur Rahman[1], Md. Mehbub Mustain Alam[1] and Md. Shakawat Hussain[2]

Abstract

The study was carried out to assess effect of gamma radiation on the acidity and vitamin C content of the *Citrus macroptera* (Satkora) and *Citrus assamensis* (Ginger lime). Irradiation with doses 0.5, 1.0, 2.0, 3.0 kGy were applied to assess the effect on the titrable acidity and vitamin C contents every one week interval for total five weeks. Both titrable acidity and vitamin C content of *C. macroptera,* and *C. assamensis* are sensitive to both gamma radiation and storage time; have decreased with increase of radiation does as well as storage time and this changes of vitamin C and titrable acidity content with gamma radiation and increasing storage period have found statistically significant.

Keywords: *Citrus macroptera* (Satkora), *Citrus assamensis* (Ginger lime), Titratable Acidity, Vitamin C and Gamma Radiation

[1]Department of Food Engineering and Tea Technology, Shahjalal University of Science and Technology, Sylhet, Bangladesh; [2]Food Technology Division, Institute of Food and Radiation Biology, Atomic Energy Research Establishment, Savar, Dhaka, Bangladesh.

*Corresponding author's email: iftekharn2003@yahoo.com (Iftekhar Ahmad)

Reviewed by Dr. Mohammad Gulzarul Aziz, Bangladesh Agricultural University, Mymensingh, Bangladesh.

Introduction

Vitamin C is one of the most important nutritional benefits of citrus fruit. Ascorbic acid (Vitamin C) is the only vitamin present in citrus fruit in amounts of major nutritional significance; one orange has 50 mg of vitamin C, which is nearly the double of the recommended daily intake.The concentration of ascorbic acid has been reported to decrease with maturity or remain constant until late in the season & then decline (Baldwin, 1993). Only 25% of ascorbic acid in the fruit is in the juice, the remainder is found in the peel, especially in the flavedo (Kefford, 1959).

Maturity standards for citrus are often based on chemical indicators of fruit flavor, including sweetness and acidity and the ratio between these components. Therefore, acidity levels have a major impact on internal fruit quality and consequently affect the time when the fruit reaches minimum market standards (Marsh *et al.,* 2000). Organic acids significantly contribute to juice acidity, citric acid being primary organic acid (70-90% of total). Citric acid is the major acid in juice, malate predominates in the flavedo and oxalate is the most abundant in albedo. Other minor organic acids found in citrus fruits are acetate, pyruvate, glutarate, formate, succinate and α-ketoglutarate (Yamaki, 1989).

Applying gamma radiation to citrus fruits at the pasteurization dose level could be the means of extending the shelf life of these commodities. The problems are related to physiological response of the citrus fruits being irradiated and the dose requirement to achieve their disinfection. The dose level for disinfection can be intolerable to some citrus fruits and can result in disorders. The feasibility of the treatments depends upon the sensibility of the host tissues. Temple mandarin has been found to be less susceptible to irradiation damage than Pineapple and Valencia oranges (Dennison *et al.,* 1966)

Ionizing radiation can safely and effectively eliminate the pathogenic bacteria from the food (Crawford and Ruff, 1996; Evangelista, 2000; Loaharanu, 1994; Sommers *et al.,* 2004), disinfest the fruits and vegetables (Moy and Wong, 2002; Fan and Mattheis, 2001; Patil *et al.* 2004; Pellegrini *et al.,* 2000), extend the shelf life of many products through ripening delay (Kilcast, 1994), inhibit the sprouting of bulbs and tubers (Aziz *et al.,* 2006; Rios and Penteado, 2003; Curzio *et al.,* 1986), and reduce or totally eliminate the parasitic microorganisms (Crawford and Ruff, 1996; Kilcast, 1994).

The present research work was conducted on *Citrus assamensis* (Ginger limes) and *Citrus macroptera,* (Satkora), two indogenous species of citrus fruits and available in Sylhet region of

Bangladesh. The objectives of the present work are to observe the effect of gamma radiation on the Titrable acidity and Vitamin C content of these two citrus fruits.

Materials and Methods

Citrus assamensis (Ginger limes) and *Citrus macroptera* (Satkora) of fresh, fully mature, large size, juicy type and free from any type of injury or deterioration were procured from the "Bandar Bazar" of Sylhet city and were taken to the Food Technology Laboratory of Institute of Food and Radiation Biology (IFRB), Atomic Energy Research Establishment (AERE), Bangladesh Atomic Energy Commission, Savar, Dhaka under ambient conditions during March.

Each sample was packed into sterilized (by 15 kGy gamma radiation) polythene paper and then sealed. The packets were well ventilated by making small holes at 0.5% area of the bag was punched with small circular holes then exposed to a Co^{60} gamma Irradiator with doses 0.5, 1.0, 2.0, 3.0 kGy. Among 25 samples of each type 5 samples were exposed to each dose and 5 samples were left as control sample. Then both irradiated and control samples were kept at room temperature (at about 22-25°C) on a clean and open surface. The acidity and ascorbic acid contents were assessed every one week interval for total five weeks duration.

Titrable acidity was determined by the method of AOAC (1990) and the estimation of ascorbic acid content were carried out by the titration result of the sample extract with 2,6-Dichlorophenol-Indophenol (Dye) (Johnson, 1948).

Mean, Standard deviation and T-test of the change of titrable acidity and vitamin C content and Analysis of variance (ANOVA) of the overall changes of titrable acidity and vitamin C content due to application of radiation doses were measured using SPSS (v.16) program.

Results and Discussion

Citurs macroptera (Satkora)

The quantity of titrable acidity of *C. macroptera* (Satkora) irradiated with different radiation doses is presented in the table 1.

Table 1. Amount of titrable acidity of satkora irradiated with different radiation doses

Weeks Doses	1st week TA (%)	2nd week TA (%)	3rd week TA (%)	4th week TA (%)	5th week TA (%)	Mean	Std. Deviation
Control (0 kGy)	0.8956	0.6307	0.5150	0.3580	N/d	0.479860	0.3323939
0.5 kGy	0.8563	0.5842	0.5087	0.4075	0.3063	0.532600	0.2090916
1.0 kGy	0.5794	0.5494	0.4363	0.3532	N/d	0.383660	0.2327432
2.0 kGy	0.5582	0.4075	0.3256	0.2568	N/d	0.309620	0.2063264
3.0 kGy	0.5150	0.3865	0.3045	0.2405	0.2321	0.335720	0.1177907

TA: Titrable Acidity,
N/d: Not done

It is found that all the values of titrable acidity obtained in the samples from 1st week to 5th week radiated with various doses are less than the titrable acidity value of fresh, non-radiated sample (0.8956% titrable acidity). And the values decrease with increase of radiation dose and storage period. So, the obtained results of present study about the change of titrable acidity of *C. macroptera* (Satkora) irradiated with different radiation doses shows a gradual decrease of titrable acidity content with the increase of radiation dose and storage time.

The table 1 shows a gradual increase of titrable acidity change with increase of radiation doses as well as with increase of storage period. So, quantity of titrable acidity of *C. macroptera* (Satkora) is inversely proportional to radiation dose and storage period. Again, the decreasing rate of titrable acidity is directly proportional to the radiation dose as well as to the storage period, i.e. the changing rate increases with increase of radiation dose and storage period.

Ladaniya *et al.* (2003) found that, doses up to 1.5 kGy cause decrease of titrable acidity of 'Nagpur' mandarin (*Citrus reticulata* Blanco), 'Mosambi' sweet orange (*Citrus sinensis* Osbeck) and 'Kagzi' acid lime (*Citrus aurantifolia* Swingle) which has similarity with our present findings.

The correlation and significance values of obtained data are measured by "Pearson Correlation (2-tailed)" method. We have taken 5% level of acceptance for measuring significance. At 5% level of acceptance, the significance values (p) below 0.05 are considered as "Significant", while above 0.05 is considered as "Insignificant". The effects of radiation doses are negatively correlated with storage time in respect of titrable acidity of *C. macroptera* and the calculated correlation co-efficient are -0.982, -0.965, -0.921, -0.971 and -0.955 for radiation doses 0, 0.5, 1.0, 2.0 and 3.0 kGy, respectively. The negative correlation is obtained due to the decrease of titrable acidity with increase of radiation doses. And all the changes of titrable acidity have found statistically significant (obtained "p" values are 0.003, 0.008, 0.027, 0.006 and 0.011 for radiation doses 0, 0.5, 1.0, 2.0 and 3.0 kGy, respectively. On the other hand, the effects of storage time are also negatively correlated with radiation doses in respect of titrable acidity of *C. macroptera* except for 5th week change and the calculated correlation co-efficient are -0.873, -0.973, -0.007, -0.896 and 0.207 for 1st, 2nd, 3rd, 4th and 5th week samples respectively. Here, the cause of negative correlation is decrease of titrable acidity with increase of storage period. The change of titrable acidity in 2nd, 3rd and 4th week have found statically significant (calculated "p" values are 0.005, 0.007 and 0.04, respectively) while changes are insignificant in 1st and 5th week (calculated "p" values are 0.054 and 0.738, respectively).

Table 2. Amount of vitamin C (mg/100g) content of *C. macroptera* (Satkora) irradiated with different doses

Weeks / Doses	1st week AA(*)	2nd week AA(*)	3rd week AA(*)	4th week AA(*)	5th week AA(*)	Mean	Std. Deviation
Control (0 kGy)	56.00	41.63	40.61	39.60	N/d	35.568000	20.9819916
0.5 kGy	50.64	33.29	30.41	27.86	25.32	33.504000	10.0261373
1.0 kGy	40.00	31.95	27.52	23.09	N/d	24.512000	15.0591358
2.0 kGy	37.06	30.50	25.09	18.01	N/d	22.132000	14.2164929
3.0 kGy	36.28	26.52	16.76	14.03	13.32	21.382000	9.8584999

AA: Ascorbic Acid or Vitamin C
N/d: Not done

In case of ascorbic acid of *C. macroptera* (Satkora), the study also shows a gradual decrease. All the obtained values of vitamin C for the samples of 1st to 5th week, radiated with different doses (0, 0.5, 1.0, 2.0 & 3.0 kGy) are less than the vitamin C content value of fresh, non-radiated sample (56.00 mg/100g). This can be seen from the Table 2 that the more the radiation dose increases, the less vitamin C content is obtained from the samples.

Here, we have compared all the values of vitamin C content obtained from different samples with the reference value of vitamin C obtained from fresh, non-radiated sample.

In case of vitamin C content, the effects of radiation doses are negatively correlated with storage time, the calculated co-efficient are -0.859, -0.884, -0.933, -0.963 -0.937 for radiation doses 0, 0.5, 1.0, 2.0, and 3.0 kGy, respectively. Except for the control sample (p=0.062), all changes in at other radiation doses are statistically significant (obtained "p" values are 0.046, 0.021, 0.008 and 0.019, respectively for 0.5, 1.0, 2.0 and 3.0kGy dose).

The effects of storage time are negatively correlated with radiation doses in respect of vitamin C content of *C. macroptera* (Satkora) except for 5th week. The calculated co-efficient values are -0.972, -0.893, -0.943, -0.927 and 0.043, respectively for 1st, 2nd, 3rd, 4th and 5th week. The changes of vitamin C content of Satkora are found significant at 5% level of acceptance for 1st to 4th week and insignificant for 5th week. The calculated "p" values for 1st to 5th weeks are 0.006, 0.042, 0.016, 0.024 and 0.945, respectively.

C. assamensis (Ginger lime)

In case of *C. assamensis* (Ginger lime) all the values of titrable acidity obtained in the samples from 1st week to 5th week radiated with various doses (0, 0.5, 1.0, 2.0 & 3.0 kGy) are less than the titrable acidity value of fresh, non-radiated sample (1.18% titrable acidity). The values decrease with increase of radiation dose and storage period. So, the obtained results of present study about the change of titrable acidity of ginger lime irradiated with different radiation doses shows a gradual decrease of titrable acidity content with the increase of radiation dose and storage time. The percentage of titrable acidity change with respect to the reference sample (fresh and non-radiated) is presented in the following table.

Table 3. Effect of gamma radiation on titrable acidity of *C. assamensis* (Ginger lime)

Weeks / Doses	1 week TA *(%)	2 week TA (%)	3 week TA (%)	4 week TA (%)	5 week TA (%)	Mean	Std. Deviation
Controlled	1.18	0.9657	N/d*	N/d	0.923	0.613740	0.5686654
0.5 KGy	1.025	0.807	0.773	N/d	0.623	0.645600	0.3884376
1.0 KGy	0.981	0.847	0.81	0.773	N/d	0.682200	0.3893632
2.0 KGy	0.98	0.91	0.88	N/d	N/d	0.554000	0.5070306
3.0 KGy	0.901	0.858	N/d	N/d	N/d	0.351800	0.4819618

Many scientists have observed the effect of irradiation on limes and other citrus. Ladaniya *et al.* (2003) found that, doses up to 1.5 kGy cause decrease of titrable acidity of 'Nagpur' mandarin lime (*Citrus reticulata* Blanco), 'Mosambi' sweet orange (*Citrus sinensis* Osbeck) and 'Kagzi' acid lime (*Citrus aurantifolia* Swingle) which has similarity with our present findings though acidity changes in our study was not statistically significant.

Here, we have taken the titrable acidity value of fresh, non-radiated *C. assamensis* as reference and its value is 1.18%. We have compared all the obtained values of titrable acidity with this reference value. It is observed from the table4 that, the values are gradually increasing in a row from left to right as well as from top to bottom in a column. This means, the decreasing rate of titrable acidity of radiated samples, stored for different period is increasing with increase of radiation dose and storage period (as radiation doses are gradually increasing from top to bottom and storage period is increasing from left to right in a row of the table 3.

Similarly, the quantity of vitamin C of *C. assamensis* irradiated with different radiation doses is presented in the following table.

Table 4. Effect of gamma radiation on vitamin C of *Citrus assamensis* (Ginger lime)

Weeks / Doses	1 week AA* (mg/100g)	2 week AA (mg/100g)	3 week AA (mg/100g)	4 week AA (mg/100g)	5 week AA (mg/100g)	Mean	Std. Deviation
Controlled	31.7135	19.80	N/d*	N/d	16.8	13.662700	13.6627159
0.5 KGy	28.66	20.00	20.00	N/d	13.34	16.400000	10.6601032
1.0 KGy	27.58	22.44	21.459	17.325	N/d	17.760800	10.5785659
2.0 KGy	27.79	21.64	12.38	N/d	N/d	12.362000	12.5473392
3.0 KGy	24.73	14.03	N/d	N/d	N/d	7.752000	11.2688318

TA*=Titrable Acidity
N/d*=Note done

The study revealed that vitamin C also changes with changes in radiation dose and storage time like titrable acidity at the same manner, that is, a gradual decrease of vitamin C content is occurred with the increase of radiation dose and storage time. The percentage of vitamin C change with respect to the reference sample (fresh and non-radiated, 31.72 mg/100g vitamin C) is presented in the table 4.

In this study, the effect of irradiation on titrable acidity of ginger lime with time is negatively correlated but the changes are not statistically significant. In this study, it is observed that changes in vitamin C content of ginger lime in the doses of 1KGy, 2KGy and 3KGy are statistically significant. Ladaniya *et al.*, (2003) found that, doses up to 1.5 kGy caused decrease of vitamin C content of 'Nagpur' mandarin lime (*Citrus reticulata* Blanco), 'Mosambi' sweet orange (*Citrus sinensis* Osbeck) and 'Kagzi' acid lime (*Citrus aurantifolia* Swingle).

The results of the study of five weeks have been statistically analyzed to do a comment that either the changes of titrable acidy or vitamin C are significant or not. The correlation and significance values of obtained data are measured by "Pearson Correlation (2-tailed)" method. We have taken 5% level of acceptance for measuring significance. At 5% level of acceptance, the significance values (p) below 0.05 are considered as "Significant", while above 0.05 is considered as "Insignificant".

The effects of radiation doses are negatively correlated with storage time in respect of titrable acidity of *C. assamensis* and the calculated correlation co-efficient are -0.408, -0.656, -0.827, -0.895, -0.873 and -0.955 for radiation doses 0, 0.5, 1.0, 2.0 and 3.0kGy, respectively. The negative correlation is obtained due to the decrease of titrable acidity with increase of radiation doses. And all the changes of titrable acidity have found statistically insignificant except for 2.0KGy (obtained "p" values are 0.496, 0.230, 0.084, 0.040 and 0.054 for radiation doses 0, 0.5, 1.0, 2.0 and 3.0 kGy, respectively). On the other hand, the effects of storage time are also negatively correlated with radiation doses in respect of titrable acidity of *C. assamensis and* the calculated correlation co-efficient are -0.873, -0.201, -0.113, -0.138 and -0.808 for 1st, 2nd, 3rd, 4th and 5th week samples, respectively. Here, the cause of negative correlation is due to decrease of titrable acidity with increase of storage period. All the changes of titrable acidity have found statistically insignificant (obtained "p" values are 0.053, 0.746, 0.857, 0.823 and 0.098 for 1st, 2nd, 3rd, 4th, and 5th week, respectively).

In case of vitamin C content, the effects of radiation doses are negatively correlated with storage time and the calculated co-efficient are -0.574, -0.143, -0.90p1, -0.973 and -0.891 for radiation doses 0, 0.5, 1.0, 2.0 and 3.0 kGy, respectively. Except for the control and 0.5 KGy samples (p=0.311and 0.143 respectively), all changes in at other radiation doses are statistically significant (obtained "p" values are 0.037, 0.005 and 0.043, respectively for 1.0, 2.0 and 3.0kGy dose).

The effects of storage time are negatively correlated with radiation doses in respect of vitamin C content of *C. assamensis*. The calculated co-efficient values are -0.906, -0.597, -0.274, -0.139 and -0.809, respectively for 1st, 2nd, 3rd, 4th and 5th week. The changes of vitamin C content of Ada lebu are found significant at 5% level of acceptance for 1st week only and it is significant for all other weeks. The calculated "p" values for 1st to 5th weeks are 0.034, 0.288, 0.655, 0.823 and 0097, respectively.

In this study, it is observed that changes in vitamin C content of ginger lime in the doses of 1 KGy, 2 KGy and 3 KGy are statistically significant. Ladaniya *et al.* (2003) found that, doses up to 1.5 kGy cause decrease of vitamin C content of 'Nagpur' mandarin lime (*Citrus reticulata* Blanco), 'Mosambi' sweet orange (*Citrus sinensis* Osbeck) and 'Kagzi' acid lime (*Citrus aurantifolia* Swingle).

Rao (1962) observed that ascorbic acid is rapidly destroyed by gamma radiation in dilute solutions. Zhang *et al.*, (2006) have found that, the loss of vitamin C of fresh-cut lettuce irradiated with 1.0 kGy was significantly (a = 0.05) lower than that of non-irradiated.

Hussain and Maxie (1974) have registered 70.2 % vitamin C losses for orange juice irradiated with 2.5 to 10 kGy. So it can be said that, the loss rate of vitamin C content increased with increase of radiation dose. This is very similar to our present finding. Benkeblia and Khali (1999) has found that gamma irradiation at 0.10, 0.15 and 0.30 kGy causes losses of 10, 13 and 20% of ascorbic acid content respectively in onion bulbs. Another study using higher radiation doses (3.0 and 4.0 kGy), resulted in 62 and 81% losses of ascorbic acid, respectively (Clark, 1959)

Increasing the radiation dosage gradually decreased the fruits vitamin C concentration of Lycium (Wen *et al.*, 2006). Evangelista (2000) has found alterations in vitamin C contents in potatoes are proportional to the irradiation dosage.

Lacroix *et al.*, (1990) have found that, the content of vitamin C of papaya and mango irradiated at 0.5 to 0.95 kGy were not significantly affected by the irradiation. In another study of the ascorbic acid content of oranges, Nagai and Moy (1985) have found no significant differences between irradiated and control fruit at dose levels up to 1 kGy. Fan and Mattheis (2001) commented that, adequate doses for insect disinfestations (normally 1-2 kGy) showed non-significant effects in vitamin C contents of citric fruits. Maxie *et al.* (1964) have found that, strawberries (*Fragaria* sp.) presented minute, non-significant decrease in vitamin C levels when submitted to 1.0-2.0 kGy doses, during 11 days of storage at 5°C. Similar observations were reported by Lopez *et al.* (1967). In our study we have found that in control and 0.5 KGy dose of radiation the changes in vitamin C is not statistically significant but in 1 KGy, 2 KGy and 3 KGy the changes are statistically significant. From the above discussion, we see that in most of the limes and citrus vitamin C content decreases with increase of doses of radiation, at the lower doses the changes may be insignificant but at the higher doses the changes are statistically significant which is too much similar to our study for ginger lime.

References

AOAC. 1990. Official Methods of Analysis. *Association of Official Analytical Chemists.* Washington D.C, USA.

Aziz, N.H., Souzan, R.M. and Aziz R. 2006. Effect of γ-irradiation on the occurrence of pathogenic microorganisms and nutritive value of four principal cereal grains. *Appl. Radiat. Isot.*, 64: 1555-1562.

Baldwin E.A. 1993. Citrus fruit. In Biochemistry of Fruit Ripening. G.B. Seymour, J.E. Taylor, and A. Tucker, Eds. Chapman and Hall, London, pp.107-149.

Benkeblia N. and Khali. M. 1999. Stability of vitamin c of irradiated Onions Allium cepa. L during storage. *J. Islamic Academy Sci.*, 9(2): 57-60.

Clark, I.D. 1959. Possible applications of ionizing radiations in the fruit, vegetable and related industries. *Int. J. Appl. Radiat. Isot.*, 6: 175-181.

Crawford, L.M. and Ruff, E.H. 1996. A review of the safety of cold pasteurization through irradiation. *Food Control.*, 7: 87-97.

Curzio O.A., Croci, C.A. and Ceci, L.N. 1986. The effects of radiation and extended storage on the chemical quality of garlic bulbs. *Food Chem.*, 21: 153-159.

Dennison, R.A., Grierson, W., and Ahmed, E.M., 1966. Irradiation of Duncan grapefruit pineapple and Valencia oranges and Temples. *Proc. Fla. State Hort. Soc.*, 79: 285–292.

Evangelista, J. 2000. Alimentos Irradiados. In: *Alimentos - um estudo abrangente.* São Paulo: Editora Atheneu, pp. 135 -169.

Fan, X. and Mattheis, J.P. 2001. 1-Methylcyclopropene and storage temperature influence responses of "Gala" apple fruit to gamma irradiation. *Postharvest Biol. Technol.*, 23: 143-151.

Hussain, A. and Maxie, E.C. 1974. Effect of gamma rays on shelf life and quality or orange juice. *Int. Biodet. Bull.,* California., 10: 81-86.

Johnson, B.C. 1948. Methods of Vitamin Determination. Burgess Publishing Co. Minneapolis, Minnesota p. 98.

Kefford, J.F. 1959. The Chemical Constituents of Citrus Fruits. *Adv. Food Res.,* 9: 285-372.

Kilcast, D. 1994. Effect of irradiation on vitamins. *Food Chem.,* 49: 157-164.

Lacroix, M., Bernard, L., Jobin, M., Milot, S., Gagnon, M. 1990. Effect of irradiation on the biochemical and organoleptic changes during the ripening of papaya and mango fruits. *Int. J. Radiation Appl. & Instr.,* Part C. Radiation Physics and Chemistry, 35: 296-300.

Ladaniya M.S., Singh, S. and Wadhawan, A.K. 2003. Response of 'Nagpur' mandarin, 'Mosambi' sweet orange and 'Kagzi' acid lime to gamma radiation. *Radi. Phy. and Chem.,* 67(5): 665-675.

Loaharanu, P. 1994. Status and prospects of food irradiation. *Food Technol.,* 48(5): 124-130.

Lopez, G.M., Rivas, G.A., Ortin, S.N. and del Val Cob, M. 1967. Preservation of food by irradiation. VI. Preliminary investigation on strawberries, in Symp. Application of Radioisotopes, Madrid, p. 21.

Marsh, K.B., Richardson, Y. and Erner, Y. 2000. Effect of environmental conditions and horticultural practices on citric acid content. Proc. Intl. Soc. Citriccult. IX Congr, pp. 640-643.

Maxie, E.C., Eaks, I.L. and Sommer, N.F. 1964. Some physiological effects of gamma irradiation on lemon fruit. *Radiat. Bot., California,* 4: 405-411.

Moy, J.H. and Wong, L. 2002. The efficacy and progress in using radiation as a quarantine treatment of tropical fruits - a case study in Hawaii. *Radiat. Phys. Chem.,* 63: 397-401.

Nagai, N.Y. and Moy, J.H. 1985. Quality of gamma irradiated California, Valencia oranges. *J. Food Sci.,* 50: 215-219.

Patil, B.S., Jairam Vanamala, J. and Hallman, G. 2004. Irradiation and storage influence on bioactive components and quality of early and late season 'Rio Red' grapefruit (*Citrus paradisi* Macf.). *Post harvest Biol. and Tech.,* 34(1): 53-64.

Pellegrini, C.N., Croci, C.A. and Orioli, G.A.. 2000. Morphological changes induced by different doses of gamma irradiation in garlic sprouts. *Radiat. Phys. Chem.,* 57: 315-318.

Rao, B.S.N. 1962. Radiolysis of Ascorbic Acid in Aqueous Solution by Gamma Radiation. Radiation Research: November 1962, Vol. 17, No. 5, pp. 683-693.

Rios, M.D.G. and Penteado, M.D.V.C. 2003. Determinação de α-tocoferol em alho irradiado utilizando cromatografia líquida de alta eficiência (CLAE). *Quim. Nova.,* 26: 10-12.

Sommers, C., Xuetong, G. and Niemira, B. 2004. Irradiation of ready-to-eat foods at USDA'S Eastern Regional Research Center-2003 update. *Radiat. Phys. Chem.,* 71: 509-512.

Wen, H.W., Chung, H.P., Chou, F.I., Lin, I.H. and Hsieh, P.C. 2006. Effect of gamma irradiation on microbial descontamination and chemical and sensory characteristic of lycium fruit. *Radiat. Phys. Chem.,* 75: 596-603.

Yamaki, Y.T. 1989. Organic acids in the juice of citrus fruits. *J. Jpn. Soc. Hort. Sci.,* 58: 587-594.

Zhang, L. Lu, Z., Lu, F. and Bie, X. 2006. Effect of γ-irradiation on quality-maintaining of fresh-cut lettuce. *Food Control,* 17: 225-228.

2

THIN LAYER CHROMATOGRAPHIC ANALYSIS OF FOOD COLORANTS FROM THREE MORPHOTYPES OF ANNATTO (*Bixa orellana* L.)

Hari Pada Seal[1]*, Mohammad Amdad Ali[1]*, Md. Usuf Ali[1], Mosammat Hasina Akhter[1] and Fowzia Sultana[2]

Abstract

This article describes a simple solvent extraction method for the extraction of colorants from the three morphotypes such as, (Morphotype-1 (M_1), Morphotype-2 (M_2), and Morphotype-3 (M_3) of Annatto (*Bixa orellana* L.) seeds, and their separation, vivid, and qualitative demonstration by thin-layer chromatography. Several solvent systems (hexane, chloroform, acetone, ethanol, and a mixed-solvent having composition of $CHCl_3/C_2H_5OH/CH_3COOH$ (80:2:1)) were applied for extraction of colored components. It was observed that a large portion of colorants was extracted by chloroform. Its effluent was deep brick red in color and transparent. Furthermore, various carrier solvent systems (Benzene-Ethyl acetate) were used to separate the components from the extracts. Carrier solvent system with the ratio of 7:3 was found as superior solvent for chloroform extracts. Three colored-spots were observed for all morphotypes. Among them, the first one was yellow colored having very low polarity and the second and third spots were both redbrick colored having medium and higher polarity respectively. In addition, for M_1 no colorless-spot was observed in low and medium polar systems, revealing that the amount of wax and gum were minimum in the extract and superior morphotype among the three.

Keywords: Solvent Extraction, Seed Coat, *Bixa orellana*, Natural Colour, R_f Value

[1]Department of Agricultural Chemistry, Faculty of Agriculture, Bangladesh Agricultural University, Mymensingh-2202, Bangladesh.

[2]Department of Poultry Science, Faculty of Animal Husbandry, Bangladesh Agricultural University, Mymensingh-2202, Bangladesh.

*Corresponding author's email: haripadaseal@gmail.com (Hari Pada Seal); amdad53@yahoo.com (Mohammad Amdad Ali)

Reviewed by Dr. Mohammad Gulzarul Aziz, Bangladesh Agricultural University, Mymensingh, Bangladesh.

Introduction

Over the past few decades, it has been a growing concern in worldwide for food quality, safety, and security. Particular attention has been drawn to colorants used in food, drug, cosmetic, and textile industry, where the safety of colorants is one of the most burning issues. While, most of the synthetic colorants tend to be perceived as undesirable and harmful; some are considered to be responsible for allergenic and intolerance reactions (Wissgott and Bortlik, 1996). In contrary, nature-derived products are considered as health friendly and of good quality.

A growing number of natural food colorants are being isolated and excepted to use commercially, such as Caramel coloring (E150) made from caramelized sugar used in cola products and also in cosmetics; Annatto, a reddish-orange dye extracted from the seed of the *Bixa orellana* L.; Cochineal (E120), a red dye derived from the cochineal insect, *Dactylopius coccus;* Pandan a green food coloring extracted from *Pandanus amaryllifolius*; Butterfly pea a blue food dye from *Clitoria ternatea.* Among them, annatto (E160b) appears as an important one for food and drug industries owing to its potential uses as a substitute of the synthetic colorant Tartrazine which is prohibited in several countries (JECFA, 1982). Annato was classified by the Food and Drug Administration in the U.S.A. as a "color additive exempt of certification" (Hallagan *et al.*, 1995). In addition, this red-orange colorant is extracted from the seeds of Annatto (*Bixa orellana* L.), can be a potential candidate for replacing the costly "Saffron" which is widely used in medicine, food and cosmetics industry. Annatto tree is frequently seen as an ornamental plant in various parts of the world. It is a tropical tree native to the Central and South American rain forest, and is vastly cultivated in warm regions of the world, such as India, Bangladesh, Sri Lanka, and Java (Wolf, 1997). The dye extracted from the seed coat of the Annatto seeds can be used in coloring cheese, soft drinks, oil, butter, sondesh, vanilla ice-cream, and soap.

Further, it can be used in some regions to color textiles (Lata *et al.*, 1990) and color fabrics and weapons (Rao *et al.*, 2002). Recently, Ali (2006) reported three morphotypes of Annatto namely, morphotype-1 (M_1), morphotype-2 (M_2), and morphotype-3 (M_3) were found in Bangladesh.

Here we reported qualitative assessment of three morphotypes of Annatto in order to develop a simple inexpensive process for extracting the colorants.

Materials and Methods

Extraction of colorants

The matured-fruit-capsule of Annatto (*Bixa orellana* L.) of three morphotypes: morphotype-1 (M_1), morphotype-2 (M_2), and morphotype-3 (M_3) were collected from Botanical Garden of Bangladesh Agricultural University (BAU). The fruit-capsules were dried in sunlight for 30 hours and seeds were separated out from the capsules. After that, the seeds were dried in sunlight for another 30 hours and preserved in a refrigerator at 20°C for further analysis. To extract the colorants, 100 g of seeds were soaked in 200 ml particular organic solvent for 30 minutes and stirred abrasively by a glass rod for one hour (Islam, 2008). Extracting solvents were selected based on their polarity, viz. hexane (Polarity: 0), chloroform (Polarity: 4.1), acetone (Polarity: 5.1), and ethanol (Polarity: 5.2) (Finar, 2000). In addition, a mixed-solvent system of $CHCl_3/C_2H_5OH/CH_3COOH$ (80:2:1) also was employed. Finally, it was filtered by Whatman Grade-114 filter paper to prepare the stock solution for Thin Layer Chromatographic (TLC) analysis.

Thin layer chromatography analysis of colorants

The chromoplate were prepared by dispersing a suspension of finely powdered silica gel adsorbent (50 g silica gel thoroughly mixed and swelled in 125 ml of water) on a glass strip (EMerck India limited, Mumbai, 400018) of suitable size (5 cm × 1 cm × 0.3 cm). A very thin uniform layer of silica was formed on glass strip with the aid of a spreader. It was then dried in an oven at 105±5 °C for 12 hours. For obtaining a chromatogram, the chromoplate was spotted with a minute drop of annatto extract with the help of a capillary tube, and then placed vertically in a solvent tank. Development of the chromatogram occurred by capillary movement of the solvent up the silica-adsorbent layer. When solvent front reached at the upper end of the TLC plate, it was marked and the plate was taken out and allowed to dry. If the components were colored, then the spots were readily located. To ascertain the presence of colorless spots, the TLC plate was transferred to iodine chamber. Both colored and colourless spots on the TLC plate were recorded and R_f values were calculated. The R_f value depends on the relative polarities of the carrier solvent and spot-components. It is defined as the ratio of the distance travelled by a component to the distance travelled by the solvent front (Finar, 2000). R_f value indicates the polarity of the spot-components in the TLC plate (Saraswathy *et al.*, 2004). Similar polarities of carrier solvent and spot-components leading higher R_f value and vice-verse.

Results and Discussion

Extraction of colorant

Colorants were extracted from the three morphotypes of Annatto seeds by employing variety of solvents (hexane, chloroform, acetone, ethanol, and a mixed solvent system), shown in Table 1. Solvents were selected based on their polarity to find out the suitable solvent for extracting the colorants. The hexane extracted effluent was transparent; however, the intensity of color was very light, indicating that only components of low polarities were extracted by hexane. When chloroform was employed, the effluent solution was very transparent and deep red brick color, shown in Fig. 1 and Table 1, suggesting that chloroform is one of the suitable solvents to extract the colorants. Furthermore, the colorants extracted by more polar solvents such as, acetone and ethanol were dark and cloudy, like emulsion. It might be due to some undesired highly polar components were also extracted along with the colorants. In addition, mixed solvents extracted effluent was transparent and brick red color, indicating that it is only selective to brick red colored pigments (Saraswathy *et al.*, 2004).

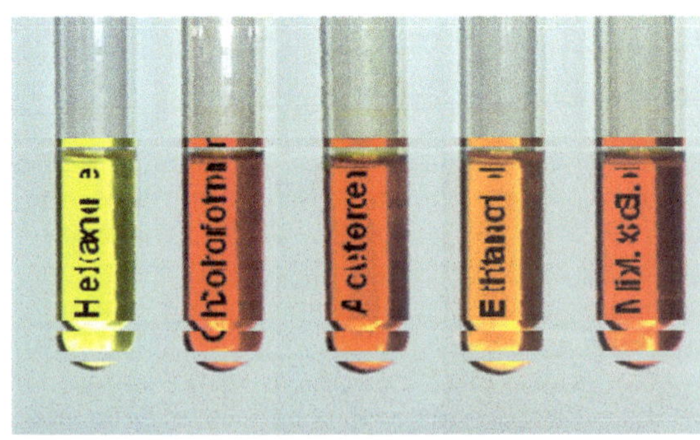

Fig. 1. Colorants extracts by solvents

Thin layer chromatographic analysis of colorant

To separate and evaluate the coloring pigments from the solvent extracted effluents, TLC was employed. Both colored and colorless spots and their R_f values were recorded (Finar, 2000; Saraswathy *et al.*, 2004).

Table 1. Physical properties of colorant extracts

Extracting solvent	Polarity	Color/Intensity	Transparency	Inferences
Hexane	0.0	Yellow/Light	Transparent	Suitable for non-polar dyes
Chloroform	4.1	Brick red/Deep	Transparent	Suitable
Acetone	5.1	Amber/Dark	Cloudy	Not suitable
Ethanol	5.2	Orange/ Dark	Cloudy	Not suitable
Mixed solvent	-	Amber/Light	Transparent	Suitable for selected colorants

Hexane extracts

Table 2 shows the R_f values of both colored and colorless spots of n-hexane extracts. The starting carrier solvent was benzene (100%), in this case, colored-spots were not developed regardless of three morphotypes. However, after exposing in I_2 vapor, M_1 and M_2 showed few spots, attributing for the colorless compounds. When mixed solvent systems of benzene/ethyl acetate (Bz/EtOAc) with the ratio of 3:2, 7:3, and 4:1 were employed, respectively, M_1 as well as M_3 showed one colored spot respectively and no colorless spot was observed, indicating that they were exclusion of colorless compounds (gum, wax, etc.). In case of M_2, two colored spots were seen: first one was yellow in color with low R_f value, and second one was brick red in color with higher R_f value respectively. It suggested that both low and high polar pigments (i.e. yellow and brick red colored constituents respectively) were present in the hexane extracted annatto. After exposing to I_2 vapor, few spots were also observed. On the basis of colored and colorless spots, and their R_f values, it can be concluded that Bz/EtOAc (3:2) and (7:3) solvent mixture were the superior carrier solvents for M_1 and M_3 morphotypes.

Table 2. TLC analysis of hexane extracted colorants

Morphotype	Eluting solvent	Value of R_f	
		Colored spot	Colorless spot
M_1	Bz = 100 %	-	0.416
M_2		-	0.602 0.504 0.416
M_3		-	-
M_1	Bz/EtOAc = 3:2	0.113	-
M_2		0.939 0.173	0.965 0.913
M_3		0.156	-
M_1	Bz/EtOAc = 7:3	0.084	-
M_2		0.940 0.110	0.983 0.840
M_3		0.067	-
M_1	Bz/EtOAc = 4:1	0.940 0.060	0.983
M_2		0.949 0.093	0.967
M_3		0.084	-

Chloroform extracts

Solvent systems of Bz/EtOAc with the ratio of 1:0, 9:1, 7:3, 1:1, and 0:1 were exploited separately as carrier solvent, presented in Table 3. Bz (100%) provided single color spot for all the morphotypes. After transferring to I_2 chamber, only M_2 showed two spots. In case of solvent systems 9:1 and 7:3, three colored spots were observed for all morphotypes. Among them, the first one was yellow colored spot with very low polarity and the second spot was red color with moderate polarity and the third one was brick red color. Most of the cases for M_1 and M_3 colorless spots were not appeared, suggesting that colorless compounds were absent in these morphotypes. These two carrier solvent systems seemed to be suitable solvent for separating the dyes from the chloroform extract. Further increasing in portion of EtOAc, as in case of 1:1 and EtOAc (100%), both color and colorless spots were observed.

Acetone extracts

Table 4 shows the R_f values for chromatogram of acetone extracts. Similar trend was observed in using Bz (100%); only colorless components were present.

Table 3. TLC analysis of chloroform extracted colorants

Morphotype	Eluting solvent	Value of R_f	
		Colored spot	Colorless spot
M_1	Bz (100 %)	0.235	-
M_2		0.227	0.504 0.390
M_3		0.219	-
M_1	Bz/EtOAc = 9:1	0.983 0.845 0.128	-
M_2		0.990 0.846 0.213	0.983 0.957 0.897
M_3		0.957 0.803 0.128	-
M_1	Bz/EtOAc = 7:3	0.929 0.835 0.144	-
M_2		0.945 0.850 0.236	0.976 0.912
M_3		0.958 0.883 0.166	0.983
M_1	Bz/EtOAc = 1:1	0.140	-
M_2		0.932 0.210	0.907 0.974
M_3		0.924 0.180	-
M_1	EtOAc (100%)	0.122	-
M_2		0.967 0.131	0.934
M_3		0.959 0.114	0.943

Table 4. TLC analysis of acetone extracted colorants

Morphotype	Eluting solvent	Value of R_f	
		Colored spot	Colorless spot
M_1	Bz (100 %)	-	-
M_2		-	0.983 0.459 0.344
M_3		-	-
M_1	Bz/EtOAc = 9:1	0.949 0.168	0.983 0.967
M_2		0.957 0.176	0.967 0.918
M_3		0.210	-
M_1	Bz/EtOAc = 7:3	0.084	-
M_2		0.924 0.101	0.983 0.840
M_3		0.067	-
M_1	Bz/EtOAc = 1:1	0.14	-
M_2		0.932 0.210	0.974 0.907
M_3		0.924 0.180	-

While mixed solvents were employed, one/two colored spots were developed. Between them one was very low polarity with yellow colored spot and other was brick red with moderate polarity. M_2 provided two colorless spots; however, M_1 and M_3 did not showed any spot under iodine vapor. On depending of the colored and colorless spots, and their R_f values, it seemed that Bz/EtOAc (9:1) solvent system was the superior carrier solvent for separating dyes from acetone extracts. In this case, TLC chromatogram of M_1 and M_2 showed two colored spots, and its R_f value was 0.949 and 0.168 for M_1; and 0.957 and 0.176 for M_2, respectively. For the M_3, one colored spot was appeared with R_f of 0.210 and did not show any spots under iodine vapor.

Ethanol extracts

For developing the chromatogram of ethanol extract, Bz (100%) and mixture of Bz/EtOAc were used as carrier solvent, shown in Table 5. As excepted, Bz did not carry any color components. With the inclusion of EtOAc, in 7:3 and 6.5:3.5 ratios two colored spots were observed for all the morphotypes. Between them one was yellow color with very low polarity and the second spot was brick red color. After exposing to iodine vapor, one spot was appeared for all morphotypes. Further increasing the portion of EtOAc, i.e., in 6:4, only single color spot was seen for all the morphotypes.

Table 5. TLC analysis of ethanol extracted colorants

Morphotype	Eluting solvent	Value of R_f	
		Colored spot	Colorless spots
M_1	Bz (100 %)	-	0.50
M_2		-	0.528 0.50
M_3		-	-
M_1	Bz/EtOAc = 7:3	0.983 0.122	0.959
M_2		0.983 0.114	0.959
M_3		0.983 0.098	0.959
M_1	Bz/EtOAc = 6.5:3.5	0.975 0.173	0.967
M_2		0.983 0.190	0.967
M_3		0.983 0.247	0.967
M_1	Bz/EtOAc = 6:4	0.459	-
M_2		0.475	-
M_3		0.508	-

Considering these chromatograms, it seemed that Bz/EtOAc with the ratio of 6:4 solvent system can be used for separating selective dye. Its chromatogram indicated that M_1, M_2, and M_3 contained one colored spots with Rf value of 0.459, 0.475, and 0.508, respectively.

Mixed-solvent of chloroform/ethanol/ acetic acid (80:2:1) extracts

Table 6 shows the R_f values of colored and colorless spots of mixed solvents extract; and Bz/EtOAc solvent systems were attempted to separate the components. In every case, only colored spots were observed in chromatogram and it was brick red color having high polarity, suggesting that extracts of mixed solvent comprised of mostly colored constituents. On the basis of separation of colored and colorless spots, and their R_f values; it may be concluded that Bz/EtOAc (1:1) solvent mixture was most suitable carrier solvent for separating the dyes from the extract of mixed solvent systems. In all case, M_1, M_2, and M_3 developed only one colored spot respectively and, their R_f value were 0.809 for M_1, 0.826 for M_2 and 0.843 for M_3.

Table 6. TLC analysis of mixed-solvents extracted colorants

Morphotype	Eluting solvent	Value of R_f	
		Colored spot	Colorless spots
M_1	Bz (100 %)	0.104	-
M_2		0.104	-
M_3		0.104	-
M_1	Bz/EtOAc= 9:1	0.295	-
M_2		0.305	-
M_3		0.315	-
M_1	Bz/EtOAc= 7:3	0.184	-
M_2		0.193	-
M_3		0.202	-
M_1	Bz/EtOAc= 1:1	0.809	-
M_2		0.826	-
M_3		0.843	-

Conclusion

Three coloring constituents in each morphotypes (M_1, M_2, and M_3) of Annatto were observed. Among the three pigments, one was associated with high R_f value, i.e., very low polarity and yellow in color. The rest of the two pigments were brick red and orange colored associated with medium and low R_f value, respectively. Moreover, after exposing in iodine vapor, most of the chromatogram indicated the presence of colorless components. Studies from R_f values of colorless components revealing that one was low polar colourless components and other two colorless compounds were of medium polarity. Furthermore, chloroform extract effluent was very deep brick red in color and transparent. On depending of the chromatogram, mixed solvent systems of Bz/EtOAc of 7:3 was superior carrier solvent for separating the dyes from the chloroform extracts. Thus, dyes separated from the extract can be used in coloring the food, medicine, cosmetic and other purpose.

References

Ali, M. 2006. Morphological variation in *Bixa orellana*. MS Thesis. Dept. Crop Bot., Bangladesh Agril. Univ., Mymensingh-2202, Bangladesh. 44p.

Finar, I.L. 2000. Organic Chemistry. Volume 2: Stereochemistry and the chemistry of Natural Products. pp. 79-80.

Hallagan, J.B., Allen, D.C. and Borzelleca, J.F. 1995. The Safety and Regulatory Status of Food, Drug and Cosmetics Colour Additives Exempt from Certification. *Food Chem. Toxicol.*, 33(6): 515-528.

Islam, S. 2008. Flower and capsule production in relation to dye yield in *Bixa orellana*. MS Thesis. Dep. Crop Bot., Agric. Univ., Mymensingh, Bangladesh. 40p.

JECFA. 1982. Evaluation of Certain Food Additives and Contaminants; Twenty-Sixth Report of the Joint FAP/WHO Expert Committee on Food Additives; Technical Report Series No. 683; WHO: Geneva. 51p.

Lata, R., Aparnathi, K.D. and Sharma, R.S. 1990. Annatto (*Bixa orellana* L.), its cultivation, preparation and usage. *Int. J. Trop. Agric.*, 8 (1): 80-88.

Rao, P.G.P., Satyanarayana, A. and Rao, D.G. 2002. Effect of Storage on the Stability of Water Soluble Annatto Dye Formulation in a Simulated Orange-RTS Beverage Model System. *Lebensmittel Wissenschaft and Technologic*, 35 (7): 617-621.

Saraswathy, S., Manavalan, R.S.A., Vadivcl, E., Manian, K. and Subramanian, S. 2004. Influence of various extraction methods on dye yield in annatto (*Bixa orellana*. L). *South Indian Hort.*, 52(1/6): 239-243.

Wissgott, U. and Bortlik, K. 1996. Prospects for New Natural Food Colorants. *Trends Food Sci. Technol.* 7: 298-302.

Wolf,M.A.1997.http://home. Braunschweig. Netsurf. Deandree. Wolf/urucum. Html.

3

EFFECTS OF STOCKING DENSITY ON GROWTH AND PRODUCTION PERFORMANCE OF INDIGENOUS STINGING CATFISH, *Heteropneustes fossilis* (Bloch)

A.H.M. Kohinoor[1]*, M.M. Khan[2], S. Yeasmine[3], P. Mandol[4] and M.S. Islam[5]

Abstract

An on-farm experiment was undertaken in nine earthen ponds to evaluate the growth and production potentials of stinging catfish shing, *Heteropneustes fossilis* for the period of six months from March to August 2010. Three stocking densities such as 1,25,000 (T_1), 1,87,500 (T_2) and 2,50,000 ha^{-1} (T_3) were tested with three replications each. Fish were fed with commercial pelleted feed containing 35% crude protein. After six months rearing, the mean harvesting weights of shing were 69.42±6.20, 58.74±3.85 and 49.50±4.52g in T_1, T_2 and T_3, respectively. Significant ($P<0.05$) highest mean harvesting weight was found in T_1. The best survival was found in T_1 (87%) among the treatments. The calculated mean production of shing (*H. fossilis*) in three treatments such as T_1, T_2 and T_3 were 7549±52, 9031±71 and 8786±60 kg ha^{-1}, respectively, which were significantly different ($p<0.05$) from each other.

Keywords: Growth, Stinging Cat fish, Stocking Densities, Earthen Pond

[1,2,3&5]Freshwater Station, Bangladesh Fisheries Research Institute, Mymensingh, Bangladesh
[4]Senior Assistant Chief, Ministry of Fisheries & Livestock, Bangladesh.

*Corresponding author's email: kohinoor41@gmail.com (A.H.M. Kohinoor)

Introduction

Among the air-breathing catfishes, stinging catfish (*Heteropneustes fossilis*) is very popular and high valued fish in Bangladesh. This fish is locally known as Shingi or Shing. It is considered to be highly nourishing, palatable and tasty and well preferred because of its less spine, less fat and high digestibility in many parts of Indian subcontinent (Khan *et al.,* 2003). The species is very high content of iron (226 mg $100g^{-1}$) and fairly high content of calcium compared to many other freshwater fishes. Due to high nutritive value the fish is recommended in the diet of sick and convalescents. Being a lean fish it is very suitable for people for whom animal fats are undesirable (Rahman *et al.,* 1982). It is a very hardy fish and can survive for quite a few hours outside the water due to presence of accessory respiratory organs. It can tolerate slightly brackish water. The fish adapts well to hypoxic water bodies and to high stocking densities (Dehadrai *et al.,* 1985). This fish was abundantly available in open water system of floodplains, canals, beel and haors of Bangladesh. But due to over exploitation and ecological changes in its natural habitats; this species have become threatened. Indiscriminate destructive practices have caused havoc to aquatic bio-diversity in Bangladesh (Hussain and Mazid, 2001). Presently, *H. fossilis* is one of the threatened fish in Bangladesh (IUCN Bangladesh, 2000). This fish has enormous aquaculture potential and it could be easily grown in ponds and small ditches. Culture of *H. fossilis* has yet been well flourished in Bangladesh due to lack of appropriate culture technique. Considering its status of threatened status, high market value and high consumer demand it is essential to develop suitable culture technique. The culture technique will helpful to prevent the fish from being extinct and at the same time this delicious tasty fish will be available for the rural and urban people.

Although the air breathing fish Shing is in high market demand, hither to very little attention has been paid to develop culture techniques. Earlier accounts on this species include seasonal morphology of gonads in relation to the biology (Azadi and Siddique, 1986; Kuddus *et al.,* 1995), induced breeding for fry production (Thakur *et al.,* 1977; Saha, 1998), food and feeding habits (Kuddus *et al.,* 1995) and nutrition (Hossain *et al.,* 1993; Anwar and Jafri, 1995). But few published literature are available on growth and production of *H. fossilis* in pond, cistern and net cages (Lipton, 1983; Haque *et al.,* 1988; Narejo *et al.,* 2005; and Khan *et al.,* 2003). Considering the lack of information on these lines, the present investigation was carried out to ascertain the optimum stocking density of *H. fossilis* in pond ecology for sustainable production.

Materials and Methods

Pond selection and preparation

The experiment was conducted for a period of six months from March to August 2010 in nine farmer's ponds of 400-600 m^2 area with a depth of 1.0-1.5m at Modhupur, Tarakanda, Mymensingh. Prior to stocking, ponds were dried and cleaned for weeds and unwanted aquatic animals. The dried ponds were left exposed to sunlight for several days and then limed at the rate of 250 kg ha^{-1}. Five days after liming, ponds were filled-up with deep tube well water up to the depth of 1 meter. After three days, ponds were fertilized with cowdung at the rate of 1,000 kg ha^{-1}.

Experimental design

Three different stocking densities of shing (*H. fossilis*) were tested in the experiment. Stocking density was maintained as treatment and which replicated thrice. Fingerlings of shing were stocked at the rate of 1,25,000, 1,87,500 and 2,50,000 ha^{-1} in T_1, T_2 and T_3, respectively.

Source of fingerlings

The fingerlings of *H. fossilis* were used in this experiment were collected from a private hatchery of Mymensingh, Bangladesh. The sizes of fingerlings were 3.26±0.55, 3.44±0.46 and 3.24±0.39g in T_1, T_2 and T_3, respectively.

Fish stocking

Fingerlings of *H. fossillis* were stocked in 01 March 2010 according to the experimental design. All ponds of various sizes were divided randomly into three treatment groups viz. group-1 (T1), group-2 (T2) and group-3 (T3). The ponds belong to Treatment-1 (T_1), Treatment-2 (T_2) and Treatment-3 (T_3) were stocked with *H. fossillis* fingerlings at the stocking density of 1,25,000 (T_1), 1,87,500 (T_2) and 2,50,000 ha^{-1} (T_3), respectively.

Post stocking management

After stocking, in order to meet up the increasing dietary demand, commercial fish feed named Saudi Bangla Fish feed (starter-2) containing 35% crude protein were applied as supplementary feed at the rate of 4-10% of standing biomass of fish twice daily at 0500 hrs and 1900 hrs daily. Proximate composition of the feed was determined following the standard methods given by Association of Official Analytical Chemists (AOAC) (1984) in the Feed and Nutrition Laboratory of Freshwater Station, Bangladesh Fisheries Research institute, Mymensingh. The amount of feed was adjusted fortnightly on the basis of sampling of experimental fish. The proximate composition of the feed is shown in Table 1. Besides this, lime was applied in all the ponds at the rate of 125 kg ha^{-1} at monthly interval.

Table 1. Proximate analyses of the feed used (% dry mater basis)

Dry mater	Protein	Lipid	Ash	Crude fiber
81.10±2.0	35.00±1.0	3.11±0.09	18.20±0.2	10.10±0.68

Fish sampling

Random samples of 50 fish from each pond were sampled fortnightly by using a seine net. The total weight was measured by using a portable balance (Tanita, Japan).

Water sampling and analysis

Water quality parameters such as water temperature, transparency, pH, dissolved oxygen (DO) and total alkalinity were determined at weekly interval between 0900 to 1000 hrs. Temperature was recorded using a Celsius Thermometer, dissolve oxygen and pH meter (Hanna pH 300) and a portable digital DO meter (MI 605, MARTINI).

Harvesting of fish

At the end of the experiment, the fishes were harvested, first by seine netting and then by draining out of the ponds. The harvested fishes were counted and weight was recorded to find out the survival rate and production, respectively. Specific growth rate (SGR) was estimated as:

SGR (% bw d^{-1}) = [In (final weight) – In (initial weight)]/culture period (days) × 100

Data analysis

Data were analysed using the statistical package, Statgraphics Version 7. ANOVA was performed on all the dependent variables to see whether the treatment had any significant effect or not.

Results

Water quality parameter

Mean values of physico-chemical parameters over the period of shing farming are presented in Table 2. The water temperature in T_1, T_2 and T_3 ranged from: 24.60- 30.7°C, 23.88-29.84°C and 24.50-30.30°C, respectively with the mean values 27.90±1.12, 27.86±1.09 and 27.49±1.11 The mean water temperature in T_1, T_2 and T_3

were not statistically significant (P<0.05). The water transparency showed significant (p<0.05) difference among the treatments. The mean values were 24.65±3.11, 26.00±3.92 and 2014±2.40 cm in T_1, T_2 and T_3, respectively. Mean transparency values differed significantly (P<0.05), increasing from T_3 to T_1. The level of pH varied from 7.10-8.20, 7.15-8.70 and 7.08- 8.60 in T1, T2 and T3, respectively.

The dissolved oxygen content in the experimental ponds ranged from 4.70 to 6.80, 4.27 to 6.70 and 3.31 to 6.90 in T_1, T_2 and T_3, respectively, with the mean values 5.32±0.14, 4.79±0.32 and 4.23±0.47 mg L^{-1}. Comparatively lower level of dissolved oxygen as observed in the treatment-3. Dissolved oxygen values (mg L^{-1}) did not differ significantly (P<0.05). Total alkalinity ranged from 85 to 128, 87 to 138 and 84 to 140 mg L-1 with mean values of 109±6.96, 113±9.58 and 117±10.29 mg L^{-1} in T_1, T_2 and T_3, respectively. These values did not show any significant difference (P<0.05) among the treatments.

The range of mean values of ammonia-nitrogen (unionized) was 0.033±0.012, 0.042±0.021 and 0.055±0.031 in T_1, T_2 and T_3, respectively. The highest ammonia-nitrogen value was 0.97 in the month of August in T_3 and the lowest value was 0.017 in the month of March in T_1. The differences among treatments were significant (P>0.05) when compared using ANOVA.

Table 2. Water quality parameter (mean ±SE) of the ponds under different treatments

Parameter	Treatment-1	Treatment-2	Treatment-3
Temperature (°C)	27.90±1.12 (24.60-30.70)	27.86±1.09 (23.88-29.84)	27.49±1.11 (24.50-30.30)
Transparency (cm)	24.65±3.11^{b}	26.00±3.92^{b}	20.14±2.40^{a}
Dissolved oxygen (mg/L)	5.32±0.14^{a}	4.79±0.32^{b}	4.23±0.47^{b}
pH	7.10-8.20	7.15 -8.70	7.08- 8.60
Total Alkalinity (mg/L)	102.00±6.96	113.00±9.58	123.00±10.29
Total ammonia (mg/l)	0.033±0.012	0.042±0.021	0.055±0.031

Figures in the same row having the same superscripts are not significantly different (P>0.05)

Growth and production

Details of stocking, harvesting, growth, FCR, SGR and production of shing (*H. fossilis*) in the three treatments during the study period are shown in Table 3. Growth of shing in ponds was investigated and the results obtained from the experiment indicated that the growth rate varied in different stocking densities. On the basis of final growth attained at harvest under T_1, T_2 and T_3 were 69.42±6.20, 58.74±3.85 and 49.50±4.52, respectively. The highest harvesting weight was obtained in T_1 and lowest in T_3. The harvesting mean weight showed significant differences (P<0.05) in T_1 followed by T_2 and T_3 when ANOVA was performed. The monthly sampling weights of fish under different stocking densities are shown in Fig. 1.

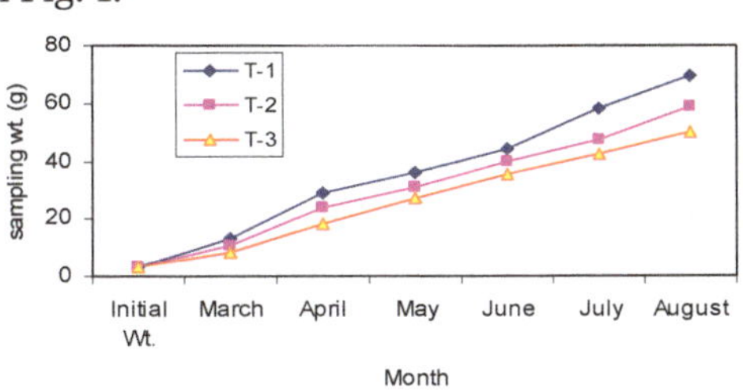

Fig. 1: Monthly growth rate (g) of Shing (*H. fossilis*) in different stocking densities

The Figure indicates that the growth rate was always higher in T_1 then followed by T_2 and T_3. The results showed that higher growth rate was observed at against lower stocking densities. The specific growth rate (% per day) of fish in different treatments varied among the treatments. Highest value was obtained in T_1 and lowest in T_3. SGR in T_1 was significantly higher than T_3, while T_2 was not significantly different from T_1, but was significantly different from T_3. FCR was significantly lower in T_1 than in T_2 and T_3. Therefore, SGR and FCR were best for fish in T_1 where lowest number of fingerlings was reared (1,25,000 ha^{-1}).

The percentage of survival as recorded in the present study was 87±3.6, 82±4.58 and 71±2.64 for T_1, T_2 and T_3, respectively. The highest survival rate was observed in T_1 and the lowest in T_3. Differences in survival rates among the treatments were found to be significant (P<0.05). The mean production of shing was 7,549±52, 9,031±71 and 8,786±60 Kg ha^{-1} in T_1, T_2 and T_3, respectively. The highest fish production was obtained in T_2 followed by T_3 and the lowest in T_1. However, production of fish differed significantly (P<0.05) among the three treatments.

Table 3. Growth, survival and production of shing (*H. fossilis*) under the three treatments

Treatment	Initial weight (g)	Harvesting mean weight (g)	Survival (%)	FCR	Production (kg ha^{-1})	SGR (%)
Treatment-1	3.26 ± 1.11^{a}	69.42 ± 6.20^{a}	87 ± 3.6^{a}	2.78	$7{,}549\pm52^{b}$	2.33^{a}
Treatment-2	3.34 ± 0.96^{a}	58.74 ± 3.85^{b}	82 ± 4.58^{a}	3.12	$9{,}031\pm71^{a}$	2.22^{a}
Treatment-3	3.24 ± 1.0^{a}	49.50 ± 4.52^{c}	71 ± 2.64^{b}	3.59	$8{,}786\pm60^{c}$	2.13^{a}

* Figures in the same column with different superscripts are significantly different ($P>0.05$)

Correlation matrix among stocking density, harvesting weight, survival, FCR, SGR (%) and production of shing is shown in Table 3 and gave a clear picture of the relationships among the parameters (Table 4). Stocking density showed a negative correlation with harvesting weight, survival and SGR. It means that if stocking density increased, then mean values of harvesting weight, survival and SGR decreased. While, harvesting weight showed positive correlation with survival but negative correlation with FCR and production. Survival rate indicated the negative correlation with FCR and production.

Table 4. Correlation matrix among stocking density, harvesting weight, survival, FCR, SGR and production of shing under grow-out system

Parameter	Stocking density	Harvesting Wt. (g)	Survival (%)	FCR	SGR	Production (Kg)
Stocking density	1.000	-	-	-	-	-
Harvesting Wt. (g)	-0.999*	1.000	-	-	-	-
Survival (%)	-0.978	0.968	1.000	-	-	-
FCR	0.996*	-0.961	-0.993*	1.000	-	-
SGR (%)	-0.998*	1.000	0.964	-0.989*	1.000	-
Production (Kg ha^{-1})	0.776	-0.804	-0.628	0.717	-0.813	1.000

* Significant difference at 5% level of probability

Discussion

Environmental parameters exert an immense influence on the maintenance of a healthy aquatic environment and production of food organisms. Growth, feed efficiency and feed consumption of fish are normally governed by a few environmental factors (Brett, 1979). The mean range of temperature (27.49-27.93°C) in the experimental ponds is within the acceptable range for farming of shing that agrees well with the findings of Haque *et al.* (1984) and Kohinoor *et al.* (1994, 2007). Boyd (1982) reported that the range of water temperature from 26.06 to 31.97°C is suitable for fish culture. Transparency was consistently higher in all the treatments, because shing did not consume plankton in adult stage. For this reason plankton abundance was found in all the ponds of treatments. The Dissolved oxygen in the morning was low in ponds stocked with high density of fish compared to ponds stocked with a low density. Similar results were observed by Saha *et al.* (1988) for rohu (*L. rohita*) and Rahman and Rahman (2003) for calash (*L. calbasu*). Fluctuation of dissolved oxygen might be attributed to photosynthetic activity and variation in the rate of oxygen consumption by fish and other aquatic organisms (Boyd, 1982). However, the level of dissolved oxygen is within the acceptable ranges in all the experiment ponds.

Roy *et al.* (2002) obtained a pH range 7.03 to 9.03 in fish ponds located in Trishal, Mymensingh. The observed pH values of water ranging from 7.3 to 9.0 indicated that the experimental ponds were suitable for fish culture (Boyd, 1982). During the study period, the pH values of pond water under different treatments were found to be alkaline. Roy *et al.* (2002) obtained a pH range 7.03 to 9.03 in fish ponds located in Trishal, Mymensingh. The observed pH values of water ranging from 7.1 to 8.6 indicated that the experimental ponds were suitable for fish culture (Boyd, 1982). The total alkalinity values of the pond water under different treatments were found to be productive level. Total alkalinity levels for natural waters may range from less than 5 mg L^{-1} to more than 500 mg L^{-1} (Boyd, 1982). Kohinoor *et al.* (1998) and Roy *et al.* (2002) found the average total alkalinity above 100 mg L^{-1} in their studies. Total alkalinity values found in the present study were within the suitable range (102 to 123). Total ammonia (including unionized fraction) was found in all the treatments because, high density if fish was stocked. The droppings of the fish might be produced more ammonia in the ponds. New (1987) reported that excessive use of feed or fertilizer caused sediments in the pond bottom which may produce ammonia in the ponds and other gases. In this experiment, this may be happened. The suitable range of Ammonia (NH_3) below 0.1 mg L^{-1} (Boyd, 1982).

Growths of fish in ponds were investigated and the results obtained from the experiment indicated that the growth rate varied in different stocking densities. Growth in terms of mean harvesting weight and SGR of *H. fossilis* was significantly higher in T_1 where the stocking density was low compared to the treatments of T_2 and T_3 although same feed was supplied in all the treatments at an equal ratio. The low growth rate of *H. fossilis* in T_2 and T_3 appeared to be related with higher densities and increased competition for food and space and an inverse relationship with the stocking density provided that space had limiting effects operate on the population (Johnson, 1965). The present results coincide with the findings of Narejo *et al.* (2005) who achieved best growth at lower stocking densities in shing farming. While Mollah (1985) reported that the lower density gave higher size and higher survival rate in *Clarias macrocephalus*. They also reported that the lower density gave larger size and higher survival rate in *H. fossilis* in cemented cistern fed on formulated feed. In the present experiment, highest harvesting mean weight and highest survival rate has been observed in lowest stocking density. The lowest stocking densities provide more space, food and less competition, which were reported by various authors like Ahmed (1982), Hasan *et al.* (1982) and Haque *et al.* (1984). The percentage of survival as recorded in the present study was 87, 82 and 81 for T_1, T_2 and T_3, respectively. Survival was found to be negatively influenced by stocking densities. It might be due to high competition and space among the fishes. The mean growth increase in this experiment was found to be comparable to the gain achieved in rice fields where *H. fossilis* was shown to grow about 25 g during 60 days culture period (Dehadrai, 1981). Lipton (1983) observed that the shing attained 30.35 g over 112 days with gross production 1242.35 g/m^2 in cage culture management. In a study, Thakur and Das, (1986) reported production range was 1642 to 7,300 kg ha^{-1} in four to eleven months culture period of *H. fossilis*. In a study, Mustafa (1991) who has reported to achieve an estimated production of 1592, 2373 and 1764 kg ha^{-1} yr^{-1}. from monoculture of mola (*A. mola*), chola punti (*P. sophore*) and colisha (*C. fasciata*), respectively. While Akhteruzzaman *et al.* (1991) reported that in monoculture condition the production of *Puntius sarana* was 1,200 kg ha^{-1} in six months. However, Khan *et al.* (2003) observed the effect of different stocking densities on production of catfish (*H. fossilis*) in earthen ponds and got the production range 2080 to 3364 kg ha^{-1}. A study conducted by Kohinoor *et al.* (2006) obtained production of 1370 to 1535 kg ha^{-1} in six months from *Mystus cavasius* in monoculture management. While in the present experiment, higher production was noticed. The reason might be due to longer culture period and higher stocking density and better management.

Therefore, the stocking density of 1,87,500 ha^{-1} for culture of shing, *H. fossilis* is advisable for farmers in commercial basis.

References

Ahmed, G.U. 1982. Intensive culture of *Labeo rohita* (Hamilton) in floating ponds with special reference to different stocking. M.Sc. Thesis. Department of Fisheries Biology and Limnology. Bangladesh Agricultural University, Mymensingh. 38 p.

Akhteruzzaman, M., Kohinoor, A.H.M., Hussain, M.G. and Shah, M.S. 1991. A study on the semi-intensive culture of *Puntius sarana* (Hamilton). *Bangladesh J. Fish.* 14 (1-2): 69-72.

Anwar, M.F and Jafri, A.K. 1995. Effects of varying dietary lipid levels on growth, feed conversion, nutrient retention and caracass composition of fingerling catfish, *Heteropneustes fossilis. Asian Fish Sci.* 8: 55-62.

AOAC Assoaiation of Official Analitical Chemists, AOAC. 1984. *In:* Howritz, W. (Ed.), Official Methods of Analysis, 14th edn. AOAC, Washington, DC. 1018 p.

Azadi, M.A and Siddique, M.S. 1986. Fecundity of catfish, *Heteropneustes fossilis* (Bloch). *Bangladesh J. Zool.* 14: 33-39.

Boyd, C.E. 1982. Water Quality Management for Pond Fish Culture. Elsevier, The Netherlands. 318 p.

Brett, J.R. 1979. Environmental factors and growth. *In:* Hoar, W.S., Randal, D.J. Brett, J.R. (Eds.) Environmental relations and behavior, Fish Physiology, vol. 6. Academic press, New York. pp. 599-677.

Dehadrai, P.V. 1981. Integration of air breathing cat fish culture with livestock and crop agriculture, *In:* Summer Institute of Farming System Integrating agriculture and livestock. 8 p.

Dehadrai, P.V., Yusuf, K.M. and Das, R.K. 1985. Package and practices for increasing production of air breathing fishes. *In:* Aquaculture Extension Manual, New Series, Information and Extension Division of CIFRI. Indian Council of Agriculture Research. pp. 1-14.

Hasan, M.R., Aminul Haque, A.K.M., Islam, M.A. and Khan, E.U.K. 1982. Studies on the effect of stocking density on the growth Nile tilapia, *Saratherodon nilotica* (Linnaeus) in floating ponds. Bangladesh. *J. Fish.* 2 (5): 73-81.

Haque, M.M., Alam, A.K.M.N. and Islam, M.A. 1988. Culture prospects of shying (*Heteropneustes fossilis)* in floating cages. *Bangladesh J. Fish.* 12: 67-72.

Haque, M.M., Islam, M.A., Ahmed, G.U. and Haq, M.S. 1984. Intensive culture of Java Tilapia (*Oreochromis mossambicus*) in floating ponds at different stocking densities. *Bangladesh J. Fish.* 7 (1-2): 55-59.

Hossain, M.A., Islam, M.N. and Ali, M.A. 1993. Evaluation of silkworm pupae meal as a dietary protein source of catfish (*Heteropneustes fossilis*). *In:* S.J. Kaaushih and P. Luguent *Heteropneustes fossilis* (Editors). Fish Nutrition in Practice. Biarritz, France. pp. 785-791.

Hussain, M.G. and Mazid, M.A. 2001. Genetics Improvement and Conservation of Carp Species in Bangladesh. Bangladesh Fisheries Research Institute, Mymensingh. 56 p.

IUCN, Bangladesh. 2000. Red book of threatened fish of Bangladesh. The world Conservation Union. Appendix 1, 61 p.

Johnson, W.E. 1965. On mechanism of self regulation of population abundance in *Oncorhynchus nerka*. Mitt. *Int. Verein. Theor. Angew.* 13: 66-87.

Khan, M.N., Saiful Islam, A.K.M. and Hussain, M.G. 2003. Marginal analysis of culture of stinging catfish (*Heteropneustes fossilis*): Effect of different stocking density in earthen ponds. *Pakistan J. Biol. Sci.* 6 (7): 666-670.

Kohinoor, A.H.M., Haque, M.Z., Hussain, M.G. and Gupta, M.V. 1994. Growth and survival of Thi punti, *Puntius gonionotus* (Bleeker) spawn in nursery ponds at different stocking densities. *J. Asia. Soc. Bangladesh Sci.* 20 (1): 65-72.

Kohinoor, A.H.M, Islam, A.K.M.S., Jahan, D.A., Zaher, M. and Hussain, M.G. 2007. Monoculture of climbing perch, Thai koi, *Anabas testudineus* (Bloch) under different stocking densities at on-farm management. *Bangladesh J. Fish. Res.* 11 (2): 173-180.

Kohinoor, A.H.M., Islam, M.L., Wahab, M.A. and Thilsted, S.H. 1998. Effect of mola (*Amblypharyngodon mola*, Hamilton) on the growth and production of carps in polyculture. *Bangladesh J. Fish.* 2 (2): 119-126.

Kohinoor, A.H.M., Momtaz Begum and Hussain, M.G. 2006. Culture potentials of gulsha (*Mustus cavasius*) in monoculture management under different stocking densities. *Bangladesh J. Fish. Res.* 8 (2): 95-100.

Kuddus, R., Kuddus, M.M.A. and Shafi, M. 1995. Feeding and breeding habits of shing fish, *Heteropneustes fossilis* (Bloch). *Dhaka Univ. J. Biol. Sci.* 5: 35-43.

Lipton, A.P. 1983. Studies on the culture of *Heteropneustes fossilis* in cages. Proc. Natl. Sem. Cage Culture. pp. 51-54.

Mollah, M.F.A. 1985. Effects of stocking density and water depth on the growth and survival of freshwater catfish (*Clarias macrocephalus*) larvae. *Indian J. fish.* 32: 1-17.

Mustafa, G. 1991. Composite culture and biology of some indigenous fishes of Bangladesh. Ph. D. Thesis. Faculty of Biological Science, Dhaka University. 299 p.

Narejo, N.T., Salam, M.A., Sabur, M.A. and Rahmatullah, S.M. 2005. Effect of stocking density on growth and survival of indigenous catfish, *Heteropneustes fossilis* (Bloch) reared in cemented cistern fed on formulated feed. *Pakistan J. Zool.* 37 (1): 49-52.

New, M.B. 1987. *Food and feeding of fish and shrimp.* Ext. Man. Aqua. Dev. Coor. Prog. ADCP/REP/87/26. FAO and UNEP. 275 p.

Rahman, M.A., Gheyasuddin, H., Rasid, M.H. and Choudhury, M.F.Z. 1982. Proximate composition and native quality of freshwater Zeol fishes of Bangladesh. *Bangladesh J. Fish.* 2-5: 34-43.

Rahman, M.R and Rahman, M.A. 2003. Studies on the growth, survival and production of calbasu (*Labeo calbasu*, Ham.) at different stocking densities in primary nursing. *Bull. Fac. Sci., Unv. Ryuyus, Jpn.* 76: 245-255.

Roy, N. C., Kohinoor, A.H.M., Wahab, M.A. and Thilsted, S.H. 2002. Evaluation of performance of Carp-SIS Polyculture technology in the rural farmer's pond. *Asian Fish. Sci.* 15: 41-50.

Saha, J.K. 1998. Studies on the induced breeding and post larval rearing of shing, *Heteropneustes fossillis* (Bloch). *Bangladesh J. Fish. Res.* 2: 10-15.

Saha, S.B., Gupta, M.V., Hussain, M.G. and Shah, M.S. 1988. Growth and survival of rohu (*Labeo rohita* ham.) fry in rearing ponds at different stocking densities. *Bangladesh J. Zool.* 16: 119-126.

Thakur, N.K and Das, P. 1986. Synopsis of biological data on Koi, *Anabas testudineus* (Bloch). *Heteropneustes fossilis* Bulletin No. 40, Central Inland Fisheries research Institute, Barrackpore, India. 47 p.

Thakur, N.K, Naser, S.A.K. and Shell, M. 1977. Spawning behaviour of an air breathing cat fish, *Heteropneustes fossilis* (Bloch). *Physiol. Behav.* 19: 53-55.

4

GROWTH AND PRODUCTION PERFORMANCES OF CROSSBRED CLIMBING PERCH KOI, *Anabas testudineus* IN BANGLADESH

A.H.M. Kohinoor*, M.S. Islam, D.A. Jahan, M.M. Khan and M.G. Hussain

Abstract

The study was performed for evaluating the growth and production performances of crossbred climbing perch (*Anabas testudineus*) during August to October 2009. The crossbred groups of fish were produced by crossing between native climbing perch and Thai climbing perch (*A. testudineus*). Growth and production performances was compared at grow out condition between crossbred groups in the two reciprocal crosses of treatments viz. T_1: Native climbing perch (♀) × Thai climbing perch (♂) and T_2: Thai climbing perch (♀)×Native climbing perch (♂. At the same time, they were compared also with their respective control F_1 pure line groups i.e. T_3: Native climbing perch (♀) × Native climbing perch (♂) and T_4: Thai climbing perch (♀) × Thai climbing perch (♂). It was evident from the results that among the treatments in grow out system, the highest mean harvesting weight was 69.25±9.01 g found in T_4 (Thai climbing perch), which was significantly different ($P<0.05$) from all other treatments. Whereas, T_3 (native climbing perch) also showed the lowest mean harvesting weight 33.38±8.74 g and showed significant differences ($P<0.05$) among all the treatments. The mean harvesting weight of both the crossbred groups i.e. T_1 and T_2 were 50.83±6.65 and 59.94±7.83 g, respectively and these results were statistically insignificant ($P>0.05$) between them but significant among all other treatments. The highest (93.60%) and lowest (83.06%) survival rates were observed in Thai koi (T_4) and native Koi (T_3) parental stocks, respectively. Wider variations in terms of productions performances were observed among the treatments, where the gross production of T_1, T_2, T_3 and T_4 were 2,753±96.67, 3,401±73.26, 1,733±168.47 and 4,051±185 kg/ha, respectively. The productions obtained from crossbred groups were at intermediate compared to their control pure line groups. Higher FCR mean value was observed in native control line (T_3), while, lower FCR mean value was observed in Thai control line (T_4).

Keywords: Growth, Crossbred, Climbing Perch Koi

Freshwater Station, Bangladesh Fisheries Research Institute, Mymensingh, Bangladesh

*Corresponding author's email: kohinoor41@gmail.com (A.H.M. Kohinoor)

Reviewed by Dr. Binay Kumar Chakraborty, Matshya Bhabon, Romna, Dhaka, Bangladesh.

Introduction

Climbing perch Koi (*Anabas testudineus*) is an important indigenous fish species of Bangladesh. The fish is very popular for its delicious taste and flavour. This species considered as a valuable item of diet for sick and convalescent. According to Saha (1971), the fish contain high values of physiologically available iron and copper essentially needed for hemoglobin synthesis. Once climbing perch was abundantly available in almost all freshwater systems of Bangladesh, however, recently population of this fish has been declining very rapidly. The reasons for such decline are many, such as ecological degradation, indiscriminate fishing, use of pesticides and fertilizers, destruction of habitats, obstruction to breeding migration, management failure, etc. In the face of diminishing natural population of climbing perch-planners, policy makers, aquaculturists, and fisheries biologists are thinking of its cultivation through intensive farming (DOF, 2002).

A substantial number of seasonal water bodies like ponds, ditches, canals etc. are presently lying fallow and not suitable for carp culture. In this context, our endemic koi fish is the promising species for culture in such waters bodies due to the presence of their accessory respiratory organs, which allow them to thrive in waters with low oxygen levels. During the past few years, natural population of these fishes is rapidly decreasing due to lack of scientific management and draining out of beels, marshes and similar natural habitat to meet up the excessive market demand and also for the production of paddy and other agricultural crops. Moreover, natural breeding grounds are under continuous destruction due to use of excessive fertilizer, insecticides and pesticides. If this situation continues, these endemic local koi will be endangered in near future. In our country, culture of these species has not yet been flourished. Among various reasons, lack of appropriate technology, paucity of fry and fingerlings, and determination of appropriate stocking density are the major setbacks for its culture in the country. The breeding technology of native Koi (*A. testudineus*) had successfully been developed in freshwater Station of the Institute (Kohinoor, 1991). But the growth rate is very slow in comparison to Thai Koi *A. testudineus*. Its slow growth and small size does not favour sustainable production per unit area in a culture system (Kohinoor *et al.*, 2009).

To overcome this situation, another fast growing climbing perch known as Thai koi (*Anabas testudineus*) has been introduced from Thailand in 2002. This strain has some special characteristics such as faster growth rate, shorter culture period, higher survival rate etc. Seed production of this species through artificial propagation technique has been developed (Kohinoor and Zaher, 2006). Introduced Thai *A. testudineus* is fast growing, however, it does not taste like Native *A. testudineus*. Although the Thai *A. testudineus* is presently used for production purposes by large number of farmers but it is not widely accepted by consumers owing to its lack of taste in comparison to native *A. testudineus*. The proposed research activities have been addressed with an objective to develop cross bred Koi through crossing between native and Thai Koi (*A. testudineus*).

Materials and Methods

Crossbred koi fry were produced through crossing between native Koi and Thai koi. Prior to breeding, 400 native Koi having mean weight 48.24±3.26 g were collected from natural sources and reared in a pond (400 m^2). While, 500 Thai Koi (68.20±4.29 g) were collected from Thailand and reared also in same size pond. Supplementary feed containing 30% crude proteins were supplied at 4-8% of estimated body weight in both native and Thai koi. During breeding season (March-May, 2009), four groups of fry were produced through induced breeding. The groups- 1, 2, 3 and 4 were assigned as tretaments- 1, 2 3 and 4, respectively. These treatments are as follows:

T_1 (Group I): Female native koi (♀) × Male Thai koi (♂)
T_1 (Group II): Female Thai koi (♀) × Male native koi (♂)
T_1 (Group III): Female native koi (♀) × Male native koi (♂)
T_1 (Group IV): Female Thai koi (♀) × Male Thai koi (♂)

Production of pure line Thai and native koi fry

Pure line Thai and native koi fry were produced in the hatchery of Freshwater Station, Bangladesh Fisheries Research Institute (BFRI) through induced breeding propagation.

Grow out system

Design of experiment

In this experiment, there were four treatments with three replications. In all the treatment groups, the stocking density was maintained at 62,500 ha^{-1}. In this experiment, four earthen ponds were used. Each pond was equally partitioned by bamboo split into three chambers (100m^2). The details of design of experiment are shown in Table 1.

Table 1. Stocking density in the four treatments

Treatment groups	Stocking density ha^{-1}
T_1: Female native koi (♀) × Male Thai koi (♂)	62,500
T_2: Female Thai koi (♀) × Male native koi (♂)	62,500
T_3: Female native koi (♀) × Male native koi (♂)	62,500
T_4: Female Thai koi (♀) × Male Thai koi (♂)	62,500

Pond preparation

The ponds were dried and cleaned for weed and unwanted animals. The dried ponds were left exposed to sunlight for seven days and then limed at the rate of 250 kg ha^{-1}. Five days after liming, underground water was supplied from a deep tube well to the ponds and filled up to the depth of 1 meter. After three days, the ponds were fertilized with cow dung at the rate of 2,000 kg ha^{-1}.

Fish stocking and management

The ponds were stocked with fish after five days of fertilization. The initial weights of fry were 1.06±0.24, 1.04±0.26, 1.09±0.30 and 1.13±0.28 g, respectively. After stocking, Feeding was done with floating pelted feed containing 30% crude protein @ 4-10 % of body weight twice daily.

Fish sampling

Sampling was done at 15 days interval and random samples of 50 *A. testudineus* fry were caught by seine net from all the treatments to assess the growth and health condition for feed adjustment.

Analyses of water quality parameters

Water samples were collected between 09:00 to 10:00 am at three days intervals of each pond surface to a depth of 20 cm (Azim *et al.*, 2001). A number of clean 500 ml black bottles marked with pond number and sampling date were used for collection of water samples. The bottles were then brought to the Water Quality and Pond Dynamics Laboratory of Freshwater Station, Bangladesh Fisheries Research Institute. Water temperature (°C) from each pond was recorded at least three spots by using a Celsius thermometer. Water transparency was determined by using a standard Seechi disc of 20 cm. diameter. Dissolved oxygen (mg/l) and pH were measured using a digital DO meter (HANNA, model HI 9142) and a direct reading digital pH meter (HANNA, model HI 9025), respectively. Total alkalinity (mg/l) was determined titrimetrically according to the standard procedure and methods (APHA, 1992). Ammonia-nitrogen was measured using a high precision HACH Kit (DR 2000).

Harvesting of fish

After 120 days of rearing, the fish were harvested from all the ponds. Primarily the harvesting of fish was performed by repeated netting using a seine net and final harvesting was done by dewatering the ponds with a submerged low lift pump. During harvest, all fishes were counted and weighed from each pond to assess the survival rate and production

Analysis of experimental data

Data were analyzed using the statistical package, Statgraphics Version 7. ANOVA was performed on all the dependent variables to find out whether treatments had any significant effect on growth. Duncan's New Multiple Range Test (DMRT) was performed to identify any significant difference among treatment means.

Results

Detailed results of the study on growth performances, survival rate, fish biomass, water quality parameters and all other aspects as recorded during the period of study are presented.

Water quality parameter

Mean levels of physicochemical parameters over 120 days of rearing of crossbred and control groups of koi are presented in Table 2.

Temperature is one of the most important physical factors, which influences the physico-chemical and biological environment of a water body. The mean values of water temperature of T_1, T_2, T_3 and T_4 were 27.70±1.32, 27.94±1.67, 27.98±1.46 and 28.19±1.59°C. The variations in temperature among the treatment means were found similar ($P<0.05$).

Water transparency varied between 26 and 34 cm when all ponds and sampling times were considered. It varied a little among the treatments with an average of 25.75±4.75, 24.83±4.12, 28.58±3.84 and 22.91±4.80 cm in T_1, T_2, T_3 and T_4, respectively. These readings were not significantly different when compared using ANOVA. Water transparency varied between 18 and 36 cm when all ponds and sampling times were considered. The water transparency did not show any significant ($P<0.05$) difference among the treatment means. The values of transparency some times varied with sampling dates which could be due to differences in abundances of plankton.

The level of pH varied from 6.89 to 8.11, 6.92 to 8.84, 7.04 to 8.2.3 and 6.86 to 8.25 in T_1, T_2, T_3 and T_4, respectively. The pH in all pond water was alkaline throughout the experimental period which might be due to regular application of lime in all the ponds at monthly interval.

The dissolved oxygen contents in the experimental ponds ranged from 4.24 to 6.80, 4.1 to 6.6, 4.11 to 7.40 and 4.20 to 9.90 mg L^{-1} in T_1, T_2, T_3 and T_4, respectively, with the mean values of 5.30±0.69, 5.36±0.65, 5.57±0.95 and 4.2±0.69 mg L^{-1}. Comparatively higher level of dissolved oxygen as observed in the experimental ponds appeared to be related to sampling time where the dissolved oxygen was monitored at about 9.00-10.00 am.

Total alkalinity ranged from 85 to 145, 92 to 142, 104 to 162 and 94 to 152 mg L^{-1} with mean values of 108.83±13.34, 113.25±14.15, 124.95±13.41 and 116.54±15.58 mg L^{-1} in T_1, T_2, T_3 and T_4, respectively. When the results of all ponds collected over the entire experimental periods were compared, there was no significant difference.

Ammonia-nitrogen is toxic to fish and above a certain level it can cause fish mortality. The range of ammonia-nitrogen was 0.021-0.80, 0.011- 0.79, 0.015-0.64 and 0.01-0.99mg L^{-1} in treatments-1, 2, 3 and 4, respectively. The highest ammonia-nitrogen value was 0.99 in the month of October in treatment-4 and the lowest value was 0.011 in the month of August in treatment-3. The differences among treatments were not significant ($P>0.05$) when compared using ANOVA.

Table 2. Mean values of water quality parameters in different treatments

Water quality parameters	Treatment			
	T_1	T_2	T_3	T_4
Temperature (°C)	27.70±1.32	27.94±1.67	27.98±1.96	28.19±1.82
Transparency (cm)	25.75±4.75	24.83±4.12	28.58±3.84	22.91±4.80
pH	6.89 to 8.11	6.92 to 8.84	7.04 to 8.23	6.86 to 8.25
Dissolved oxygen (mg L^{-1})	5.30±0.69	5.36±0.65	5.57±0.95	4.2±0.69
Total alkalinity (mg L^{-1})	108.83±13.34	113.25±14.15	124.95±13.41	116.54±15.58
NH_3-N (mg L^{-1})	0.29 ± 0.15	0.26 ± 0.13	0.19±0.11	0.34±0.17

Growth and production

Details of stocking, growth, survival, harvesting and production of koi under four treatments are presented in Table 3. It was evident from the results that among the treatment groups in grow-out system, the highest harvesting mean weight 69.25±11.35 g was found in T_4 (Thai climbing perch) which was significantly different ($P<0.05$) from all other treatments. Whereas, T_3 (native climbing perch) also showed the lowest harvesting mean weight (33.38±8.7 g), which was significantly different ($P<0.05$) from all other treatments. The harvesting mean weight of both the crossbred groups i.e. T_1: Native koi (♀) × Thai koi (♂) and T_2: Thai koi (♀) × Native koi (♂) were 50.83±6.65 and 59.94±7.83 g, respectively and these results were not statistically significant ($P>0.05$) between them. The growth of crossbred was at intermediate between the pure line Thai koi and native koi. It is remarkable that, the growth rate of both crossbred groups were significantly higher ($P<0.05$) than that of pure line native koi, while growth rate of both crossbred group of koi showed significantly lower ($P<0.05$) than that of pure line Thai stock. The growth patterns of koi in different months of different treatments are shown in Figure 1, where a clear distinguishable difference in weight was observed in every month among the treatments.

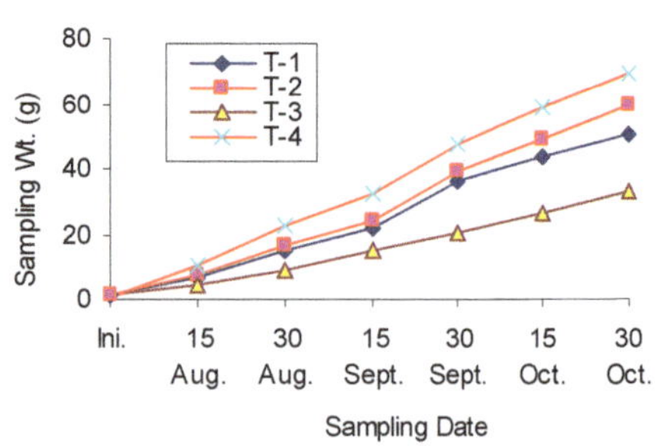

Fig. 1: Sampling Wt. (g) Koi in different treatments

The mean survival rate of koi was found to vary in different treatments. The highest (93.6%) and lowest (83.06%) survivals were observed in Thai koi (T_4) and native Koi (T_3) parental stocks, respectively. The differences among the treatment means of survivals were found to be insignificant ($P>0.05$). The survival of cross bred group i.e. T_1 and T_2 were 86.67 and 90.80%, respectively, where the difference between the treatments are apparent but were not statistically significant. The survival rate of Thai Koi (T_4) showed significant differences with cross bred groups and native koi i.e. T_1, T_2 and T_3.

The specific growth rate (SGR) of koi at different treatments was observed to be 3.23, 3.40, 2.90 and 3.52 for T_1, T_2, T_3 and T_4, respectively. The SGR of native koi in T_3 was significantly ($P<0.05$) different from T_1, T_2 and T_4, whereas, T_1 and T_2 did not show any significant difference ($P>0.05$) between them.

However, the gross production of T_1, T_2, T_3 and T_4 were 2,753±96.67, 3,401±73.26, 1,733±168.47 and 4,051±185 kg/ha, respectively. Wider variations in terms of production performances were observed among the treatments. These results when tested statistically, showed significant differences among the treatments. The highest and lowest production was observed in T_4 (Thai koi) and T_3 (native koi), respectively. The production obtained from cross bred groups of koi showed higher values over native koi, but Thai koi showed higher production over crossbred koi. The production variation between crossbred groups was not significant ($P>0.05$). Production of cross bred groups showed 58-96% higher than that of native koi while 16-33% less than Thai koi.

Table 3. Mean weight, survival and production of Koi under different treatments

Treatment	Stocking Density ha^{-1}	Mean initial Wt. (g)	Mean Harvesting Wt. (g)	Mean survival (%)	Mean FCR	Mean SGR (%)	Mean Production (kg ha^{-1})
T_1	62,500	1.06±0.24	50.83±6.65^b	86.67^b	3.35^c	3.23^b	2,753±96.67^b
T_2	62,500	1.04±0.26	59.94±7.83^b	90.80^b	3.20^c	3.40^b	3,401±73.26^b
T_3	62,500	1.09±0.30	33.38±8.74^d	83.06^c	4. 25^a	2.90^c	1,733±168.47^c
T_4	62,500	1.13±0.28	69.25±9.01^a	93.60^a	2.90^b	3.52^a	4,051±185^a

* Dissimilar superscript indicates significant difference at 5% level of probability

Discussion

All the water quality parameters of the experimental ponds were found to be within the acceptable ranges for aquaculture and there was no abrupt change in any parameter of the pond water. In this experiment, the variations in water temperature among the treatment means were found similar and were within the suitable range of growth of fish in tropical ponds (Rahman *et al.*, 1982; Roy *et al.*, 2002; Begum *et al.*, 2003, Kohinoor *et al.*, 2007; Mondol *et al.*, 2010). Boyd (1982) reported that the range of water temperature from 26.06 to 31.97°C is suitable for fish culture. Normally, the transparency value was low in this experiment because usually koi did not effectively consume plankton (Nargis and Hossain, 1987; Singh and Samual, 1981). Boyd (1982) also recommended a transparency between 15-40 cm as appropriate for fish culture. The pH in all pond water was alkaline throughout the experimental period which might be due to regular application of lime in all the ponds at monthly interval. According to Swingle (1967), pH from 6.5 to 9.0 is suitable for pond fish culture and pH more than 9.5 is unsuitable. Different authors have reported a wide variations in pH from 7.18 to 7.24 (Kohinoor *et al.*, 1998), 7.03 to 9.03 (Roy *et al.*, 2002), 6.8 to 8.20 (Begum *et al.*, 2003) and 7.50 to 8.20 (Chakraborty *et al.*, 2005) in fertilized fish ponds and found the ranges to be productive.

Rahman *et al.* (1982) measured dissolved oxygen range of 3.8 to 10.7 mg L^{-1} from a lake situated at Mymensingh. Wahab *et al.* (1994) recorded dissolved oxygen values ranging from 5.10 to 7.15 mg L^{-1}, while Kohinoor *et al.* (2007) recorded dissolved oxygen values ranging from 4.12 to 6.80 mg L^{-1}. In another study, Mondol *et al.* (2010) also recorded dissolved oxygen ranging from 6.51 to 6.65 mg L^{-1} during their experiment in the pond of Bangladesh Agricultural University Campus, Mymensingh. The variations in total alkalinity in all the treatments were found in productive range for aquaculture ponds (Boyd, 1982; Wahab *et al.*, 1995; Kohinoor *et al.*, 1998 and Kohinoor, 2009). Dewan *et al.* (1991) recorded total ammonia-nitrogen (NH_3-N) values of 0.50 to 6.20 mg L^{-1}, which was less than the present study. Ahmed (1993) found NH_3-N values 0.03 to 0.04 mg L^{-1}, while Azim *et al.* (1995) recorded mean total ammonia-nitrogen of 0.10 mg L^{-1}. Mondol *et al.* (2010) recorded total ammonia- nitrogen range from 0.098 to 0.167 mg L^{-1} in the research ponds of Bangladesh Agricultural University Mymensingh.

The productivity of Thai koi was much more higher than native and cross bred groups. While, the production of crossbred groups were higher than native koi. The mean harvesting weight and survival rate of Thai Koi was higher than all other treatments. For this reason, Thai Koi showed highest production among the treatments. While the mean harvesting weight and survival of crossbred groups exhibited higher values than native koi. Therefore, the production of crossbred groups also higher than native koi but lower than that of Thai koi. The survival rate of all treatments was fairly high. Among the treatments, the lowest survival was observed in native koi (T_3).The main factors that might have attributed to the high survival were the proper stocking of healthy seed, freedom from predation, favourable ecological conditions (physicochemical factors) and also proper feeding etc. High survival rate of climbing perch was reported by many authors (Rao, 1971, Kohinoor *et al.*, 2007, 2009 and Kim Van Van and Vo Qui Hean, 2009). These indicated that climbing perch is an excellent candidate for aquaculture perhaps more so than other culture species.

Thakur and Das (1986) reported that Koi (*Anabas testudineus*) production was 1,800 kg ha^{-1} in India by applying supplementary feed (rice bran, mustard oil cake and fish meal) with the stocking density of 60,000 ha^{-1} in 170 days. They also stated that by applying the above feed, achieved 702 kg ha^{-1} over a period of 330 days, where the stocking density was 1,25,000 ha^{-1}. Earlier study conducted by Akhteruzzaman (1988) evaluated the production potentials of native koi (*Anabas testudineus*) in monoculture management at the density of 16,000 ha^{-1} and obtained a production of 450 kg ha^{-1} in five months rearing with supplementary feed consisted of rice bran (50%), mustard oil cake (30%) and fish meal (20%).

In a trial conducted by Kohinoor *et al.* (2007) investigated on the effects of stocking density on the growth and production of Thai Koi (*Anabas testudineus*) in Freshwater Station, Bangladesh Fisheries Research Institute at three stocking densities (50,000 to 1,25,000 ha^{-1}) by applying supplementary feed (35% crude protein). They observed growth was increased in the treatment with lower stocking density and obtained production 6480 to 6617 kg ha^{-1} in 150 days culture period.

Kohinoor *et al.* (2009) assessed the production potentials of native koi (*Anabas testudineus*) at different stocking densities (50,000 to 62,250 ha^{-1}) in Freshwater Station, Bangladesh Fisheries Research Institute, Mymensingh. They found the highest total net yield of 1,916 ha^{-1} over a period of 5 months culture period by applying supplementary feed containing 35% crude protein. The production obtained in the cross bred group was lower than these mentioned results of Thai koi by different authors. Whereas, the production of native Koi was lower over cross bred groups these also mentioned by different authors (Akhteruzzaman, 1988; Thakur and Das, 1986; Kohinoor *et al.,* 2009).

Conclusion

In native koi, yellowish colour found in the abdomen region, which also observed to some extent in the crossbred group (T_1). Body shapes of cross bred koi (T_1) was more or less same as native koi, while the production of both the two groups of crossbred Koi were comparatively higher over native koi. Organoleptic testing of the four groups of koi revealed that the crossbred koi showed superior taste performances compared to Thai koi. Therefore, in terms of taste in associated with production values, crossbred koi can be a good culture species in Aquaculture Avenue.

References

Akhteruzzaman, M. 1988. A study on the production of koi fish (*Anabas testudineus*) under semi-intensive culture system. *J. Zool.*, 3: 39-43.

APHA. 1992. *Standard Methods for the Examination of Water and Waste Water.* American Public Health Association, Washington DC.

Azim, M.E., Talukder, G.S., Wahab, M.A., Haque, M.M. and Haq, M.S. 1995. Effect of liming and maintenance of total hardness levels on fish production in fertilized ponds. *Progress. Agric.*, 6 (2): 7-14.

Azim, M.E., Wahab, M.A., Van Dam, A.A., Beveridge, M.C.M., Huisman, E.A. and Verdegem, M.C.J. 2001. Optimization of stocking ratios of two Indian major carps, rohu (*Labeo rohita* Ham.) and catla (*Catla catla* Ham.) in a periphyton based aquaculture system. *Aquaculture*, 203:33-49.

Begum, M., Hossain, M.Y., Wahab, M.A. and Kohinoor, A.H.M. 2003. Effects of iso-phosphorus fertilizers on water quality and biological productivity in fish pond. *J. Aqua. Trop.*, 18 (1): 1-12.

Boyd, C.E. 1982. *Water quality management for pond fish culture.* Elsevier Sci. Publ. CO. Amsterdam-Oxford- New York. 318 p.

Chakraborty, B.K., Miah, M.I., Mirza, M.J.A. and Habib, M.A.B. 2005. Growth, yield and returns to *Puntius sarana* (Hamilton) Sharpunti, in Bangladesh under semi intensive aquaculture. *Asian Fisheries Science,* 18: 307-322.

Dewan, S., Wahab, M.A., Beveridge, M.C.M., Rahman, M.H. and Sarker, B.K. 1991. Food selection, electivity and dietary overlap among planktivorous Chinese and Indian major carps fry and fingerlings grown in extensively managed, rain-fed ponds in Bangladesh. *J. Aquaculture and Fisheries Management*, 22: 277-294.

DOF. 2002. *Fisheries fortnight souvenir.* Department of Fisheries, Dhaka, Bangladesh, 87p.

Kim Van van and Vo Qui Hean. 2009. Intensive nursing climbing perch (*Anabas testudineus* (in hapas using pellet feed at different protein levels. *J. Sci. Dev.* (Eng. Tsn.), 7: 239-242.

Kohinoor, A.H.M and Zaher, M. 2006. Breeding of exotic koi (*Anabas testudineus*) at BFRI. *Fisheries Newsletter*, 14 (1): 1-2.

Kohinoor, A.H.M., Akhteruzzaman, M., Hussain, M.G. and Shah, M.S. 1991. Observations on the induced breeding of koi fish, Anabas testudineus (Bloch) in Bangladesh. *Bangladesh J. Fish.,* 14(1-2): 73-77.

Kohinoor, A.H.M., Islam, A.K.M.S., Jahan, D.A., Zaher, M. and Hussain, M.G. 2007. Monoculture of climbing perch, Thai koi, *Anabas testudineus* (Bloch) under different stocking densities at on-farm management. *Bangladesh J. Fish. Res.*, 11(2): 173-180.

Kohinoor, A.H.M., Islam, M.L., Wahab, M.A. and Thilsted, S.H. 1998. Effect of mola (Amblypharyngodon mola Ham.) on the growth and production of carps in polyculture. *Bangladesh J. Fish. Res.,* 2(2): 119-126.

Kohinoor, A.H.M., Jahan, D.A., Khan, M.M., Ahmed, S.U. and Hussain, M.G. 2009. Culture potentials of climbing perch, *Anabas testudineus* (Bloch) under different stocking densities at semi-intensive management. *Bangladesh J. Fish. Res.*, 13(2): 115-120.

Mondal, M.N., Shahin, J., Wahab, M.A., Asaduzzaman, M. and Yang, Y. 2010. Comparison between cage and pond production of Thai Climbing Perch (*Anabas testudineus*) and Tilapia (*Oreochromis niloticus*) under three management systems. *J. Bangladesh Agril. Univ.*, 8(2): 313–322.

Rahman, M.S., Chowdhury, M.Y., Haque, A.K.M.A. and Haq, M.S. 1982. Limnological studies of four ponds. *Bangladesh J. Fish.*, 2-5(1-2): 25-35.

Rao, S.V. 1971. The Indian climbing perch, Koi. *Science Report*, 8 (2): 63.

Roy, N.C., Kohinoor, A.H.M., Wahab, M.A. and Thilsted, S.H. 2002. Evaluation of performance of Carp-SIS Polyculture technology in the rural farmer's pond. *Asian Fisheries Science*, 15: 41-50.

Saha, K.C. 1971. Fisheries of West Bengal. West Bengal Government Press, Alipore, West Bengal, India.

Shingh, K.P and Samuel, P. 1981. Food and feeding habits and gut contents of *Anabas testudineus* (Bloch). *Matsya.*, 7: 96-97.

Swingle, H.S. 1967. Standardization of chemical analyses for waters and pond muds. *FAO Fish. Rep.*, 4(4): 397-421.

Thakur, N.K and Das, P. 1986. Synopsis of biological data on Koi, *Anabas testudineus* (Bloch). Bulletin No. 40, *Central Inland Fisheries research Institute*, Barrackpore, India. 47p.

Wahab, M.A., Ahmed, Z.F., Islam, M.A. and Rahmatullah, S.M. 1995. Effect of introduction of common carp, *Cyprinus carpio* (L), of the pond ecology and growth of fish in polyculture. *Aquaculture Research*, 26: 619-628.

Wahab. M.A., Ahmed, Z.F., Haq, M.S. and Begum, M. 1994. Compatibility of silver carp in the poly culture of cyprinid fishes. *Progress. Agric.*, 5 (2): 221-227.

5

ASSESSMENT OF POND WATER QUALITY OF THAKURGAON SADAR UPAZILA FOR FISH PRODUCTION

M.J. Islam[1*], A. Akter[1], M. Kamrujjaman[2], A. Siddiqa[1], S. Aktar[1]

Abstract

During dry season, physico-chemical properties of waters from 30 ponds of Thakugaon Sadar Upazila, Thakurgaon, Bangladesh were analyzed for its quality and suitability for aquaculture. The variation in the physico-chemical parameters of the aquaculture ponds above or below standard values has potential effects on the health and productivity of aquaculture. Overall, we found that the pond water were acidic to neutral in nature (pH varied from 6.0 to 7.2) and could be suitable for aquaculture. The dissolved oxygen (DO) concentration was suitable for fish production but more DO level should be present for all aquatic life especially for fish production. Chemical oxygen demands (COD) of all pond waters were within the permissible limits for fish production. The temperature values were remained within the standard values in all the aquaculture ponds. Pond water samples contained Ca^{2+}, Mg^{2+} and Na^+ as the dominant cations and HCO_3^- and Cl^- were the dominant anions. All samples were within 'soft' class regarding hardness. Based on Zn^{2+}, Mn^{2+}, Fe^{3+}, SO_4^{2-}, NO_3^- and Cl^- all pond water samples were within the 'safe' limit for fish production during dry period.

Keywords: Pond, Water Quality, Fish Production

[1]Department of Agricultural Chemistry, Hajee Mohammad Danesh Science and Technology University, Dinajpur-5200, Bangladesh

[2]Department of Mathematic and Physics, Hajee Mohammad Danesh Science and Technology University, Dinajpur-5200, Bangladesh

*Corresponding author's email: jahid@mail2world.com (M.J. Islam)

Introduction

Water quality describes the chemical, physical, and biological aspects of water, generally in terms of suitability for aquaculture. Water quality is one of the most overlooked aspects of pond management until it affects fish production. Several important variables influence water quality for fish including water temperature, dissolved oxygen, phytoplankton, photosynthesis, pH, carbon dioxide, ammonia, alkalinity and hardness. Each water quality factor interacts with and influences other parameters, sometimes in complex ways. What may be toxic and causes mortalities in one situation can be harmless in another. At present time, surface waters of Bangladesh are polluting by various ways (Islam *et al.*, 2009).

Water quality would be determined by the weathering of bedrock minerals, atmospheric processes of evapotranspiration and deposition of dust and salt by wind, by the natural leaching of organic matter and nutrients from soil, by hydrological factors that lead to runoff, and by biological processes within the aquatic environment that can alter the physical and chemical composition of water. The ionic composition of surface water is governed by exchanges with the underlying geology of the drainage basin and with atmospheric deposition. Global average concentrations of the four major cations (Ca^{2+}, Mg^{2+}, Na^+, and K^+) and the four major anions (HCO_3^-, CO_3^{2-}, SO_4^{2-}, and Cl^-) in surface water tend to approach patterns in which Ca^{2+} concentrations dominate the cations and HCO_3^- and/or CO_3^{2-} concentrations dominate the anions (Wetzel, 2001). The consequences of eutrophication for humans are bad taste and odour events in public water supplies, and production of cyanobacterial toxins that can threaten animal and human health (Ongley, 1996). Elevated levels of heavy metals in aquatic systems have resulted from a number of land use activities including agriculture, urbanization, impoundments, mining, and industrial activities. Heavy metals effects include reduced growth rates, impaired reproduction, and sometimes death. Bioconcentration and bioaccumulation of these elements in the food chain can put terrestrial consumers, including humans, at risk. In humans, prenatal exposure to high mercury levels, particularly in fish-eating populations, has been associated with developmental problems related to the central nervous system (WHO, 2004). However, it is essential to assess the quality of surface water as well as its suitability for fish production. Therefore, the objectives are to assess the water quality of ponds of Thakurgaon sadar upazila for fish production.

Materials and Methods

Water samples were collected from 30 ponds of Thakurgaon Sadar Upazila between May and June 2012. Table 1 indicates the sampling details of the ponds during the study period. Samples were collected in 10 liter plastic bottles that had been cleaned with hydrochloric acid (1:1) and then rinsed with tap water followed by rinsing with distilled water. Before collecting each sample, plastic bottles were rinsed 3 to 4 times with sample. All reagents used in chemical analysis were of analytical grade. Samples were analyzed in Department of Agricultural Chemistry, Hajee Mohammad Danesh Science and Technology University, Dinajpur.

Table 1. Information regarding sampling ponds of the Thakurgaon Sadar Upazila under the District of Thakurgaon, Bangladesh; location, depth, and the duration of uses

Sample No.	Sampling sites	Depth (m)	Duration of uses (year)
1-4	Akcha	2.5±0.58	44.25±9.94 (30-52)
5-8	Akhanagor	2.75±0.75	42.50±6.45 (35-50)
9-12	Begunbari	2.50±0.50	59.25±2.50 (52-70)
13-16	Chilarong	2.50±0.75	50.75±4.57 (45-56)
17-20	Gorea	2.75±0.50	36.25±8.54 (25-45)
21-24	Jogonnatpur	2.50±0.70	30.75±10.04 (20-36)
25-28	Jamalpur	2.50±0.50	25.00±2.16 (24-28)
29-30	Mohammadpur	2.75±0.75	32.50±10.60 (25-40)

For assessing the suitability classes for fish cultivation, we measured pH, EC (electrical conductivity), DO, COD_{Mn}, alkalinity, TDS (total dissolved solids), Ca^{2+}, Mg^{2+}, Na^+, K^+, Zn^{2+}, Cu^{2+}, Mn^{2+}, Fe^{3+}, CO_3^{2-}, HCO_3^-, SO_4^{2-}, NO_3^- TN (total nitrogen), NO_2^-, NH_3, PO_4^{3-}, TP (total phosphate) and Cl^-. Temperature was measured by thermometer. The pH (HANNA pH 211) and electrical conductivity (EC) were determined electrometrically (APHA, 1998). DO was measured by DO meter (Milwaukee MW 600). For COD_{Mn}, 30-mL samples were treated with 5 mN $KMnO_4$ in 1% NaOH for 1 hour at 100°C in an autoclave. A total dissolved solid (TDS) was measured by drying and weighing method. K^+ and Na^+ were estimated by flame emission spectrophotometry. Ca^{2+} and Mg^{2+} were analyzed by complexometric titration. Alkalinity was determined by measuring the amount of acid (hydrochloric acid) needed to bring the sample to a pH of 4.5. Chloride was estimated by argentometric titration (APHA, 1998). SO_4^{2-} was determined turbidimetrically. CO_3^{2-} and HCO_3^- were analyzed titrimetrically. Zn^{2+}, Cu^{2+}, Mn^{2+} and Fe^{3+} were analyzed by atomic absorption spectrophotometry (APHA, 1998) in the Soil Resources Development Institute, Dinajpur, Bangladesh.

The dissolved inorganic phosphate (DIP) i.e., PO_4^{3-} was analyzed after GF/F filtration (0.45 μm) by applying the molybdenum blue method at 880 nm, according to the American Public Health Association (APHA, 1998). The total phosphorus (TP) was analyzed from the unfiltered sample as the DIP after persulfate digestion and determined after the Mo-blue method, according to APHA (1998). The total N (TN) was determined by the cadmium-reduction method after persulfate digestion, using a flow-injection auto analyzer (SKALAR). The nitrate level was determined with an autoanalyzer by the Cd-reduction method. The Nessler method (Drum *et al.*, 2002) was used for measurement of ammonia. Ammonia-N was measured by colorimetric nesslerization method after distillation of samples. Nitrite-N was determined by colorimetric method after formation of a reddish purple azo dye produced at pH 2.0 to 2.5 by coupling diazotized sulfanilamide with N-(1-naphthyl)-ethylenediamine dihydrochloride (NED dihydrochloride). Nitrate-N was determined by colorimetric method in the ultraviolet range. Calculation of hardness was performed by the following equation of $H_T = 2.5 \times Ca^{2+} + 4.1 \times Mg^{2+}$ (Freeze and Cherry, 1979) where concentrations of ionic constituents are in mg/L.

Results and Discussion

Physico-chemical characteristics of water samples

Summary results from our survey of pond waters are shown in Tables 2, 3 and 4. The water temperatures ranged from 21.5 to 27.6°C and the pH of the samples ranged from 6.0 to 7.2 (Table 2). DO concentrations varied from 6.4 to 8.9 mg O_2/L. In this investigation, the CODMn values of the water samples ranged from 3.2 to 7.9 mg O_2/L ((Table 2). Higher CODMn values indicate that waters of these ponds were to some extent polluted with non-biodegradable chemical pollutants. The EC and TDS ranged from 50 to 490 µS/cm and 32 to 274 mg/L, respectively.

Table 2. pH, EC, DO, TDS, CODMn, alkalinity and hardness of surface water of Thakurgaon Sadar Upazila during the study period

	Temp (°C)	pH	DO mgO_2/L	COD_{Mn} mgO_2/L	EC µS/cm	TDS mg/L	Alkal mg/L	Hardn mg/L
Min	21.5	6.0	6.4	3.2	50	32	18	28
Max	27.6	7.2	8.9	7.9	490	274	88	212
Mean	25.3	6.5	7.7	5.6	145	89	32	70
SE(±)	0.3	0.1	0.1	0.3	16	9	3	8

Phosphate (PO_4^{3-}) level of water samples was in between 0.011 and 0.270 mg/L (Table 3). Nitrite-nitrogen values ranged between 0.01 and 0.10 mg/L while nitrate levels in surface water varied from 3.64 to 10.50 mg/L. On the other hand extensive epidemiological data support the current guideline values for NO_3^--N of 10 mg/L (WHO, 2004). Higher concentrations of NO_3^--N and trace amount of ammonia nitrogen were noticed in all ponds. As most of the study areas were in intensive irrigation, the fertilizers used for agriculture may be the source for the elevated concentration of nitrate in a few locations (Chandna *et al.*, 2010). During the present study, major chemical contents such as NO_3^--N, NO_2^--N and phosphorus values were within suitable level from pollution point of view. The TN: TP ratios ranged from 10~98:1. A recent mathematical model has suggested that the origin of Redfield's N: P = 16 (Redfield *et al.*, 1963) could be the consequence of the most advantageous ratio of protein: RNA under optimal growth conditions for microbes and algae (Loladze and Elser, 2011).

Concentrations of Na^+ and K^+ ranged from 0.12 to 1.82 meq/L and 0.06 to 0.87 meq/L, respectively. K^+ concentrations were generally lower than Na^+ concentrations. Ca^{2+} and Mg^{2+} were major cations in pond water and ranged from 0.20 to 3.89 meq/L and 0.85 to 4.25 meq/L, respectively (Table 4). An appreciable amount of HCO_3^- was present in all water samples, though CO_3^{2-} was negligible in most cases. The concentrations of HCO3- were from 0.85 to 6.20 meq/L while Cl^- and SO_4^{2-} concentrations ranged from 0.55 to 2.10 meq/L and 0.03 to 1.67 meq/L, respectively (Table 3). The order of the relative abundance of major cations, expressed in percent of meq/L, was $Mg^{2+} > Ca^{2+} > Na^+ > K^+$ while that of the anions was $HCO_3^- > Cl^- > SO_4^{2-}$. The concentration of earth alkalis elements (Ca and Mg) represents 75% of total cations; this shows a high rock/water interaction in dry season. Fe^{3+}, Cu^{2+}, Zn^{2+} and Mn^{2+} concentrations varied from 0.11 to 0.23 mg/L, 0.004 to 0.047 mg/L, 0.033 to 0.659 mg/L and 1.00 to 1.42 mg/L, respectively (Table 4). Hardness varied from 28 to 212 mg/L (Table 2).

Suitability based on pH, alkalinity and hardness for fish production

In the study area, the pH range recorded from 6.0 to 7.2. The pH of 6.0 to 8.5 is acceptable for most pond life (Alabaster and Lloyd, 1980). Fish can become in stress in water with a pH ranging from 4.0 to 6.5 and 9.0 to 11.0. Long term conditions above 9.0, can cause kidney damage to the fishes. When pH rises over 11, the gills, lens and cornea of fish eyes are destroyed (Jhingran, 1988). Lower pH increases the toxicities of hydrogen sulphide (H_2S), copper and other heavy metals to fish.

Alkalinity found to range from 18 to 88 mg/L. Alkalinity is water's ability to resist changes in pH and is a measure of the total concentration of bases in pond water including carbonates, bicarbonates, hydroxides, phosphates and borates. A suitable range of alkalinity is 50 to 300 ppm (APHA, 1998; Boyd and Tucker, 1992). In the study area, hardness ranged from 28 to 212 mg/L. Calcium and magnesium are essential to fish for metabolic reactions such as bone and scale formation. Total alkalinity was high and total hardness was low, pH was risen to extremely high levels during periods of rapid photosynthesis. Desirable level of total hardness and total alkalinity for fish culture generally was fallen within the range of 20-300 ppm. Total alkalinity and total hardness raised by liming (Joseph *et al.*, 1993).

Suitability based on temperature, DO, TDS and heavy metals for fish production

In the study area, the temperature recorded to range from 21.5-27.6°C. However, the ideal range of temperature is between 20°C-25°C and the acceptable range is 2°C-30°C (Alabaster and Lloyd, 1980). In study area it was recorded that water temperature increased in the pond water,, the toxicity of ammonia was increased and the amount of dissolved oxygen decreased. Water temperature influenced the onset of fish spawn, aquatic vegetation growth and the biological demand for oxygen in ponds. In addition, plants and animals used more oxygen due to increased respiration rates. These factors commonly resulted in less available oxygen for fish during the study period.

The value of DO of the study area varied from 6.4 to 8.9 mg/L (Table 2). This range was suitable for fish culture in the study area (Jhingran, 1988). He found that growth and production is optimum at more than 5 mg/L in pond culture. Above 5 mg/L, almost all aquatic organisms can survive indefinitely, provided the allowable limits of other environmental parameters. Higher concentrations of phosphorus, nitrogen and iron can occur in the deeper portions of ponds when anoxic conditions convert bound and solid forms in sediments into soluble forms thus released into the water column.

Total dissolved solids (TDS) ranged from 32 to 274 mg/L (Table 2). The high amount of dissolved solids in water increases the water density; it influences osmoregulation of freshwater organisms and reduces solubility of gases. In the study area, Fe content varied from 0.11 to 0.23 mg/L and was within the limit of the Department of Environment (DOE, 2003). The concentration of Cu and Mn of the study area were also within the limit of DOE (2003). The content of Zn ranged from 0.033 to 0.659 mg/L. The maximum recommended concentrations of Fe, Cu, Zn and Mn are 2.0, 0.5, 5.0 and 5.0 mg/L, respectively (DOE, 2003).

Suitability based on ammonia, nitrite, nitrate, chlorine and phosphorus concentrations

In the study area, the concentrations of ammonia varied from 0.23 to 2.72 mg/L (Table 3). According to Alabaster and Lloyd (1980), ammonia poisoned fish congregates close to the water surface, gasp for air and are restless. In some cases, hemorrhages occur mainly at the base of the pectoral fins. Fishes are able to withstand levels of unionized ammonia of up to 0.6 to 2 mg/L for only short periods (Alabaster and Lloyd, 1980).

In the study area, the concentrations of nitrite ranged from 0.01 to 0.10 mg/L. Nitrite can be deadly, particularly to the smaller fishes, in concentrations as low as 0.25 ppm. Vamos and Szollozy (1974) documented that in nitrite poisoned fish, a brownish colour of blood on the gills is indicative of the increase of methaemoglobin, as nitirite bound to haemoglobin giving rise to methaemoglobin that reduces the oxygen transporting capacity of the blood.

Concentrations of nitrate from zero to 200 ppm are acceptable. The NO_3^- concentration in the water samples of the study area ranged from 3.64 to 10.50 mg/L (Table 3) which were slightly higher than the limit of 5 mg/L (UCCC, 1974) that may be due to continuous uses of nitrogenous fertilizers for agricultural field crops.

Table 3. TN: TP and anionic constituents of pond water of Thakurgaon Sadar Upazila during the study period

	NH_3 Mg/L	NO_2^- mg/L	NO_3^- mg/L	TN mg/L	Cl^- Meq/L	HCO_3^- meq/L	SO_4^{-2} meq/L	TP mg/L	PO_4^{-3} mg/L	TN: TP
Min	0.23	0.01	3.64	5.65	0.55	0.85	0.03	0.081	0.011	10
Max	2.72	0.10	10.50	14.25	2.10	6.20	1.67	0.84	0.27	98
Mean	1.24	0.04	6.66	9.05	1.03	2.71	0.31	0.27	0.05	49
SE(±)	0.10	0.004	0.28	0.38	0.06	0.27	0.07	0.04	0.01	4.76

Even in very small concentrations of chlorine ($<$ 0.5 mg/L), it burns the edges of the gills of fish and kills the beneficial nitrifying bacteria in the pond. The Cl concentration was much lower than the DOE standard. All the chloride (Cl^-) values were within the permissible limits of 200 mg/L (Gupta, 2005). The buccal spasm hinders respiration, so that the fish suffocate and ultimately dies (Bohl, 1989).

The total phosphorus content of the study area ranged from 0.081 to 0.840 mg/L (Table 3), these high levels of phosphate originated from municipal wastewater discharges from nearby locality or from sediments of ponds. The increase in nutrients of water was spurted in algal productivity i.e., eutrophication. Bird droppings, natural run-off, fall of leaves and twigs from surrounding vegetation, continuous discharge of domestic waste, bathing, washing were principal sources of N, P, K in the wastewater of the study area.

Correlations among the parameters

The correlation matrix of 14 parameters, for the 30 samples in the study area is shown in Table 5. There were few significant relationships among the measured concentrations. The high correlations between Cl^- and Ca^{2+}(r=0.86), Cl^- and Mg^{2+} (r=0.80), alkalinity and hardness (r=0.89), and Cl^- and Na^+ (r=0.54) indicating that they most likely derive from the same source of water (Table 5). For most of the ions however, there were no significant correlation between them. The high correlation between EC and TDS (r=0.99) reflects the interdependency of these measurements as general measures of the amount of total dissolved solutes.

Table 4. Cationic composition of pond water samples of Thakurgaon Sadar Upazila

	Ca^{2+} Meq/L	Mg^{2+} meq/L	Na^+ meq/L	K^+ meq/L	Cu^{2+} mg/L	Zn^{2+} mg/L	Mn^{2+} mg/L	Fe^{3+} mg/L
Min	0.20	0.85	0.12	0.06	0.004	0.033	1.00	0.11
Max	3.89	4.25	1.82	0.87	0.047	0.659	1.42	0.23
Mean	1.21	1.98	0.70	0.38	0.028	0.136	1.25	0.15
SE(±)	0.15	0.16	0.08	0.05	0.002	0.04	0.02	0.01

Table 5. Correlation matrix of the physico-chemical characteristics of pond water

	Temp	EC	TDS	Cl	HCO_3	SO_4	NO_3	Ca	Mg	Na	K	TP	Alka	H_T
Temp	1.00													
EC	-0.04	1.00												
TDS	-0.03	0.99	1.00											
Cl	-0.04	0.42	0.39	1.00										
HCO_3	0.19	0.33	0.34	-0.05	1.00									
SO_4	0.20	0.09	0.09	0.09	0.15	1.00								
NO_3	-0.01	0.14	0.16	0.04	-0.04	-0.26	1.00							
Ca	0.09	0.30	0.28	0.86	0.18	0.28	0.09	1.00						
Mg	-0.24	0.52	0.50	0.80	0.17	0.06	-0.12	0.58	1.00					
Na	0.34	0.40	0.37	0.54	0.29	0.47	0.07	0.57	0.27	1.00				
K	0.31	0.15	0.12	0.45	0.15	0.20	0.03	0.53	0.07	0.44	1.00			
TP	0.02	0.17	0.14	0.15	-0.13	-0.02	0.17	0.18	-0.06	0.13	0.33	1.00		
Alka	-0.05	-0.06	0.01	0.25	0.16	0.17	0.34	0.42	0.03	0.31	0.26	0.10	1.00	
H_T	0.02	-0.02	-0.04	0.37	0.28	0.22	0.34	0.54	0.19	0.45	0.26	0.07	0.89	1.00

Values are significant at p =0.05 where r ≥ 0.36

Conclusions

Nitrate shows an elevated concentration in most of the ponds. Fertilizer application in agricultural fields and decay of organic matter are the probable origin of these nitrate concentration. These anions and cations concentrations might be helpful for algal production. Based on different water quality parameters, we concluded that all the pond water samples of the Thakurgaon Sadar Upazila, Bangladesh were suitable for fish production. These results illustrate the necessity of improving our knowledge of pond water quality in the study area. Especially for fish production, continued monitoring of physical, chemical and biological indicators of water quality is needed.

Acknowledgements

The author is grateful to the Director, Research and Training Division and Chairman of the Department of Agricultural Chemistry, Hajee Mohammad Danesh Science and Technology University, Dinajpur-5200, Bangladesh for providing some research grants to conduct this research.

References

Alabaster, J.S. and Lloyd, R. 1980. Water quality criteria for freshwater fish. Butterworths. 297 p.

APHA. 1998. Standard methods for the examination of water and wastewater. 20th edn., American Public Health Association, 1015 Fifteenth Street, NW, Washington, DC. pp. 3-103.

Bohl, M. 1989. Optimal water quality-basis of fish health and economical production. Current trends in fish therapy. Deutsche Veterinarmedizinische Gesellschaft e.v., Giessen. pp. 18-32.

Boyd, C.E. and Tucker, C.S. 1992. Water quality and pond soil analyses for aquaculture. Auburn University. 183 p.

Chandna, P., Khurana, M.L., Ladha, J.K., Mehla, R.S. and Punia, M. 2010 Factors affecting the concentration of nitrates in groundwater in the rice-wheat cropping system of India: A geo-spatial approach. Poster presented at the ASA/CSSA/SSA annual meetings. Long Beach, USA. 705 p.

DOE. 2003. A compilation of environmental laws of Bangladesh. Department of Environment pp. 212-214.

Drum, D.A., Bauman, S.L. and Shugar, G.J. 2002. Environmental field testing and analysis ready reference handbook. McGraw-Hill. 800 p.

Freeze, A.R. and Cherry, J.A. 1979 *Groundwater*, Prentice Hall Inc. Englewood Cliffs, New Jersey 07632. pp. 84-187.

Gupta, P.K. 2005. Methods in environmental analysis: water, soil and air. Agrobios Jodhpur, India. pp. 13-73.

Islam, M.S., Mahmud, M.S., Chowdhury, N. and Shamsad, S.Z.K.M. 2009. Water and sediment anality of Fay's Lake, Chittagong. *Dhaka Univ. J. Bio. Sci.* 18: 147-158.

Jhingran, V.G. 1988. Fish and fisheries of India. Hindustan Publishing Corporation, Delhi, India. pp. 79-100.

Joseph, K.B., Richard, W. and Daniel, E.T. 1993. An introduction to water chemistry in fresh water aquaculture. NRAC Fact sheet No. 170-1993. pp. 15-23.

Loladze, I. and Elser, J.J. 2011. The origins of the Redfield nitrogen to-phosphorus ratio are in a homoeostatic protein-to rRNA ratio. *Ecol Lett.* 14: 244-250.

Ongley, E.D. 1996. Control of water pollution from agriculture: FAO irrigation and drainage paper 55. Food and Agriculture Organization of the United Nations: Rome. 53 p.

Redfield, A.C., Ketchum, B.H. and Richards, F.A. 1963. The influence of organisms on the composition of sea-water. *In:* Hill MN (ed) The sea, vol 2. Wiley, New York. pp. 26-77.

UCCC. 1974. Guidelines for interpretations of Water Quality for Irrigation. Technical Bulletin, University of California Committee of Consultants, Cafornia, USA. pp. 20-28.

Vamos, R. and Szollozy, G. 1974. There is noit a danger of ammonia intoxication of fish if there is enough oxygen in water. Halaszat, 20, No. 4: 124 (in Hungarian).

Wetzel, R.G. 2001. Limnology. Lake and River Ecosystems. 3rd Edn, Academic Press: San Diego, USA. pp. 169-186.

WHO. 2004. *Guidelines for Drinking Water Quality*, Third Edition. Volume 1: Recommendations. World Health Organization, Geneva. 402 p.

CHEMICAL FRACTIONATION OF CU AND ZN AND THEIR IMPACTS ON MICROBIAL PROPERTIES IN SLIGHTLY CONTAMINATED SOILS

Liu Aiju[1]*, Wang Honghai[2], Gao Peiling[1], Xu Hong-zhi[3]

Abstract

Chemical fractionation of Cu and Zn in bulk soil and its effects on soil microbial properties were determined in Cu and Zn contaminated soils (Cu: 35.57~46.37 $mg \cdot kg^{-1}$, Zn: 74.33~127.20 $mg \cdot kg^{-1}$) sampled from an agricultural field in outskirts of Zibo, China during the month of September, 2011. A sequential extraction technique (SET) was used for metals chemical fractionation analysis in soils and a correlation analysis was applied to determinate the effects of metal on soil microbial properties. Chemical speciation showed that Cu and Zn were mostly present in the residual fraction and their concentrations in the most labile fraction (acid soluble fraction) were the lowest in the investigated soils. However, the correlation analysis indicated that the labile forms of Cu/Zn, such as its acid soluble, reducible or oxidizable fractions, were usually significantly negatively correlated with the tested microbial activities at 0.05 or 0.01 probability levels. These results indicate that the metal labile fractions could exert an inhibitory effect on the soil microbial parameters even in the minor contaminated soils.

Keywords: Chemical Fractionation, Heavy Metals, Microbial Biomass, Soil Enzymes

[1]School of Recourse and Environmental Engineering, Shandong University of Technology, 12 Zhangzhou Road, Zibo 255049, China
[2]Zibo entry-exit inspection and quarantine bureau, 63 Liuquan Road, Zibo 255033, China
[3]Analysis & Testing Center, Shandong University of Technology, 12 Zhangzhou Road, Zibo 255049, China

*Corresponding author's email: liuajnju@yahoo.com.cn (Liu Aiju)

Introduction

Soil microbial biomass and enzyme activities are considered to be good bio-indicators reflecting natural and anthropogenic disturbances, and evaluating soil enzyme activities is one of the cheapest and easiest techniques that can be used to evaluate soil pollution (Hinojosa *et al.,* 2004; Khan *et al.,* 2007). A considerable body of information has been accumulated in the last century on the inhibition of heavy metals on soil microbial activities and the factors that affect this inhibition (Karaca *et al.*, 2010). Nevertheless, an enormous disparity of results is evident from the studies, in particular such as the concentration to which metal is toxic (Ge *et al.,* 2000; Li *et al.,* 2001). These discrepancies are due to factors, which modify the toxicity of the metals and to the differences in sensitivity of soil enzymes (Giller *et al.,* 2009; Karaca *et al.*, 2010).

Beside the environmental factors, microbial toxicity of heavy metals had been attributed to the two factors: metal concentration and bioavailability. The available forms of a metal are significant when attempting to understand metal toxicity, and its available forms are related to its chemical forms in the soil (Wang *et al.,* 2007a, b). Several workers studied the correlation between metal forms and microbial activities in long-term contaminated soils (Wang *et al.,* 2007a, b; Bhattacharyya *et al.,* 2008), but not much information is available on the relationship of different metal fractions and soil microbial activities in slightly contaminated soils with trace metals.

Present work aimed to study chemical fractionation of Cu and Zn in bulk soil and its effects on soil microbial biomass and enzymes activities in minor Cu and Zn contaminated soils.

Materials and Methods

Study area and samples collection

The study sites are situated at the agricultural fields grown with wheat and corn in the outskirts of Zibo, China (36°48′ N, 118°03′E). Four sites were selected, of which the site 1, 2 and 4 lie in district of Zibo city, the site 3 is near to industrial area.

Three surface soil samples from each of the 4 selected study sites were collected during the month of September, 2011. Altogether 12 soil samples were brought to the laboratory in the sealed polythene bags. The soils were passed through a 2 mm sieve and dispersed well. A portion of the soil samples was air-dried for chemical analysis. The others were store at 4°C for microbial analysis, which was completed within two weeks.

Physico-chemical analysis of soils

The pH was determined in soil: water @ 1:2.5 suspensions. Organic carbon (OC) and total metal contents were determined by the method of Lu (1999). Soil texture was determined with a laser particle size analyzer (Model: Rise-2028, China).

Fractionations of metals

Chemical fractionation of trace metal Cu and Zn in soil was carried out following a sequential extraction technique (Quevauviller *et al.*, 1997). To explain the method briefly, 1.0 g of soil samples (in dry weight) were sequentially extracted with 0.11 M CH_3COOH (16 h shaking, 120 rpm) for acid soluble fraction, with 0.1 M $NH_3{\cdot}HCl$ (pH = 1.5, 16 h shaking, 120 rpm) for reducible fraction, with 8.8 M H_2O_2 (pH=2) at room temperature (1 h, shaking, 120 rpm) and 85 □(1 h, shaking,120 rpm), and then added 1M NH_4COOH (pH = 2, 16 h shaking, 120 rpm) for oxidizable fraction, and finally with concentrated HNO_3 at 105°C for residual faction. As concentrations in the extracts were measured by ICP-MS (7500ce, Agilent, USA) and the results were expressed on a moisture-free basis.

Microbiological analysis

Microbial biomass carbon (MBC) content was determined by the fumigation extraction method (Gregorich *et al.,* 1990) using K_{EC} (K_{EC} is the calibration factor) factor of 0.38 (Joergensen, 1995). Urease, Dehydrogenase and arylsulphatase activities of soils were estimated following the methods suggested by Tabatabai and Bremner (1970, 1972); Trevors (1984); García *et al.* (1993) respectively. These measurements were performed using the field moist soils. However, the parameters were expressed on a moisture-free basis.

Statistical analysis

The significance of differences among the parameters was tested by two-way analysis of variance (ANOVA) followed by Tukey's test. Pearson correlation analysis and partial correlation analysis were performed to investigate the forms of metals suppressing the soil microbial activities in the soil. All statistical analyses were carried out with the program SPPS 17.0 for Windows.

Results and Discussion

Physico-chemical properties of soil

Table 1 shows the properties of the sampled soils. The soils were slightly alkaline in nature as the pH values ranged from 7.39 to 7.54. Organic carbon contents of the soil samples were ranged from 7.76% to 8.70%, which are somewhat lower than or comparable with other agriculture soils in China (Meng *et al.*, 1996; Dai *et al.*, 2009) and the clay contents were ranged from 45.53% to 48.87%. No differences were found between the investigated sites on the tested soil properties.

Table 1. Soil characteristics and heavy metals contents at each sampling sites

Sample sites	pH	OC (g/kg)	Clay (%)	Cu ($mg{\cdot}kg^{-1}$)	Zn ($mg{\cdot}kg^{-1}$)
1	7.52±0.06a*	8.30±0.11a	45.53±2.30a	41.23±1.50a	83.33±1.53a
2	7.54±0.09a	8.02±0.14a	48.87±1.89a	35.17±2.76b	121.01±5.31b
3	7.44±0.01a	8.50±0.10a	57.73±3.02a	46.37±0.95c	127.20±4.58b
4	7.39±0.03a	8.76±0.59a	46.90±1.31a	42.20±0.95a	74.33±3.61c
Background values[b]	--	--	--	20.4±10.05d	60.6±18.88d

Mean values followed by the same letter are not significantly different according to ANOVA and multiple comparisons with Tukey test ($p < 0.05$); [b] the background values of soils (Luo, 1994).

Total amount of Cu and Zn in soils ranged from 35.17 to 46.37 $mg{\cdot}kg^{-1}$ and from 74.33 to 130.67 $mg{\cdot}kg^{-1}$, respectively, and both were highest in the site 3. However, in the present study, soil concentrations of Cu and Zn are generally lower than those reported in other Cu/Zn-contaminated sites (Li *et al.,* 2008), although they are significantly higher than their soil background values.

Chemical speciations of Cu and Zn in soils

Soil Cu and Zn were present in various forms from easily leachable to recalcitrant ones due to interactions with various soil components. Consequently, the total concentrations cannot provide a precise index representing the influence of Cu and Zn on soil microorganisms and enzyme activities (Kunito *et al.,* 2001; Sun *et al.,* 2006). Distribution of Cu and Zn forms in the investigated soils is given in Fig. 1.

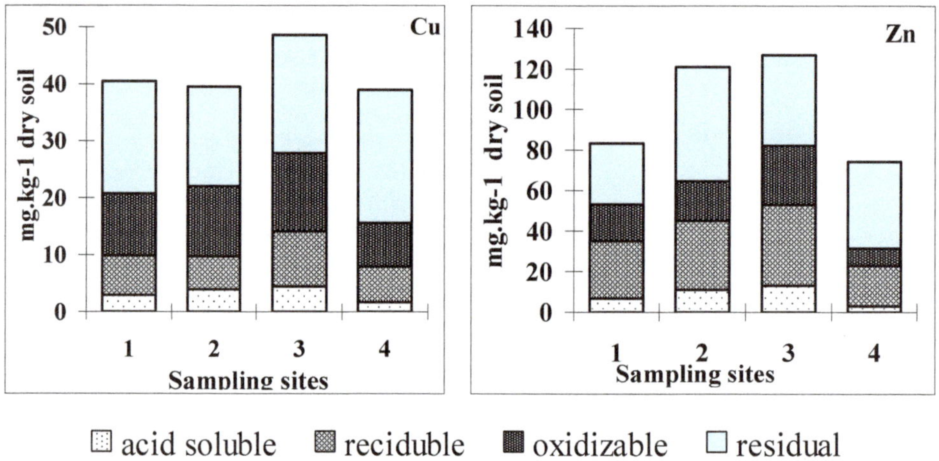

Fig. 1. Form patterns of Cu and Zn in the studied soils

It was shown that Cu and Zn were mainly present residual fractions in the investigated soils, and their concentrations in the most labile fraction (acid soluble fraction) were the lowest. In addition, the oxidizable Cu fraction and the reducible Zn fraction possessed the highest level among their three bioavailable fractions (acid soluble-, reducible- and oxidizable fraction), respectively. These are consistent with those of other reports (Bhattacharyya *et al.*, 2008, Kunito *et al.*, 2001; Saeki and Okazaki, 1993; Saeki *et al.*, 1993). The high percentages of reducible-Zn and oxidizable-Cu might indicate that the labile fractions of soil Cu or Zn would potentially increase in the future by reductive dissolution of Fe (III)/Mn (IV) oxides and mineralization of organic moieties under an appropriate condition in the investigated area (Lagomarsino *et al.*, 2011). Finally, in all the investigated sites, the sum of three bioavailable fractions of Cu/Zn, was highest in the site 3 and the lowest in the site 4. These trends are consistent with that of their total amount (Table 1). Some studies also reveal that total metal encompasses all metal forms including water soluble and exchangeable ones in soils (Bhattacharyya *et al.*, 2008).

Soil microbial activities

Soil enzymes and microbial biomass carbon are used to estimate the adverse effects of various pollutants on soil quality (Dick, 1997). Some significant variations in soil mcirobial activities between the soils were indicated in this study (Table 2).

Table 2. Microbial biomass carbon and enzyme activities in each soil

Parameter	1	2	3	4
MBC ($mg\cdot kg^{-1}$)	17.32±0.75a*	16.82±0.26a	14.85±0.33c	19.18±0.39d
Urease (mmol NH4$kg^{-1}\cdot 5h^{-1}$)	1.84±0.04a	1.94±0.08b	1.05±0.08c	1.99±0.02b
Dehydrogenase (mmol INTF $kg^{-1}\cdot 2h^{-1}$)	7.83±0.21a	7.41±0.20a	7.05±0.14b	7.86±0.15a
Arylsulfatase (mmol p-nitrophenol $kg^{-1}\cdot h^{-1}$)	43.21±1.60a	37.35±1.21b	36.03±0.90b	41.71±0.92a

*Mean values followed by the same letter are not significantly different according to ANOVA and multiple comparisons with Tukey test ($p < 0.05$)

The MBC in the tested soils ranged from 104.85 to 119.18 $mg\cdot kg^{-1}$, being lowest in the site 3 (Table 2). The tested soils enzymes also showed the significantly lower activities in soils from the site 3 than that from other sites. This indicated the inhibitory effect of metals in the contaminated soils on the production of the enzymes, which were conformed by the Pearson-correlation analysis results showed in Table 3. Others, activities of the studied soil enzymes and soil MBC were all almost significantly and positively correlated with soil pH and OM at 0.05 or 0.01 level of statistical significance (Table 3). Therefore, beside of Cu and Zn, soil organic matter and pH are also key factors for microbial growth and enzyme synthesis in the investigated soils (Karaca *et al.*, 2010).

Table 3. Pearson-correlation analysis results between studied parameters

	Urease	Dehydrogenase	Arylsulphatase	MBC
A1	-.919**	-.873**	-.741**	-.959**
B1	-.365	-.628*	-.664*	-.445
C1	-.903**	-.890**	-.776**	-.968**
Cu	-.647*	-.838**	-.806**	-.735**
A2	-.808**	-.739**	-.652*	-.904**
B2	-.676*	-.659*	-.594*	-.870**
C2	-.410	-.404	-.374	-.132
Zn	-.455	-.674*	-.657*	-.606*
pH	0.839**	0.820**	0.809**	0.670*
OM	0.760**	0.889**	0.807**	0.880**

*,** Correlation is significant at 0.05 and 0.01 levels, respectively; the same below

Influence of metals on soil microbial parameters

In the present study, partial correlations analysis, under the control of pH and OM, were conducted to distinguish the effects of trace metals on the microbial parameters from the soil properties. The total metal amount and its bioavailable values (acid soluble-, reducible- and oxidizable-fractions) were used for the analysis. The results obtained were showed in Table 4.

Table 4. Partial correlations of metals forms and microbial properties under the control variables of pH and OM

	Urease	Dehydrogenase	Arylsulphatase	MBC
A1	-.871**	-.857**	-.341	-.926**
B1	.522	.156	-.602*	.378
C1	-.858**	-.883**	-.436	-.935**
Cu	-.257	-.517	-.540	-.552
A2	-.815**	-.844**	-.479	-.926**
B2	-.799*	-.805**	-.547	-.946**
C2	.389	.503	.470	.634*
Zn	-.436	-.714*	-.911**	-.638*

As shown in Table 4, the labile Cu/Zn forms, such as its acid soluble, reducible and oxidizable fractions, were significantly negatively correlated with the studied enzyme activities at 0.05 or 0.01 probability levels. According to Bhattacharyya *et al.* (2008), the MBC and soil enzymes activities were significantly and negatively correlated with the metal labile forms as well as with its total metal concentrations due to the colinearity effect because total metal encompasses all the metal forms. However, no significant inhibition of the total Cu amounts on MBC and enzyme activities except of Arysulfatase were found in the partial correlation analysis (Table 4), although the adverse effects of the total Zn amounts on dehydrogenase, arysulfatase and MBC is increased in the partial correlation analysis (Table 4). These results indicate that labile metal forms are the most important factors regulating MBC and enzyme activities in heavy metal contaminated soils (Karaca *et al.*, 2010).

The Cu and Zn concentrations in the present study (Table 1) were quietly lower than the lowest observed adverse effect concentrations (LOAECs) of Cu (150 $mg \cdot kg^{-1}$) and Zn at Ludington (McGrath *et al.*, 1995). However, the significant inhibition of their forms on the tested microbial parameters was also found in the investigated soils. This may be related to that there are wide variations in the concentrations because various soil properties may also influence the LOAECs (Baath, 1989; McGrath *et al.*, 1995).

Conclusions

The trace elements Cu and Zn were slightly enriched in the investigated soils according to the soil back ground values, and they were mostly present in the residual fractions in bulk soil. Moreover, the correlation analysis between chemical speciation and soil microbial parameters also revealed that the labile forms of Cu and Zn could exert an inhibitory effect on soil Urease, Dehydrogenase and soil microbial biomass in the investigated field. The results signify the important role of the labile metal fractions exerting inhibitory effect on the microbial parameters even though they occupy only a small part of the total metal concentration.

Acknowledgements

The study was supported by the National Natural Science Foundation of China (41001145), and the Ph.D Foundation of Shandong University of Technology (406033).

References

Baath, E. 1989. Effects of heavy metals in soil on microbial processes and populations (a review). *Water Air & Soil Pollution*. 47: 335-379.

Bhattacharyya, P., Tripathy, S. and Chakrabarti, K. 2008. Fractionation and bioavailability of metals and their impacts on microbial properties in sewage irrigated soil. *Chemophere*. 72: 543-550.

Dai, W.H., Huang, Y. and Wu, L. 2009. Relationships between soil organic matter content (SOM) and pH in top soil of zonal soils in china. *Acta pedologica snica*. 146 (5): 851-860.

Dick, R.P. 1997. Soil enzyme activities as integrative indicators of soil health. *In:* Pankhurst C E, Doube B M, Gupta V V S R. (Eds.), Biological Indicators of Soil Health. CAB International, New York. pp. 121–156.

García, C., Hernandez, T. and Costa, F. 1993. The dehydrogenase activity of soils as an ecological marker in processes of perturbed system regeneration. *In:* Gallardo J F. (Ed.), Proceedings of the Eleventh International Symposium of Environmental Biogeochemistry, Salamanca, Spain. pp. 89-100.

Ge, Y., Murray, P. and Hendershot, W.H. 2000. Trace metal speciation and bioavailability in urban soils. *Environ. Pollution*. 107: 137-144.

Giller, K.E., Witter, E. and McGrath, S.P. 2009. Heavy metals and soil microbes. *Soil Biol. Biochem*. 41 (10): 2031-2037.

Gregorich, E.G., Wen, G. and Voroney, R.P. 1990. Calibration of rapid direct chloroform extraction method for measuring soil microbial biomass C. *Soil Biol. Biochem*. 22: 1009–1011.

Hinojosa, M.B, Carreira, J.A. and Garcia-Ruiz, R. 2004. Soil moisture pre-treatment effects on enzyme activities as indicators of heavy metal-contamiated and reclaimed soils. *Soil Biol. Biochem*. 36: 1559–1568

Joergensen, R.G. 1995. Microbial biomass. *In:* Alef, K, Nannipieri P (Eds), Methods in applied soil microbiology and biochemistry. Academic Press, London. pp. 382-386.

Karaca, A., Cetin, S.C. and Turgay, O.C. 2010. Effects of heavy metals on soil enzyme activities. I. Sherameti and A. Varma (eds.): Soil Heavy Metals, *Soil Biol*. 19: 237-262. DOI 10.1007/978364202436811.

Khan, S., Cao, Q. and Hesham, A.E.L. 2007. Soil enzymatic activities and microbial community structure with different application rates of Cd and Pb. *J. Environ. Sci*. 19: 834–840.

Kunito, T., Saeki, K. and Goto, S. 2001. Copper and zinc fractions affecting microorganisms in long-term sludge-amended soils. *Bioresourse & Tech*. 79: 135–146.

Lagomarsino, A., Mench, M., Marabottini, R., Pignataro, A., Grego, S., Renella, G. and Stazi, S.R. 2011. Copper distribution and hydrolase activities in a contaminated soil amended with dolomitic limestone and compost. *Ecotoxicology and Environmental Safety*. 74 (7): 2013-2019.

Li, L.F., Zeng, X.B. and Bai, L.Y. 2008. Accumulation of copper and zinc in soils under different agricultural and natural field (Chinese). *Acta Ecologica Sinica*. 29 (9): 4372-4380 .

Li, X., Poon, C.S. and Liu, P.S. 2001. Heavy metal contamination of urban soils and street dusts in Hong Kong. *Applied Geochemistry*. 16: 1361-1368.

Lu, R.K. 1999. Chemical analysis of agricultural soil. Beijing: Agricultural Science and Technology Press of China.

Luo, M.J. 1994. Shandong Soil. Beijing: Agriculture Press of China.

McGrath, S.P., Chaudri, A.M. and Giller, K.E. 1995. Long-term effects of metals in sewage sludge on soils, microorganism and plants. *J. Indus. Microbiol. Biotech*. 14: 94-104.

Meng, Q.H., He, M.R. and Wang, Q.Z. 1996. Research on the soil condition and its nutrient limitation factor in high yield farmland of the main soil types of Shandong Province. *J. Soil Sci*. 27 (6): 256-258.

Quevauviller, P., Rauret, G. and López-Sánchez, J.F. 1997. Certification of trace metal extractable contents in a sediment reference material (CRM 601) following a three-step sequential extraction procedure. *Science of Total Environment*. 205: 223-234.

Saeki, K. and Okazaki, M. 1993. Heavy metal accumulations in a semenclosed hypereutrophic system: lake Teganuma, Japan. Part 2. Heavy metal accumulations in sediments. *Water air & soil pollution*. 69: 79-91.

Saeki, K., Okazaki, M. and Matsumoto, S. 1993. The chemical phase changes in heavy metals with drying and oxidation of the lake sediments. *Water Research*. 27: 1243-1251.

Sun, S.J., Xu, J., Dai, S.G. and Han, X. 2006. Influences of copper speciation on toxicity to microorganisms in soils. *Biomedical Environ. Sci*. 19: 409-413.

Tabatabai, M.A. and Bremner, J.M. 1970. Arylsulphatase activity of soils. *Soil Science Society of America Proceedings*. 34: 225-229.

Tabatabai, M.A. and Bremner, J.M. 1972. Assay of urease activity in soils. *Soil Biol. Biochem*. 4: 479-487.

Trevors, J.T. 1984. Dehydrogenase activity in soil: a comparison between the INT and the TTC assay. *Soil Biol. Biochem.* 16: 673-674.

Wang, Y., Shi, J., Lin, Q., Chen, X. and Chen, Y. 2007a. Heavy metal availability and impact on activity of soil microorganisms along a Cu/Zn contamination gradient. *J. Environ. Sci.* 19: 848-853.

Wang, Y., Shi, J., Lin, Q., Chen, X. and Chen, Y. 2007b. The influence of soil heavy metals pollution on soil microbial biomass, enzyme activity, and community composition near a copper smelter. *Ecotoxicology and Environmental Safety.* 67: 75-81.

7

STUDY ON BIOFORTIFICATION OF RICE BY TARGETED GENETIC ENGINEERING

Sumon M. Hossain* and A.K.M. Mohiuddin

Abstract

Micronutrient malnutrition is a major health problem in Bangladesh and also in many other developing countries, where a diversified diet is not affordable for the majority. In the present world- one, out of seven people suffers from hunger. Yet, there is a stealthier form of hunger than lack of food: micronutrient malnutrition or hidden hunger. While often providing enough calories, monotonous diets (of rural poor) frequently fail to deliver sufficient quantities of essential minerals and vitamins. Due to micronutrient deficiencies different characteristic features have been observed to the victims. Various estimates indicate that over two-thirds of the world population, for the most part women and children specially, pre-school children are deficient in at least one micronutrient. This can have devastating consequences for the life, health and well being of the individuals concerned (like premature death, blindness, weakened immune systems etc). Genetic engineering approach is the upcoming strategy to solve this problem. Genetically engineered biofortified staple crops specially, rice that are high in essential micronutrients (Fe, Zn, vitamin A) and adapted to local growing environments have the potential to significantly reduce the prevalence of micronutrient deficiencies specially to the rural poor.

Keywords: Malnutrition, Biofortification, Rural Health, Genetic Engineering, Iron, Zinc, Vitamin A

Department of Biotechnology and Genetic Engineering, Mawlana Bhashani Science and Technology University Tangail-1902, Bangladesh

*Corresponding author's email: sumonmeister@gmail.com (Sumon M. Hossain)

Introduction

Globally, malnutrition is a matter of concern and erratically increase day by day, which including both overt nutrient deficiencies as well as diet-related chronic diseases (e.g., heart disease, cancer, stroke and diabetes) and is responsible for world huge population deaths than any other cause addressing for over 20 million mortalities annually (Kennedy *et al.*, 2003; WHO and FAO, 2003). Malnutrition also contributes to increase morbidity, disability, stunt mental and physical growth and reduce national socio-economic development (WHO and FAO, 2003). Micronutrient malnutrition are mainly affected over two billion people mostly among resource-poor families in developing countries like Bangladesh with iron, iodine, zinc and vitamin-A deficiencies most prevalent (Kennedy *et al.*, 2003). Over five millions pre-school children deaths occur due to the micronutrient malnutrition every year (Anonymous, 2007). World leading global economists have identified investing in strategies to reduce micronutrient malnutrition as the most cost effective manner (Anonymous, 2008).

Micronutrients are nutrients required by humans and other living things throughout life in small quantities to orchestrate a whole range of physiological functions, but which the organism itself cannot produce and body meet this valuable nutrient from diet.

Dysfunctional food systems that cannot supply all the nutrients and health promoting factors that are required for human life in sustainable ways. However, food systems that feed the disadvantaged are very complex (Sobal *et al.*, 1998). Therefore, dysfunctions in numerous interacting factors can result in inadequate supplies of nutrients reaching the most vulnerable populations (World Bank, 2007). Importantly, because food systems are dependent on agricultural products as their source of most nutrients, agricultural systems must be contributing to this worldwide quandary in public health (Welch, 2001). Unfortunately, agricultural systems have never been explicitly designed to promote human health mostly focusing on increased profitability for farmers and agricultural industries. Agriculture meets the challenge of feeding the world's poor during the "Green Revolution" focusing primarily on three staple crops – rice, wheat and maize. These crops provided enough energy to prevent widespread famines in many developing nations like Bangladesh. An unforeseen consequence of that agricultural revolution was the rapid rise in micronutrient malnutrition in many developing nations that adopted the cropping systems that prevented large scale starvations as well as malnutritional diseases (Welch and Graham, 1999).

Ten leading causes of illness and diseases in low income developing countries (Table 1).

Table 1. Ten leading cause of illness and diseases in low income developing countries (WHO, 2002)

Risk Factors	Ratio (%)
Under Weight	14.9
Unsafe Sex	10.2
Unsafe Water	5.5
Indoor Smoke	3.7
Zinc Deficiency	3.2
Iron Deficiency	3.1
Vitamin A Deficiency (VAD)	3.0
Blood Pressure	2.5
Tobacco	2.0
Cholesterol	1.9

Here we see that deficiency of Iron, Zinc and Vitamin A are the leading cause of illness and death in the low income developing countries.

Mentioned before that, several strategies are pursued to defeat micronutrient malnutrition worldwide, notably the nutritional enhancement of rice and other cereals through fortification, traditional or conventional breeding and genetic engineering (Fig. 1).

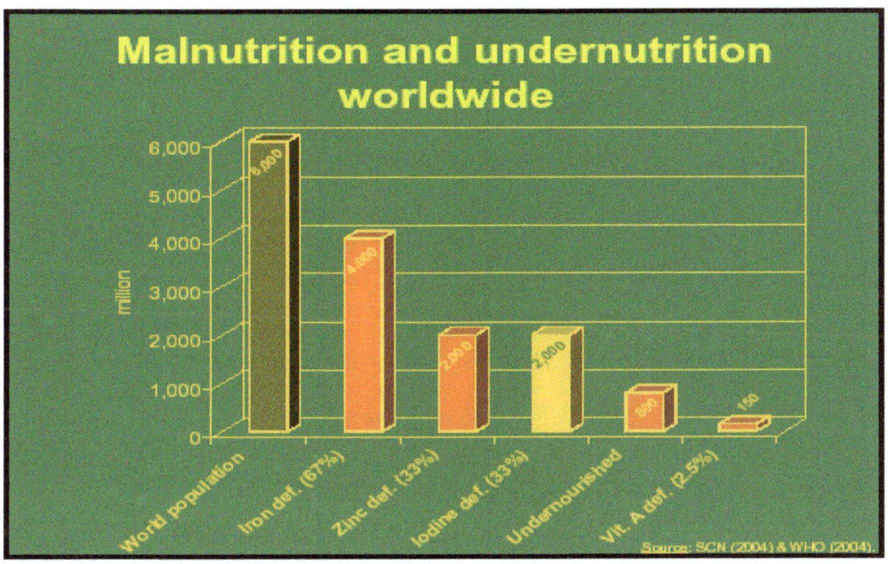

Fig. 1. Malnutrition and under nutrition worldwide [Source: UN/SCN (2004) and WHO (2004)]

Biofortification (developing food crops that fortify themselves) is a method of increasing nutritional value by adding the essential minerals and vitamins, is the first agricultural tool now being employed to address micronutrient malnutrition worldwide.

According to **International Center for Tropical Agriculture,** (CIAT/IFPRI, 2002) *"The goal of the proposed program is to improve the health of poor people by breeding staple food (Rice) crops that are rich in micronutrients, a process referred to here as "biofortification."*

According to **Sally Brooks, "***Bio-fortification is the enrichment of staple food crops with essential micronutrients. Considerable attention and resources are being directed towards the biofortification of rice- the world's most important food crop."*

Rice is the predominant staple food in at least 35 countries in the world, where micronutrient malnutrition is widely spread in the developing countries among them. Rice endosperms are the most important source of human food and which can provide as much as 80-85% in the daily calorie intake to the 2.5-3.0 billion people, half of the world's populations (Meng *et al.*, 2005)

Biofortification of rice is a method of increasing nutritional value by adding the essential minerals and vitamins into the rice endosperm. This can be done either through conventional selective breeding, or through genetic engineering. Biofortification of rice differs from ordinary fortification because it focuses on making rice foods more nutritious as the rice plants are growing, rather than having nutrients added to the rice foods when they are being processed. This is an improvement on ordinary fortification when it comes to providing nutrients for the rural poor, who rarely have access to commercially fortified foods. As such, biofortification of rice is seen as an upcoming strategy for dealing with deficiencies of micronutrients in the developing world. In Bangladesh it has been rarely seen in the rural primary school where the government freely supplies fortified biscuits to the pre-school children to meet the essential nutrients.

Genetic engineering, or rDNA (ribosomal DNA), is a technique that offers still greater speed and reach because it moves specific genes with desired traits from a source organism one which does not have to be a related organism directly into the living DNA of a target organism. The *transgenic* trait is added without normal biological reproduction, but once in the plant it becomes inheritable through normal reproduction. Scientists first developed this technique in the laboratory in 1973 and have been using it to transform agricultural crop plants since the 1980s. Once a useful gene has been identified (which can require a major research project and many years), it is attached to both marker and promoter genes and then inserted into a plant, usually using a non-viable virus called *Agrobacterium* as a carrier. GE produces plants that are known as transgenic, or less precisely as GMOs.

This review will focus on genetic engineering approaches to improve nutritional quality of rice: increasing the amount of bioavailable iron, zinc and introducing the β-carotene biosynthetic pathway in the rice endosperm, Because conventional breeding approach has some limitations: it is a long term process requiring variety of breeding activities and huge resources. In addition, it is uncertain whether this strategy will effectively work after all the long-term efforts. The breeding steps include at least (1) identification of a useful genetic variation and the most promising parents, (2) long-term crossing and back-crossing activities, (3) stability of the target traits (e.g. high grain Fe, Zn concentrations) across the different environments that feature huge variation in soil and climatic conditions, and finally (4) adaptation of the newly developed biofortified genotypes over a range of crop and soil management practices applied in the target regions or countries. The acceptance of biofortified crops by producers is a further issue that needs a special attention. Most importantly, newly developed genotypes should be able to extract sufficiently large amounts of Fe, Zn from potentially Fe, Zn-deficient soils and accumulate it in whole grain at sufficient levels for human nutrition.

Overview of the proposal

A new paradigm for agriculture in the 21st century was proposed (Welch and Graham, 2000) that views agriculture as an instrument for public health and focuses attention on the role of agriculture in delivering nutrients to humans and animals in balanced amounts that can sustain maximal physical and mental activity of the humans who are simultaneously the drivers of the food system and its dependents. This is known as the productive, sustainable, and nutritious food systems paradigm for agriculture and public health.

"It is time that IRRI lets go of the illusion that genetic engineering is an upcoming tool that would be benefited farmers and humankind. IRRI should start focusing its attention and resources on tools that would actually work to ensure that rice developers, growers and consumers have access to non-GE rice that is free of corporate control," said Daniel M. Ocampo, Sustainable Agriculture Campaigner of Greenpeace Southeast Asia.

Rationale and significance

The biofortification of rice strategy seeks to take advantage of the consistent daily consumption of large amounts of rice by all family members in Bangladesh (because rice is our staple food), including women and children who are most at risk for micronutrient malnutrition. As a consequence of the predominance of rice staples in the diets of the poor people, this strategy implicitly targets low-income households. After the one-time investment is made to develop seeds that fortify themselves, recurrent costs are low and germplasm may be shared internationally (conventional breeding). It is this multiplier aspect of plant breeding across time and distance that makes it so cost-effective. Once in place, the biofortified rice crop system is highly sustainable. Nutritionally improved varieties will continue to be grown and consumed year after year, even if government attention and international funding for micronutrient issues fades.

Table 2. Global mortality in children under 5 years of age in 2004 (Source: Black *et al.*, 2008)

Micronutrient Deficiency	Deaths
Vitamin A	6,66,771
Zinc	4,53,207
Iron	20,854
Iodine	3,619

From the above chart, Vitamin A deficiency (VAD) is a serious form of malnutrition to the children that weakens the immune system and may cause blindness. Several measures address

VAD and have shown positive results. Genetically modified rice containing beta-carotene is a new approach in an early experimental stage.

Biofortification of rice provides a truly feasible means of reaching malnourished populations in relatively remote rural areas, delivering naturally fortified rice to people with limited access commercially-marketed fortified rice, which are more readily available in urban areas. Biofortification and commercial fortification of rice, therefore, are highly complementary.

The ultimate goal of biofortification of rice strategy is to reduce mortality and morbidity rates related to micronutrient malnutrition and to increase food security, productivity, and the quality of life for poor populations of developing countries like Bangladesh where 40% people living under poverty (http://www. indexmundi. com) because of, there poverty most of them are eaten only two times in a day. So, they need more and more nutrients to meet the normal physical activity as well as for leading their healthy life especially for the pre-school children. After all it is the ongoing process that children are the future asset of a country.

Biotechnology *Vs* Biofortification

Biofortification capitalizes on the consistent daily intake of food staples, thus indirectly targeting low-income households who cannot afford a more diverse diet especially for the rural poor. After the initial investment of developing fortified staple crops, no extra costs are needed, and also making this strategy very sustainable and beneficiary. Furthermore, the genetically improved varieties can be shared internationally. The rice endosperms should be biofortified specially that rice variety, whose has huge yield in local territory because targeting for biofortification of rice is not for high yielding but also improved the rice quality especially for micronutrients. Biofortified rice endosperms are also likely to have an indirect impact in agriculture and environment, as a higher trace mineral content in seeds that confer better protection against pests, herbicides, diseases, and environmental stresses (like PH, temperature etc), thereby increasing yield (Welch and Graham, 2004). Biofortification is not a panacea in itself but a very important complement to dietary variety and to supplementation.

Biofortified rice can be developed by genetic engineering or conventional breeding methods, provided there is sufficient genetic variation in crop populations for the desired trait (such as high protein content). In staple grains such as rice, improvement of some complex traits such as vitamin A is not possible using conventional breeding strategies, as there are no natural rice varieties rich in this vitamin. All plants produce pro-vitamin A, but only in the green organs of the plant and not in the starch-storing part of the seed. Conventional breeding is also very difficult in vegetatively propagated varieties (such as cassava and potatoes), due to the scarcity of genetically well-defined breeding lines (Welch and Graham, 2004). In addition, conventional breeding can change important traits of the crops desired by consumers, such as taste. Agricultural biotechnology methods, and in specific genetic engineering (GM), represent therefore a very valuable, complementary strategy for the development of more nutritious rice.

Micronutrient deficiency and its impact

A prerequisite for rural health is a diversified diet with sufficient micronutrients. These micronutrients include trace minerals and vitamins. Iron deficiency anaemia and vitamin A deficiency blindness are two major forms of disorders caused by micronutrient malnutrition (FAO, 1997).

Anaemia which caused by deficiency of Iron is far the most common micronutrient deficiency worldwide. It is also highly prevalent in developed countries both in women of childbearing ages (Milman and Kirchhoff, 1992) and in children (Dallman *et al.*, 1980). Severe anaemia in pregnant women is estimated to be responsible for up to 42% of the half a million deaths associated with childbirth each year (Seymour and Klein, 1996).

Vitamin A deficiency affects about 30 countries (FAO, 1997), including some 16 million children in the developing world. The early stages of vitamin A deficiency are characterized by impaired dark adaptation, which will progress, if uncorrected, to nyctalopia (night blindness), xerophthalmia and total blindness. In Southeast Asia, it is estimated that a quarter of a million children go blind each year because of vitamin A deficiency (Sommer, 1988).

The importance of micronutrient deficiencies is well known and documented (Table 3). According to the World Health Organization (WHO), "More than 2.0-2.5 billion people in the world today are estimated to be deficient in key vitamins and minerals, particularly vitamin A, iodine, iron and zinc. Most of these people live in low income countries and are typically deficient in more than one micronutrient" (WHO, World Food Programme, & UNICEF, 2007).

Table 3. Micronutrient deficiencies and their estimated impacts

Iron	Highest risk of iron deficiency occurs during rapid growth and nutritional demand, such as in children, adolescence and pregnancy. Also associated with maternal death, impaired physical and cognitive development, increased risk of morbidity in children, and reduced work productivity in adults. Can be addressed through fortification of wheat products. WHO estimates 2 billion people are anemic, and this is frequently exacerbated by infectious diseases. Malaria, HIV/AIDS, hookworm infestation, schistosomiasis, and tuberculosis contribute to a high prevalence of anemia in some areas. Efforts to increase iron intake must be accompanied by efforts to control infectious disease.
Zinc	Associated with reduced immune status in neonates and children. Preliminary research shows that additional zinc can reduce incidence of diarrhea and pneumonia in children and improves maternal health. One estimate shows zinc as close to iron deficiency in contribution to the global burden of disease. Can be provided through supplements.
Vitamin A	Associated with reduced immune system functioning and a higher incidence of illness and complications from illness and also associated with blindness and increased risk of disease and death for small children and pregnant women. Can be addressed through supplements, which are now estimated to reach children at least once a year in 40 countries. The UN Standing Committee on Nutrition (UN/SCN) estimates that 140 million children and 7 million pregnant women are VA deficient, primarily in Africa and South/Southeast Asia. In 1998, WHO, UNICEF, Canadian International Development Agency, USAID, and the Micronutrient Initiative launched the VA Global Initiative. This provides support to countries in delivering VA supplements.
Iodine	Associated with brain damage. Easily mitigated with iodized salt. While incidence has declined dramatically in recent years due to the universal adoption of salt iodization starting in 1993, WHO estimates that 54 countries still have some iodine deficiency.
Folate	Deficiency results in serious disorders, including neural tube defects (NTD) such as *spina bifida* in infants, with an estimated 2,00,000 severe birth defects every year and megaloblastic anemia. Adequate dietary folate intake can prevent onset of these conditions. Deficiency also associated with increased risk of maternal death and complications in birth; can be addressed through fortification of wheat products.

Note: Data from De Steur *et al.* (2010); Storozhenko *et al.* (2007); Shekar *et al.*, (2006); UNICEF/MI (2004); UN/SCN (2004); and WHO (2004).

From the Table 3, deficiencies of iodine, vitamin A, and iron are the most prevalent for global public health in terms of numbers of individuals in a position to secure large benefits; folate and zinc have important but more limited or less understood impacts in the body. The World Bank estimates that $180 billion in lost productivity globally could be prevented with only $4 to $5 billion invested in deficiency prevention, using currently available technologies (Behrman *et al.*, 2004). It is now recognized that micronutrient deficiencies are persistent public health problems that require direct intervention during the process of economic development.

Iron and Zinc biofortificaion

Iron (Fe) and zinc (Zn) are essential micronutrients in both plant and animal kingdom. Amplification of the natural iron stores is a further strategy to increase iron accumulation in the rice endosperms. Both plants and animals use ferritin (organic form) as the form of iron storage during developmental stage. Plant ferritin is composed of 24 subunits (Lobreaux *et al.*, 1992), which together can store up to 4000 iron atoms in the central cavity of the ferritin complex. A recent revaluation demonstrated that ferritin as a good bioavailable iron source (Murray-Kolb *et al.*, 2003).

Now-a–days many reports investigated the possibility of accumulating iron in plants by ferritin gene transformation strategies. Several methods are being used for gene transformation, such as:

- ✓ *Agrobacterium* mediated gene transfer
- ✓ Chemically mediated gene transfer
- ✓ Electroporation
- ✓ Microinjection
- ✓ Microprojectile bombardment method

The well developed strategies is the *Agrobacterium*-mediated gene transfer among all others gene transformation method, because *Agrobacterium tumefaciens* is a soil born natural genetic engineer. The transformation of dicotyledons by *Agrobacterium* is well established. However, transformation methods based on the use of *Agrobacterium* are still preferred in many instances because of the following properties:

- ✓ Easy to handle
- ✓ Higher efficiency
- ✓ More predictable
- ✓ Pattern of foreign DNA integration, and
- ✓ Low copy number of integration.

This vector was the discovery that the T-DNA and the Vir region could be separated onto two different plasmids.

For Example, Transgenic tobacco plants with a high iron-content have been produced via *Agrobacterium*-mediated gene transfer of the soybean ferritin cDNA under control of a CaMV 35S promoter (Goto *et al.*, 1998).

Goto and his colleagues (1998) reported an outstanding correlation between ferritin levels and iron accumulation in ferritin-transformed tobacco leaves, raising the possibility that higher iron concentration could be achieved by increasing ferritin expression. In rice, the same ferritin gene expression, driven by a constitutive promoter, led to an increase in iron content of the vegetative parts of rice but not in the seeds (Drakakaki *et al.*, 2000). Goto *et al.* (2000) reported that transgenic lettuce, constitutively expressing ferritin, grew larger and faster than the control. Moreover, he reported the endosperm-specific expression of soybean ferritin in rice seeds with maximum three-fold increase in one of the transformants (Goto *et al.*, 1999).

Recently, an extremely high expression of soybean ferritin driven by the rice glutelin promoter was reported in rice endosperm (Qu *et al.*, 2005), with levels of the transgenic protein up to 13-fold higher than previously reported (Goto *et al.*, 1999). Unfortunately, the maximum iron concentration in the seeds of the new lines was only about 30% higher compared with previously transformants and about three-fold higher than in the non-transformed control seeds. The authors (Qu *et al.*, 2005) observed that the mean iron concentration in leaves of high ferritin expressing lines decreased to less than half of the non-transformed plants, and the plants showed chlorosis after flowering, even on iron-rich medium. Qu and collaborators concluded that accumulation of iron in the seeds of hyper-expression ferritin rice did not only depend on the expression level of exogenous ferritin but may also be limited by Fe uptake and transport.

Two plant genes help to mobilize and store Iron and Zinc in rice

Recently succeeded in increasing the iron content in polished rice by transferring two plant genes into an existing japonica rice variety.

- ➢ One gene encodes nicotianamine synthase, the enzyme that produces nicotianamine. Nicotianamine chelates iron temporarily and facilitates its transport in the rice plant. Nicotianamine synthase is expressed under a constitutive promoter
- ➢ Second gene encodes the protein ferritin, which functions as a storage depot for up to four thousand iron atoms per protein molecule in both plants and humans. Since the ferritin gene is expressed under the control of a specific promoter.
- ➢ These two genes allow the rice plant to absorb more iron from the soil, transport it in the plant, and store it in the rice kernel.
- ➢ A third gene encoding phytase was also engineered into this rice line. Phytase degrades phytate which can store phosphate and bind divalent cations like iron and thus inhibits their absorption in the intestine.
- ➢ The pmi (*Escherichia coli* phosphomannose-isomerase) gene was used as a selectable marker
- ➢ starting material and gene transfer is mediated by *Agrobacterium tumefaciens* strain LBA4404
- ➢ Primary transformants is obtained after transformation and selection on mannose.
- ➢ Plants were regenerated and analysed by Southern blot.
- ➢ The genetically engineered lines expressing nicotianamine synthase, ferritin, and phytase (NFP-line) contain up to a 6.3-fold increase of iron in the endosperm of polished kernels.
- ➢ Overexpression of the *NAS* gene in rice would enhance the synthesis of NA (Nicotinamine) and DMA (2′-deoxymugineic acid), and thus increase Fe and Zn concentrations in seeds.

Higuchi *et al.* (2001) introduced the *HvNAS1* gene into rice under the control of the *35S* promoter. The resultant transgenic lines produced two to threefold greater concentrations of NA compared to non-transgenic lines. DMA production and secretion also increased under Fe-sufficient conditions. Laser ablation inductive coupled plasma mass spectroscopy (LA-ICP-MS) confirmed the accumulation of additional iron in the endosperm of NFP lines by imaging micro-X-ray fluorescence spectroscopy.

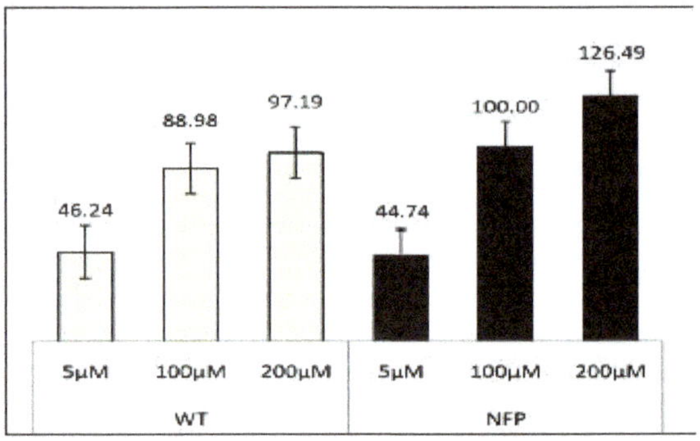

Fig. 2. Iron concentration in leaves (WT vs NEP line)

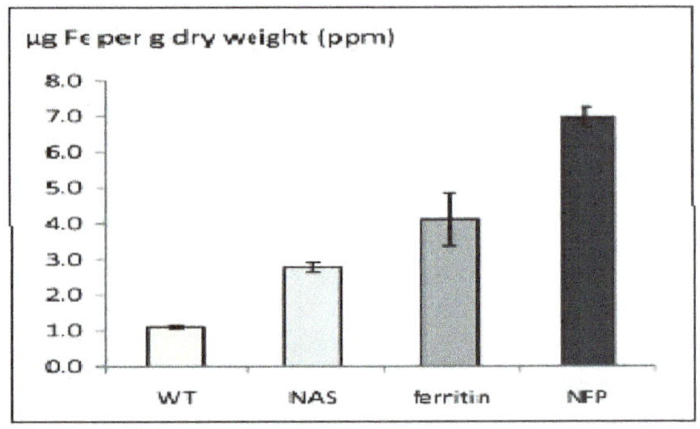

Fig. 3. Iron concentration in endosperm

Note: Fig. 2. Concentration of iron in leaves of wild type (WT) and NEP rice as a function of external iron concentration (Johnathan Napier *et al.,* 2010)
Fig. 3. Concentration of iron in polished grains of wild type (WT), NAS, Ferritin, and NEP rice on 200 micro meteriron supplied in the medium (Johnathan Napier *et al.*, 2010).

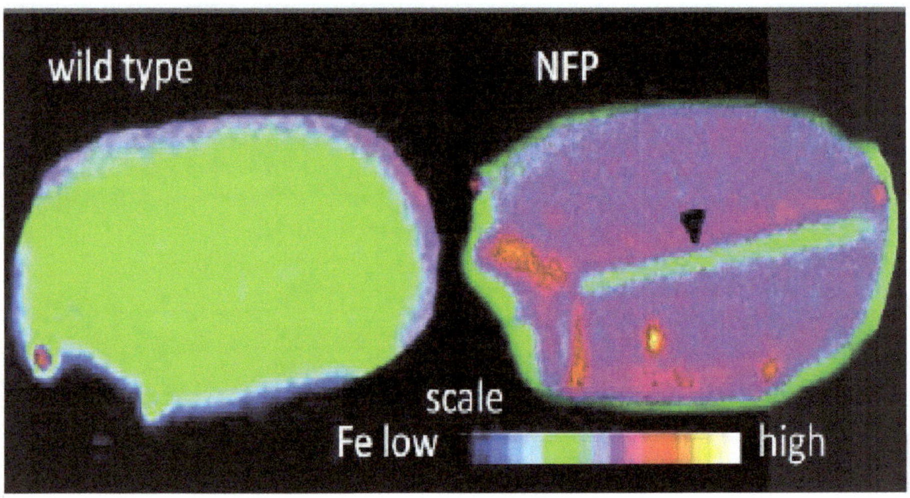

Fig. 4. Iron localization in endosperm grains^{-1} (Wild type vs NEP line)

Note: Fig 4. Element-sensitive mapping of iron concentration in a representative WT and NEP grain, acquired using micro-X-ray fluorescence spectrometry. The color indicates iron concentration. The low iron in NEP line representation the laser ablation track, which is out of focus and therefore iron concentration appears to be reduced along this line (arrowhead).

Introducing β-carotene into the rice endosperm: (Golden rice)

Introducing β-carotene into rice was developed about 20 years ago. Orange in color, it was genetically engineered to be rich in beta carotene, a nutrient that our body can easily converts its into Vitamin A. Scientists hoped that beta carotene containing rice would save lives and the eyesight of a quarter of a billion kids worldwide not getting enough vitamin A in their diets, children who typically live in less developed countries where ordinary rice is a dietary staple. The breakthrough of the so-called **'golden rice'** made apparent that it was possible to engineer not only single proteins but to redirect a complete pathway with several involved enzymatic functions with no apparent pleiotropic effect. German professors Ingo Potrykus and Peter Beyer invented the 'golden rice' by genetic engineering at Switzerland's Federal Institute of Technology for the poor in developing countries with public money.

Early critiques of this transgenic biofortification were disappointed that 'golden rice' contained only low amounts of β-carotene and that it was a Japonica variety, which is not eaten in the areas of the most spread provitamin A deficiency. Meanwhile, this provitamin A biofortification has been transformed also into Indica rice lines (Datta *et al.,* 2003), the varieties widely consumed in Asia. Indica varieties IR64 and MTL250 were obtained (Hoa *et al.,* 2003) avoiding the antibiotic selection system, using a positive selection strategy (Lucca *et al.,* 2001). The polished indica rice seeds contained lower level of β-carotene compared with the previous Japonica rice variety (Ye *et al.,* 2000).

Gene source

Golden rice (Vitamin A) is an example of a GM crop developed for its nutritional value. Golden rice contains PSY gene from daffodil and the CRTI gene from the bacterium *Erwinia uredovora*, both expressed only in the rice endosperm (Ye *et al.*, 2000). Replacing PSY with genes from maize and rice increased the level of β-carotene by 23 times in Golden Rice (Paine *et al.*, 2005) which can be converted by the body into vitamin A. This can help alleviate symptoms of vitamin A deficiency.

Notwithstanding the fact that rice plants synthesise β-carotene in vegetative tissues, located only in the outer cell layers of its seeds but not in the grain, which, however, are removed before storage by polishing. Processed rice is therefore full free of provitamin A.

β-Carotene synthesis in the endosperm of rice grains requires phytoene synthase to produce the intermediate phytoene from geranylgeranyl diphosphate (GGPP). Synthesis of β-carotene from phytoene needs three more plant enzymatic steps: phytoene desaturase (PDS), ζ-carotene desaturase (ZDS) and lycopene β-cyclase. The *Erwinia uredovora* enzyme is able to perform the functions of both desaturases (for an overview of the pathway and the enzymes involved, (Burkhardt *et al.*, 1997). A first step towards the production of β-carotene in staple crops has been reported in rice with the constitutive and endosperm-specific expression of a recombinant daffodil phytoene synthases cDNA (Burkhardt *et al.*, 1997). The transgene driven by the constitutive promoter CaMV 35S produced a high infertility rate of the transgenic plants.

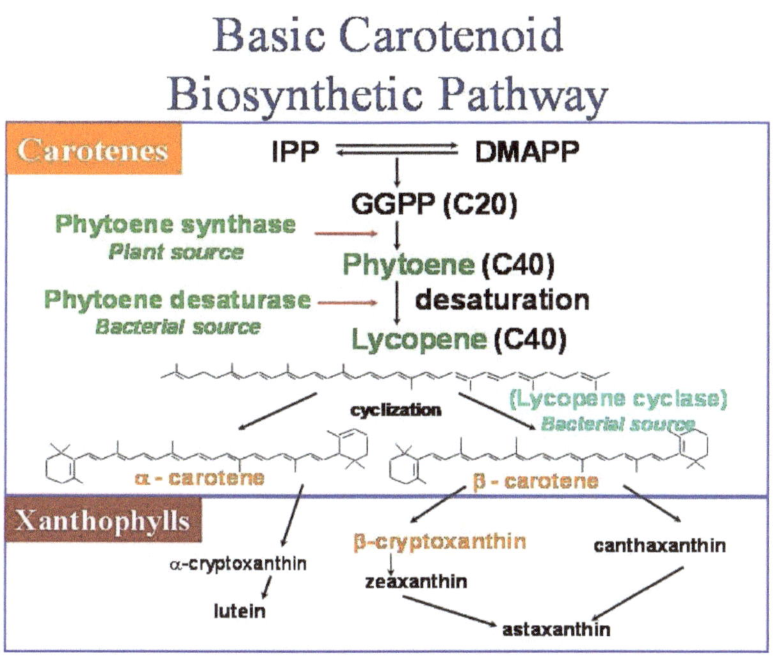

Fig. 5. Caroteniod synthesis

- To obtain the complete provitamin A biosynthetic pathway only in rice endosperm, Ye *et al.* (2000) have introduced the genes encoding the three enzymes, necessary to synthesize β-carotene form GGPP, into Japonica rice variety. Phytoene synthase (PSY) gene from daffodil and the phytoene desaturase (CRTI) gene from the bacterium *Erwinia uredovora*, both expressed only in the rice endosperms.
- A systematic test with *psy* cDNAs from alternative plant sources particularly rich in carotenoids indicated that the best strategy was probably to use the phytoene synthase (*psy*) from maize (Paine *et al.*, 2005) instead of that from daffodil.
- The combination of this *psy* cDNA with the *Erwinia uredovora* carotene desaturase under the control of an endosperm-specific promoter produced transgenic rice seeds with approximately 23-fold higher provitamin A content than the original Japonica golden rice variety.
- Transformation of tobacco (*Nicotiana tabacum*) BY2 cells and Arabidopsis (*Arabidopsis thaliana*) plants is performed using the binary vector pSToff containing CRTI under the control of the constitutive CaMV 35S promoter, and PSY under the control of Gt1 promoter and hpt marker gene.

- The resulting vector is then transformed into the *Agrobacterium tumefaciens* strain GV3101.
- After that this *Agrobacterium tumefaciens* when infect the target rice plant its modified T-DNA is introduced into the rice genome, and the rice plant get the desire gene and to accumulate the β-carotene inside the rice endosperm.

This new development is an exciting advance, and a further support that golden rice could complement existing efforts to end blindness and other diseases caused by vitamin A deficiency. The golden rice obtained from these will allow extended human nutritional work for investigation on bioavailability and carotenoid stability in rice food preparation with various forms of processing, storage and cooking.

Fig. 6. β-carotene containing rice (yellow color) Vs normal rice (white color)

Conclusion

Biotechnology can help to address many challenges in the developing world but the political issue is always facilities this work to weak all over the world. The real challenges to solve micronutrient malnutrition diseases in the developing countries as well as are reduction of poverty and diseases and the increase of infrastructure and education system. All these issues are important and should continue to be addressed, as they have been in the past. In developing countries, rice biofortification is increasingly recognized as an effective approach to improve the micronutrient status of large parts of the populations like India, Myanmar, and Indonesia etc. improving nutritional quality of rice grains through genetic modification in the rice genome. Therefore, even a small improvement in nutritional content of rice seeds can have a dramatic impact on human health especially to the rural poor.

So, the review suggested that genetic engineering can be used as a promising approach for the biofortification of important staple crops focus to rice to address micronutrient deficiencies and improve human health.

Future Prospect

Bangladesh is an agriculture based developing country. It has a huge population, among them most of living under poverty and being suffered from basic nutrient in their diet.

Biofortification in rice is a relatively new field, while increasing essential micronutrients is our goal. Using this technique (genetic engineering) we would like to produce GM rice by introducing the folic acid biosynthesis gene into the rice plant. Folate is a β-group vitamin, also called folacin, critical for normal cellular function, insufficient intake causes megaloblastic anemia and there are strong linkages to cardiovascular diseases, various cancers and cognitive decline. It is also vital raw materials for red blood cell production in the blood and also has brain function.

Folic acid can be found in the leafy green vegetable, yeast, mushrooms, papaya etc. After over expressing two *Arabidopsis thaliana* genes of the pterin and para-aminobenzoate branches of the folate biosynthetic pathway from a single locus into the rice genome then, when the people eat the GM rice they will meet the sufficient folic acid that is crucial for the preventing the above diseases.

If Bangladesh would approve folate enriched rice as in the future, several policy issues should be addressed, such as labeling, intellectual property rights and trade or price policy issues. Furthermore, the implementation of folate biofortification programs should initially reach its target group, women of childbearing age from poor, rural areas. Also the procedures to commercialize the rice and distribute the seeds (through farmers or other channels; voluntary or mandatory) should be clarified and supported. And even if the Bangladeshi policy would accept the production of biofortified staple crops, it should be taken into account that biofortification alone will only alleviate micronutrient malnutrition. A combination of interventions will be probably the most feasible and effective strategy to combat folate deficiency.

Acknowledgements

The authors express their sincere gratitude to Prof. Dr. Md. Shahidul Haque, Dept. of Biotechnology, Bangladesh Agricultural Univerity for their critically going through the manuscript. Special thanks due to Md. Abdul Khaleque, Israt Nadia, Tanusree Dey, Hriday Banik and Rifat Sultana for their valuable help in the preparation of the manuscript.

References

Anonymous. 2007. Global childhood malnutrition. *The Lancet*. 367: 1459.

Anonymous. 2008. The Copenhagen Consensus 2008 - Results. The Copenhagen Consensus Center, 1-6. (http://www.copenhagenconsensus.com)

Behrman, J., Alderman, H., and Hoddinott, J. 2004. *Hunger and malnutrition* (Copenhagen Consensus Challenge Paper). Copenhagen, Denmark: Author. Bishai, D., & Nilubola, R. 2002. The history of food fortification in the United States: Its relevance for current fortification efforts in developing countries. *Economic Development and Cultural Change*. 51 (1): 37-53.

Black, R.E., Allen, L.H., Bhutta, Z.A., Caulfield, L.E., De Onis, M., Ezzati, M., Mathers, C. and Rivera, J. 2008. Maternal and child under nutrition: *global and regional exposure and health consequences. Lancet*. 371: 243–260.

Burkhardt, P.K., Beyer, P., Wu¨nn, J., Kloeti, A., Armstrong, G.A., Schledz, M., V. Lintig, J, and Potrykus, I. 1997. Transgenic rice (*Oryza sativa*) endosperm expressing daffodil (*Narcissus pseudonarcissus*) phytoene synthase accumulates phytoene, a key intermediate of provitamin A biosynthesis. *Plant Journal*. 11: 1071–1078.

CIAT/IFPRI. 2002. (International Center for Tropical Agriculture/International Food Policy Research Institute). Biofortified crops for improved human nutrition. *Program proposal*. PP. III

Dallman, P.R.S.I., Imes, M.A. and Stekel, A. 1980. Iron deficiency in infancy and childhood. *American J. Clin. Nutr*. 33: 86–118.

Datta, K., Baisakh, N., Oliva, N., Torrizo, L., Abrigo, E., Tan, J., Rai, M., Rehana, S., Al-Babili, S., Beyer, P. Potrykus, I., Datta S.K. 2003. Bioengineered "golden" indica rice cultivars with b-carotene metabolism in the endosperm with hygromycin and mannose selection systems. *Plant Biotechnol J*. 1: 81–90.

De Steur, H., X. Gellynck, S. Storozhenko, G. Liqun, W. Lambert, D. Van Der Straeten, J. Viaene. 2010. Willingness-to-Accept and Purchase Genetically Modified Rice With High Folate Content in Shanxi Province, China. 54: 118-125.

Drakakaki, G., Christou, P.and Stoger, E. 2000. Constitutive expression of soybean ferritin cDNA in transgenic wheat and rice results in increased iron levels in vegetative tissues but not in seeds. *Transgenic Res*. 9: 445–452.

FAO. 1997. Preventing micronutrient malnutrition: a guide to food-based approaches. Why policy makers should give priority to food-based strategies.

Goto, F., Yoshihara, T. and Saiki, H. 1998. Iron accumulation in tobacco plants expressing soyabean ferritin gene. *Transgenic Res*. 7: 173–180.

Goto, F., Yoshihara, T., Shigemoto, N., Toki, S. and Takaiwa, F. 1999. Iron fortification of rice seed by the soybean ferritin gene. *Nature Biotechnol*. 17: 282–286.

Goto, F., Yoshihara, T. and Saiki, H. 2000. Iron accumulation and enhanced growth in transgenic lettuce plants expressing the iron-binding protein ferritin. *Theor. Appl. Genet*. 100: 658–664.

Higuchi, K., Watanabe, S., Takahashi, M., Kawasaki, S., Nakanishi, H., Nishizawa, N.K., and Mori, S. 2001. Nicotianamine synthase gene expression differs in barley and rice under Fe-deficient conditions. *Plant J*. 25: 159–167.

Hoa, T.T., Al-Babili, S., Schaub, P., Potrykus, I. and Beyer, P. 2003. Golden Indica and Japonica rice lines amenable to deregulation. *Plant Physiol*. 133: 161–169.

http://www.indexmundi.com/bangladesh/population_below_poverty_line.html

Johnathan Napier, Noemi Ruiz-Lopez, Tianbi Li, Richard Haslam and Olga Sayanova. 2010. The accumulation of novel omega-3 fatty acid in transgenic plants. ISB News Report. pp.1-3

Kennedy, G., Nantel, G. and Shetty, P. 2003. The scourge of "hidden hunger": global dimensions of micronnutrient deficiencies. *Food Nutr. Agr.* 32: 8-16.

Lobreaux, S., Yewdall, S.J., Briat, J.F. and Harrison, P.M. 1992. Amino-acid sequence and predicted three-dimensional structure of pea seed (*Pisum sativum*) ferritin. *Biochem J.* 288 (3): 931–939.

Lucca, P., Hurrell, R. and Potrykus, I. 2001. Genetic engineering approaches to improve the bioavailability and the level of iron in rice grains. *Theor. Appl. Genet.* 102: 392–397.

Meng, F., Wei, Y. and Yang, X. 2005. Iron content and bioavailability in rice. *J. Trace Element Med. Biol.* 18: 333–338.

Milman, N. and Kirchhoff, M. 1992. Iron stores in 1359, 30- to 60 years old Danish women: evaluation by serum ferritin and haemoglobin. *Ann Hematol.* 64: 22–27.

Murray-Kolb, L.E., Welch, R., Theil, E.C. and Beard, J.L. 2003. Women with low iron stores absorb iron from soybeans. *American J. Clin. Nutr.* 77: 180–184.

Paine, J.A., Shipton, C.A., Chaggar, S., Howells, R.M., Kennedy, M.J., Vernon, G., Wright, S.Y., Hinchliffe, E., Adams, J.L., Silverstone, A.L. and Drake, R. 2005. Improving the nutritional value of golden rice through increased pro-vitamin A content. *Nature Biotechnol.* 23: 482–487.

Qu, L.Q., Yoshihara, T., Ooyama, A., Goto, F. and Takaiwa, F. 2005. Iron accumulation does not parallel the high expression level of ferritin in transgenic rice seeds. *Planta*, (doi: 10.1007/s00425-005-1530-8), available at http://www.springerlink.com/media/3DGH VGQWVQW6D4V5PN5U/Contributions/G/1 /4/X/G14X2411H716P336_html/fulltext.ht ml.

Shekar, M., Heaver, R. and Lee, Y. (Eds.). 2006. *Repositioning nutrition as central to development: A strategy for large scale action* (Directions in Development Series). Washington, DC: World Bank.

Seymour, J.T. and Klein, T. 1996. Hungry for a new revolution. *New Scientist*. pp. 34–37

Sobal, J., Khan, L.K. and Bisogni, C. 1998. A conceptual model of the food and nutrition system. *Soc. Sci. Med.* 47: 853-863.

Sommer, A. 1988. Avoidable blindness. *Aust. N Z J. Ophthalmol.* 16: 31–35.

Storozhenko, S., De Brouwer, V., Volckaert, M., Navarrete, O., Blancquaert, D., Zhang, G.F., Lambert, W. and Van der Straeten, D. 2007. Folate fortification of rice by metabolic engineering. *Nature Biotechnol.* 25: 1277-1279.

UN/SCN (United Nations Systems Standing Committee on Nutrition). 2004. *5th report on world nutrition situation: Nutrition for improved development outcomes*. Geneva, Switzerland. pp. 27-30.

UNICEF/MI. 2004. Vitamin & Mineral Deficiencey. A Global Progress Report. Micronutrient Initative; Ottawa, Canada [http://www.micronutrient.org/reports/]

Welch, R.M. 2001. Micronutrients, agriculture and nutrition; linkages for improved health and well being. *In:* Perspectives on the Micronutrient Nutrition of Crops. Singh K, Mori S, Welch RM eds. Jodhpur, India: Scientific Publishers. pp. 247-289.

Welch, R.M. and Graham, R.D. 1999. A new paradigm for world agriculture: meeting human needs - Productive, sustainable, nutritious. *Field Crops Res.* 60: 1-10.

Welch, R.M. and Graham, R.D. 2000. A new paradigm for world agriculture: productive, sustainable, nutritious, healthful food systems. *Food Nutr. Bull.* 21: 361-366.

Welch, R.M. and Graham, R.D. 2004. Breeding for micronutrients in staple food crop from human nutrition perspective. *J. Exp. Bot.* 55: 353-364.

WHO (World Health Organization). 2002. The World Health Report 2002, Geneva.PP.8-10

WHO (World Health Organization). 2004. *Iodine status worldwide: WHO global database on iodine deficiency*. Geneva, Switzerland. p. 2

WHO and FAO. 2003. Diet, nutrition and the prevention of chronic diseases. Report of a joint WHO/FAO expert consultation. WHO Technical Report Series 916.

WHO, World Food Programme & UNICEF. 2007. *Preventing and controlling micronutrient deficiencies in populations affected by an emergency*. Geneva, Switzerland: WHO. From http://www. who. int/nutrition/publications/WHO_WFP_U NICEFstatement.pdf. p. 1

World Bank. 2007. From agriculture to nutrition: pathways, synergies and outcomes. Report 40196-GLB. Washington, D.C.: The International Bank for Reconstruction and Development / The World Bank. pp. 1-83.

Ye, X., Al-Babili, S., Kloti, A., Zhang, J., Lucca, P., Beyer, P. and Potrykus, I. 2000. Engineering the provitamin A (beta-carotene) biosynthetic pathway and (carotenoid free) rice endosperm. *Science*. 287: 303-305.

8

GROWTH AND YIELD PERFORMANCE OF SHINGI, *Heteropneustes fossilis* AND KOI, *Anabas testudineus* IN BANGLADESH UNDER SEMI-INTENSIVE CULTURE SYSTEMS

B. K. Chakraborty[1*] and N.N. Nur[2]

Abstract

Production potential of shingi, *Heteropneustes fossilis* and koi, *Anabas testudineus* in polyculture were assessed at a stocking density of 2, 47,000 and 3, 70,500 ha^{-1}, respectively in treatment T_1 and T_2. Monoculture of *H. fossilis* and *A. testudineus* was designed at a stocking density of 2, 47,000 ha^{-1} in treatment T_3 and T_4, respectively. Culture period of shingi was 120 days and koi was 100 days in all treatments. All fingerlings were of the same age at stocking, with a mean weight of 2.54±0.08 g for shingi and 0.50±0.01 g for koi, respectively. Commercial fish feeds (30.0% crude protein) supplied at the rate of 100 to 5% of total biomass twice daily. Fish production in treatment T_1, T_2, T_3 and T_4 were 18,803±111, 12,388±115, 10,042±5 kg ha^{-1} day^{-120} and 22,176±7 kg ha^{-1} day^{-100}, respectively. The feed conversion ratio (FCR) was significantly ($P<0.05$) lower in T_4 than that of the other three treatments. The net financial benefits incurred from treatment T_1, T_2, T_3 and T_4 were Bangladeshi Taka 17,65,769; 6,691; 15,83,990 and 16,29,409 BDT ha^{-1}, respectively. The mean differences of gross yields and net benefits among different treatments were significant ($P<0.05$). The polyculture technology of shingi and koi, and monoculture technology of koi may help to meet the dietary needs and improve the socio-economic status of the people of Bangladesh.

Keywords: Growth, Yield Performance, Shingi, Koi

[1]Mathsya Bhabon, Department of Fisheries, Dhaka-1000, Bangladesh
[2]Programme Manager, NETZ, Dhaka, Bangladesh
*Corresponding author's email: bborty@gmail.com (B.K. Chakraborty)

Introduction

Polyculture or composite fish culture is the system in which fast growing compatible species with different feeding habits are grown in the same pond (Jhingran, 1975). Polyculture management technique is based on the relationship between fishes at different levels of the food chain and environment. The outcome of fish production from polyculture systems depends on the species combinations and their stocking densities. Polyculture is one of the major culture techniques that have been used traditionally in Bangladesh, where carp polyculture is practised in the farmer's ponds for rapid growth and maximum production. However, there is no information on polyculture practice of *H. fossilis* (Heteropneustadae) and *A. testudineus* (Anabantidae) in Bangladesh. The local name of *H. fossilis* and *A. testudineus* is shingi and koi (IUCN, 2000). So, this polyculture culture technology is a completely new one in South East Asia (Chakraborty and Mirza, 2008).

Heteropneustes fossilis (Bloch) is an indigenous stinging catfish of South-East-Asia, which is locally known as Shingi or Shing in different parts of Bangladesh. It is not only recognised for its excellent taste and market value but is also highly sought after for its nutritional and medicinal benefits. The species has high content of iron (226 mg 100 g^{-1}) and fairly high content of calcium compared to many other freshwater fishes (Saha and Guha, 1939). Due to its high nutritive value the fish is recommended in the diet of the sick and the convalescents (Singh Kohli and Goswami, 1989). Being a lean fish it is very suitable for people for whom animal fats are undesirable (Rahman *et al.*, 1982). The perch fish *Anabas testudineus* (Bloch) is one of the important fresh water fish of Bangladesh which is locally known as koi in different places of Bangladesh. Shingi and koi are commonly found in open waters (streams, lakes, floodplains and beels), paddy fields and swamps of Bangladesh and its preferred habitats are heavily-vegetated, stagnant waters. They are very hardy fish and can survive for a few hours out of the water due to the presence of accessory respiratory organs. Indiscriminate destructive practices have caused havoc to aquatic bio-diversity (Hussain and Mazid, 2001) in Bangladesh. International Union of Conservation of Nature (IUCN, 2000) enlisted *H. fossilis* and *A. testudineus* in the "not threatened" category in Bangladesh. The native species are threatened now due to poorly planned water management policy for irrigation, over exploitation, illegal fishing and various ecological changes in its natural habitat (Chakraborty, 2010). Considering the importance of these species form the nutritional, economic and biodiversity point of view, appropriate culture technologies for *H. fossilis* and *A. testudineus* are needed to meet the dietary demand and ultimately more of these tasty fishes will be available for the rural people of Bangladesh. For large scale production of these fishes, comprehensive information on culture technologies is required.

Therefore, the present study was conducted to find the maximum growth, yield and economic performance of shingi and genetically improved koi in various culture systems. The present experiment was also done to expand a practical and economically viable methodologies for mass production of these two species under controlled grow out culture practices. The polyculture technology of shingi and koi, and monoculture technology of koi can help to meet the dietary needs and improve the socio-economic status of the people of Bangladesh.

Materials and Methods

Study area and experimental design

The research was carried out at the private rearing ponds of three Fish Farms in Sadar, Gouripur and Gaffargaon Upazilla of Mymensingh district, Bangladesh. Foyaz Uddin Farm, Gouripur and Nazrul Islam Shabajpur, Sadar, Mymensingh was designated as treatment T_1 and T_2 for polyculture, and Babul Fish Farm, Basutia and Rubel Fish farm, Doulatpur, Gaffargaon was designated as treatment T_3 and T_4 for monoculture with three replicates each.

The experiment was conducted for a period of 100 days for koi and 120 days for shingi from May to August, 2010 in 12 ponds. The ponds were rectangular with an area of 0.30±0.11 ha and an average depth of 1.10±0.12 m. The growth performance of *H. fossilis* and *A. testudineus* was evaluated under four treatments and was replicated thrice and the pond dyke was covered with netting.

Pond preparation and fertilization

The ponds, with well-designed inlet and outlet system, were drained, cleared of aquatic vegetation and exposed to sunlight. After drying, lime ($CaCO_3$, 250 kg ha^{-1}) was spread over the pond bottom. All the ponds were filled with ground water. Six days subsequent to liming, the ponds were fertilized with organic manure (cow dung @ 6,175 kg ha^{-1}). Lime application (50 kg ha^{-1}) was maintained fortnightly to control water quality in all the treatments.

Stocking of fingerlings

Treatment T_1 and T_2 were stocked with shingi and F_2 population of (hormone 17 α methyltestosterone treated) koi. Treatment T_3 was stocked with only shingi and treatment T_4 was stocked with only F_2 population of koi (Table 1). Stocking density was planned as shown in Table 1. The fingerlings of shingi were stocked 15 days before stocking koi. The experimental ponds were stocked with shingi and koi with an initial length of 3.05±0.01 cm and weight of 2.54±0.11 g, and 0.75±0.01 cm and 0.50±0.02 g, respectively. The same stocking density of 0.247 million ha^{-1} was maintained in treatments T_1 T_3 and T_4, while stocking density of 0.3705 million ha^{-1} was maintained in treatments T_2. Culture period of shingi was recorded 120 days in treatment T_1, T_2 and T_3 and Culture period of Koi was recorded 100 days in treatment T_1, T_2 and T_4.

Table 1. Details of species combination and stocking density (individual ha^{-1}) of four treatments

Sl. No.	Name of the species	Stocking densities treatment wise ha^{-1}			
		T_1	T_2	T_3	T_4
1.	*Heteropneustes fossilis*	74,100	1,23,500	2,47,000	-
2.	*Anabas testudineus*	1,72,900	2,47,000	-	2,47,000
	Total	2,47,000	3,70,500	2,47,000	2,47,000

Supplementary feeding

Saudi Bangla and Mega feed were used in all treatments. Proximate composition of these commercial feeds was analysed according to AOAC (1995) method and nitrogen free extract (NFE) by subtraction (Castell and Tiews, 1980). Proximate composition (% dry matter) of the supplementary floating Mega feed (crude protein, crude lipid, crude fibre, ash, moisture and nitrogen-free extract) of experimental feeds was 30.0%, 03.0%, 10.0%, 17.0%, 12.0% and 28.0% and sinking Saudi Bangla feed (crude protein, crude lipid, crude fiber, ash, moisture and nitrogen-free extract) of experimental feeds was 30.0%, 06.0%, 07.0%, 18.0%, 12.0% and 27.0%, respectively. Feeds were supplied at the rate of (100-60)% for the 1st 15 days, (50-30)% from 16-30 days, (30-20)% from 31-45 days, (20-10)%

from 46-60 days, (10-8)% from 61-75 days , 6% from 76-90 days and 5% from 91-120 days twice daily in the morning and afternoon commencing from the first day of stocking. Daily ration was adjusted by estimating the standing crop once every fortnight by random sampling of the stock. Fifteen species of shingi and fifteen species of koi were sampled by using a seine and cast net.

Water quality parameters

Physico-chemical parameters of pond water were monitored fortnightly between 0900 and 1000 hr. Water temperature was recorded using a Celsius thermometer and transparency (cm) was measured by using a Secchi disc of 20 cm diameter. Dissolved oxygen and pH were measured directly using a digital electronic oxygen meter (YSI, Model 58, USA) and an electronic pH meter (Jenway, Model 3020, UK). Total alkalinity was determined by titrimetric method (Clesceri *et al.*, 1989).

Water recycling

Low-lift-pump was used to recycle surface water and shallow tube-well was used for adding under ground water in the pond (8000 L day^{-1}). This method mitigated pollution from excretory product of individuals and maintained water quality suitable for the experimental fish and primary productivity.

Estimation of growth, survival, production and feed utilisation

Fishes were sampled by using seine and cast net. The length (cm) and weight (g) of individual fish was recorded separately with the help of a measuring scale and a portable sensitive balance. Weight of each species was measured separately to assess the health condition of fish and their growth. Twenty individuals of both *H. fossilis* and *A. testudineus* species from each pond were sampled fortnightly until they attained marketable size. Growth in terms of weight, average daily growth (ADG), specific growth rate (SGR), food conversion ratio (FCR), and percentage survival and mean values (±SD) for each parameter were computed. ADG and percentage survival were followed according to De Silva (1989). SGR was calculated according to Brown (1957) and Ricker (1979), and FCR was calculated according to Brown (1957) and Gangadhara *et al.* (1997). The marketable size koi and shingi were harvested after 100 and 120 days, respectively by repeated netting, followed by draining or drying the ponds. The number of individuals were counted and weighed. Survival percentage and production (individuals ha^{-1}) of fingerlings were then calculated and compared among the treatments.

Analysis of experimental data

The data were analysed through one way analysis of variance (ANOVA) using MSTAT followed by Duncan's New Multiple Range test to find out whether there was any significant difference among treatment means (Duncan, 1955; Zar, 1984). The results were calculated and expressed as mean ±SD. A simple cost-benefit analysis was done to estimate the net benefits from different treatments.

Results

Water quality parameters

Mean levels of physico-chemical parameters over the 120 days culture of *H. fossilis* and *A. testudineus* are presented in Table 2. The mean water temperature, pH and dissolved oxygen in treatments T_1, T_2 ,T_3 and T_4 were not significantly ($P>0.05$) different during the study period. Mean Secchi disk transparency and total alkalinity differed significantly ($P<0.05$) among different treatments. Despite these variations, water quality parameters in all the experimental treatments were within the normal range for fish culture.

Table 2. Physico-chemical parameters of experimental ponds under four treatments

Parameters	Treatment T_1	Treatment T_2	Treatment T_3	Treatment T_4
Temperature (°C)	25.98±4.21 (14.18-32.40)	25.94±7.06 (14.15-32.40)	25.81±7.24 (14.20-32.40)	25.92±5.20 (14.85-32.30)
Transparency (cm)	28.48±2.19^{c} (23.80-39.50)	30.30±3.64^{b} (24.40-41.40)	36.24±4.16^{a} (24.10-45.20)	25.54±3.06^{d} (22.10-33.40)
pH	8.12±0.30 (7.70-8.50)	8.08±0.33 (7.80-8.60)	8.17±0.31 (7.70-8.55)	8.20±0.30 (7.60-8.60)
Dissolved oxygen (mg L^{-1})	4.72±0.66 (3.80-6.12)	4.34±0.70 (3.50-6.10)	4.05±0.88 (3.20-5.43)	4.40±0.68 (3.20-5.83)
Total alkalinity (mg L^{-1})	160.60±6.24^{a} (148.40-174.20)	153.23±16.69^{d} (150.60-188.20)	155.81±8.85^{c} (146.50-181.40)	156.43±7.80^{b} (146.20-171.40)

Figures with different superscripts in the same row varied significantly ($P<0.05$).
Figures in the parenthesis indicate range.

Growth and production of fish

The results of growth and production of fishes in terms of gain in weight in the four treatments are presented in Table 3. The results showed that there was significant difference in growth variation in each treatment and continued until the final harvest.

Table 3. Survival and production of *H. fossilis* and *A. testudineus* species as obtained under four treatments during 120 and 100 days study

Parameters	Fish species	Treatments			
		T_1	T_2	T_3	T_4
Initial length (cm)	*H. fossilis*	3.05±0.01	3.05±0.01	3.05±0.01	-
	A. testudineus	0.75±0.01	0.75±0.01	-	0.75±0.01
Final length (cm)	*H. fossilis*	18.02±3.42^a	12.88±3.86^b	18.17±3.88^a	-
	A. testudineus	15.71±2.40^a	10.02±3.11^b	-	15.88±2.61^a
Net length (cm)	*H. fossilis*	14.97±3.02^a	9.83±3.66^b	15.12±3.38^a	-
	A. testudineus	14.96±2.36^a	9.27±3.01^b	-	15.13±2.41^a
Initial weight (g)	*H. fossilis*	2.54±0.08	2.54±0.08	2.54±0.08	-
	A. testudineus	0.50±0.01	0.50±0.01	-	0.50±0.01
Final weight (g)	*H. fossilis*	50.14±3.22^a	30.24±3.91^c	46.44±3.11^b	-
	A. testudineus	98.84±3.02^a	56.62±3.66^b	-	98.92±3.82^a
Net weight gain(g)	*H. fossilis*	47.60±3.11^a	27.70±3.44^c	43.90±3.05^b	-
	A. testudineus	98.34±3.01^a	56.12±3.32^b	-	98.42±3.08^a
Average daily gain (g)	*H. fossilis*	0.40±0.01^a	0.23±0.01^c	0.37±0.03^b	-
	A. testudineus	0.98±0.04^a	0.56±0.02^b	-	0.98±0.02^a
SGR (% bw·day^{-1})	*H. fossilis*	3.14±0.04^a	2.63±0.03^c	3.06±0.02^b	-
	A. testudineus	5.29±0.02^a	4.73±0.04^b	-	5.29±0.02^a
FCR	*H. fossilis* *A. testudineus*	2.0±0.04^c	2.8±0.06^a	2.6±0.04^b	1.9 ±0.04^d
Species wise survival rate (%)	*H. fossilis*	87.08±2.51^b	64.70±4.88^c	87.55±2.02^a	-
	A. testudineus	90.55±2.24^a	70.77±3.41^b	-	90.76±2.07^a
Average survival rate (%)	*H. fossilis* *A. testudineus*	88.82±2.45^b	67.74±4.29^d	87.55±2.02^c	90.76±2.07^a
Species wise production	*H. fossilis* (kg ha^{-1} day^{-120})	3328.50±5.66^b	2416.30±6.42^c	10337.36±5.44^a	-
	A. testudineus (kg ha^{-1} day^{-100})	15474.08±6.14^b	9897.23±6.88^c	-	22175.60±7.35^a
Total production (kg ha^{-1})		18802.58±111.06^b	12388.25±115.22^c	10042.56±5.44^d	22175.60±7.35^a

Values in the same row having the same superscript are not significantly different ($P>0.05$).
Shingi and Koi fish were harvested after day^{-120} and day^{-100}

H. fossilis (Shingi)

During the investigation, mean final length and weight of *H. fossilis* was recorded as 14.97±3.02 cm, 9.83±3.66 cm and 15.12±3.38 cm; and 50.14±3.22 g, 30.24±3.91 g and 46.44±3.11 g in treatments T_1, T_2 and T_3, respectively (Fig. 1 and 2). The increase in weight of *H. fossilis* was the highest in T_1 followed by T_3 and T_2, respectively. The initial weight of fingerlings, stocked in all the ponds did not vary significantly ($P<0.05$). The fish in T_1 treatment showed the highest gain in weight as compared to the polyculture treatment T_2 and monoculture treatment T_3. However, the mean final weight of individuals in different treatments were significantly different ($P<0.05$). SGR in treatment T_1 was significantly higher ($P<0.05$) than in T_3 and T_2. The highest survival rate was also observed in T_3 and the lowest in T_2. There was a significant variation ($P<0.05$) in the survival rate of *H. fossilis* among different treatments. The mean productions of *H. fossilis* were 3,328.5±5.66, 2,416.30±6.42 and 10,337.36±5.44 kg ha^{-1} day^{-120} in treatments T_1, T_2 and T_3, respectively. The total production of *H. fossilis* differed significantly ($P<0.05$) among the treatments (Table 3). Production was higher in monoculture treatment T_3 and lower in polyculture treatments T_1 and T_2.

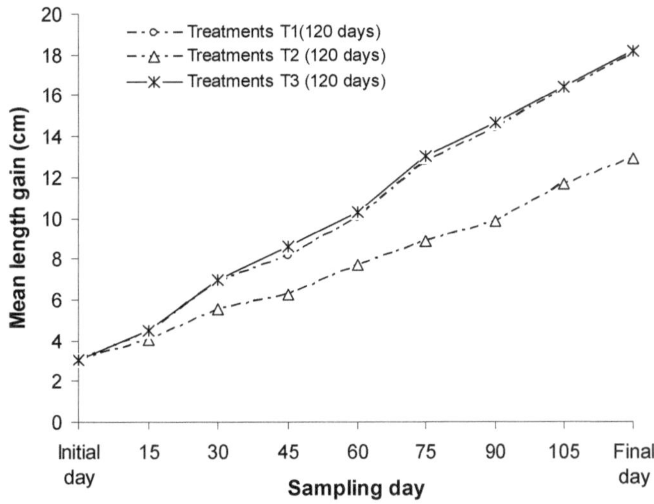

Fig.1. Fortnightly mean length (cm) gain of shingi under different treatments over a period of 120 days

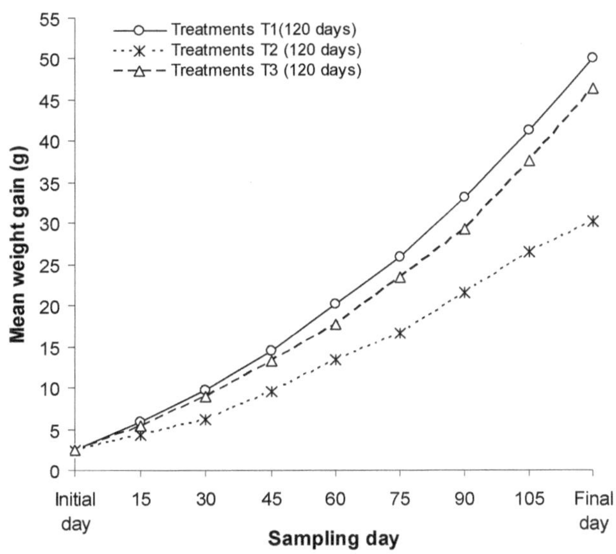

Fig.2. Fortnightly mean weight (g) gain of shingi under different treatments over a period of 120 days

A. testudineus (Koi)

During the experimental period, the final mean length and weight of *A. testudineus* was found to be 14.96±2.36 cm, 9.27±3.01 cm and 15.13±2.41 cm, and 98.34±3.02 g, 56.62±3.66 g and 98.92±3.82 g in treatments T_1, T_2, and T_4, respectively (Fig. 3 and 4). Growth and production parameters of *A. testudineus* are shown in Table 3. The initial weight of fingerling, stocked in all treatments did not differ significantly. The fish in monoculture treatment T_4 showed the highest gain in weight as compared to the polyculture treatments T_1 and T_2. However, the mean final weight of *A. testudineus* in treatment T_1 and T_4 were not significantly different ($P>0.05$). SGR in polyculture treatment T_1 and monoculture treatment T_4 was the same but significantly higher than in another polyculture treatment T_2 ($P<0.05$). The highest survival rate was also observed in monoculture treatment T_4 and the lowest in polyculture treatment T_2. There was a significant variation ($P<0.05$) in the survival rate of *A. testudineus* individuals among different treatments. The mean productions of *A. testudineus* were 15,474.08±6.14, 9,897.23±6.88 and 22,175.6±7.35 kg ha^{-1} day^{-100} in treatments T_1, T_2 and T_4, respectively. Production was higher in monoculture treatment T_4 and lowest in polyculture treatment T_2. The growth rate was the same in treatment T_1 and T_4 but lower in treatment T_3. The production of *A. testudineus* also differed significantly ($P<0.05$) among the three treatments.

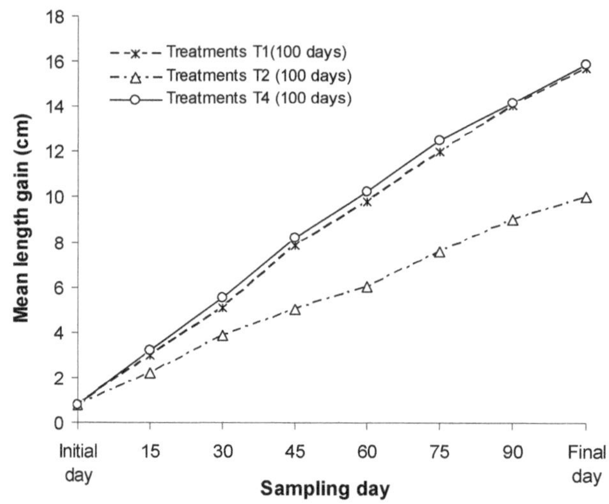

Fig. 3. Fortnightly mean length (cm) gain of koi under different treatments over a period of 100 days

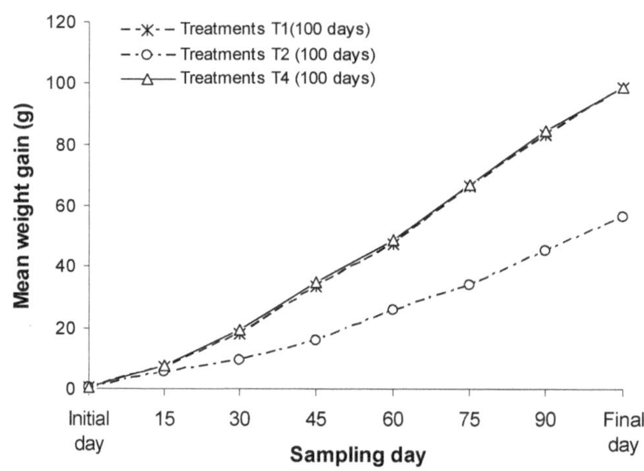

Fig. 4. Fortnightly mean weight (g) gain of koi under different treatments over a period of 100 days

Total production

In the study, FCR was significantly lower in monoculture treatment T_4 than in polyculture treatment T_1 followed by the monoculture treatment T_3 and polyculture treatment T_2. Best FCR was recorded in monoculture treatment T_4. Total net production of fish as recorded in treatments T_1, T_2, T_3 and T_4 were 18,802.58±111.06, 12,388.25±115.22, 10,042.56±5.44 and 22,175.6±7.35 kg ha^{-1}, respectively. The production of fish was higher in monoculture treatment T_4, where only F_2 population of *A. testudineus* was stocked at 2, 47,000 ha^{-1} and lowest production was recoded in monoculture treatment T_3, where only *H. fossilis* was stocked at 2, 47,000 ha^{-1}. Second highest production was recorded in polyculture treatment T_1, where *H. fossilis* and *A. testudineus* were stocked respectively at 74,100 and 1,72,900 no. ha^{-1} and an intermediate production was obtained in polyculture treatment T_2, where *H. fossilis* and *A. testudineus* were stocked respectively at 1,23,500 and 2,47,00 ha^{-1}. The production levels in all treatment differed significantly ($P<0.05$).

Economic analysis

A simple cost-benefit analysis was performed to estimate the amount of profit that had been generated from these four types of culture operations. The results of the analysis are shown in Table 4. The cost of production in treatment T_2 was consistently significantly higher than those of treatments T_4, T_1 and T_3. Highest net benefit was obtained in treatment T_1 (BDT. 17,65,769) followed by T_4 (BDT. 16,29,406), T_3 (BDT. 15,83,990) and T_2 (BDT. 6,691). Second and third net return benefit was found to be in the monoculture treatments T_4 and T_3. Treatment T_2 appeared to give poor net return levels and differed significantly ($P<0.05$) from T_1, T_3 and T_4. Although the higher production was recorded in monoculture of *A. testudineus* (treatment T_4), the higher net benefit was found in polyculture treatment T_1 due to reason of high market price of *H. fossilis*.

Table 4. Cost and return of fish production under a polyculture management of *H. fossilis* and *A. testudineus* over a period of 120 and 100 days

Item	Amount BDT ha^{-1} day^{-120}				Remarks
	Treatment T_1 (Tk)	Treatment T_2 (Tk)	Treatment T_3 (Tk)	Treatment T_4 (Tk)	
Total return (TR)	3453527^a	1813585^d	3213619^c	3326340^b	Price related with size and weight
a. Variable cost:					
1. Price of fingerlings	321100	494000	494000	247000	
2. Feed	1024912	961208	806286	1100710	(BDT 25.00 kg^{-1})
3. Fertilizer	10369	10369	10369	10369	(BDT 9.00 kg^{-1})
4. Human labour cost	48000	48000	48000	48000	(BDT 200.00 day^{-1})
5. Chemicals	12470	11580	12870	13750	
6. Miscellaneous	100250	100250	100250	100250	
Total Variable cost (TVC)	1517101	1625407	1471775	1520079	
b. Fixed cost :					BDT 200.00 $dec.^{-1}$
1. Pond rental value (day^{-120})	18947	18947	18947	18947	according to local rate, Mymensingh
Interest of operating capital	151710	162540	138907	157907	10% interest according to BKB, Bangladesh
Total fixed cost (TFC)	170657	181487	157854	176854	
Total cost (TC= TVC+TFC)	1687758^b	1806894^c	1629629^d	1696933^a	
Gross margin (GM= TR-TVC)	1936426^a	188178^b	1741844^d	1747261^c	
Net return (TR-TC)	1765769^a	6691^d	1583990^c	1629409^b	

Values with different superscripts in the same row varied significantly ($P<0.05$).
1.00US$ = 75.00 BDT (Bangladeshi Taka).
BKB= Bangladesh Krishi (Agricultural) Bank.
Sale price of shingi and koi in treatment T_1 was BDT 350.00 kg^{-1} and 150.00 kg^{-1}, and in T_2 BDT 300.00 kg^{-1} and 100.00 kg^{-1}, respectively, in T_3 (only shingi) BDT 320.00 kg^{-1} and in T_4 (only koi) BDT 150.00 kg^{-1}.

Discussion

Growth, feed efficacy and feed consumption of fish are normally governed by a few environmental factors (Brett, 1979). The temperature, water transparency, pH, dissolved oxygen and total alkalinity of the experimental ponds were within the acceptable range for fish culture that agrees well with the findings of Boyd (1979), Wahab *et al.* (1994) and Chakraborty *et al.* (2005). Higher total alkalinity level in the water of the experimental ponds might be due to higher amount of lime applied during pond preparation and frequent liming every 15 days interval during the study period (Boyd, 1982; Jhingran, 1991).

In this experiment, similar supplementary feeds are supplied for the growth of *H. fossilis* and *A. testudineus,* as explained by Haque and Barua (1989). The crude protein levels (30.0% dry weight) in supplementary feeds were very near the dietary protein of 31% for the optimal growth of *Labeo rohita* (De Silva and Gunasekera, 1991). Growth in terms of weight, weight gain and SGR of individuals of *H. fossilis* and *A. testudineus* was significantly higher in T_1 (polyculture), where the combined stocking density of the two species was the same with the stocking density of *H. fossilis* monoculture in T_3 as well as the stocking density of *A. testudineus* monoculture in T_4, although the same food was supplied in all the treatments at an equal ratio. The low growth rate of *H. fossilis* and *A. testudineus* in treatment T_2 appeared to be related with higher densities and increased competition for natural food and space. The inverse relationship between the stocking density and the growth rate is a space limiting effect on the population (Johnson, 1965).

During the experimental period, ecological factors, pond preparation and good management practices, feed quality, healthy fingerlings, and stocking rate influenced the high performance in the survival rate of *H. fossilis* and *A. testudineus* (Choudhury *et al.*, 1978). Chiu *et al.* (1989) and De Silva and Davy (1992) stated that digestibility plays an important role in lowering the FCR value by efficient utilization of food. The FCR value in monoculture treatment T_4 was significantly lower than those of polyculture treatments T_1 and T_2, and monoculture treatment T_3. Digestibility, in turn, depends on daily feeding rate, frequency of feeding, and type of food used. There was a general decrease in FCR for the population of treatment T_4 than that of the treatment T_1, T_3 and T_2 which is supported by Pechsiri and Yakupitiyage (2005). The FCR values of different treatments were acceptable and indicated better food utilization, which is agreed by Das and Ray (1989), Reddy and Katro (1979) and Islam (2002). Significantly, higher survival was recorded in monoculture treatment T_4 and polyculture treatment T_1, where, the stocking density was lower than treatment T_2. The reason for reduced survival rate in treatment T_2 might be due to higher stocking density of individuals as well as competition for natural food and space in the water area of pond which is supported by Tripathi *et al.* (1979), Haque *et al.* (1994) and Chakraborty *et al.* (2005).

In the present investigation, the amount of supplementary feeds given in different treatments was based on the number of fingerlings stocked and amount of feed provided per individual was kept at the same level. In this experiment, at higher stocking densities, presence of abundant food substances could produce a comparative interaction among the population causing a stressful situation (Houde, 1975). Hence, the observed poor growth at higher stocking densities could be due to space limiting effect, stressful situation caused by supplementary feed, some variations in environmental parameters and less availability of natural food. The results in the present experiment are very similar to the study of Ameen *et al.* (1984); Vijayakumar *et al.* (1998); Usmani *et al.* (2003) and Chakraborty *et al.* (2005).

In the present study, a significant higher production (22,175.6±7.35 kg ha^{-1} day^{-100}) was recorded in treatment T_4 than those of treatment T_1, T_2 and T_3, respectively. Despite this, consistently higher net benefits (BDT. 17,65,769 ha^{-1}) were obtained from treatment T_1 than those of treatment T_4, T_3 and T_2. The higher net benefit in treatment T_1 was obtained due to stocking of *H. fossilis* (high market value fish), although the second higher production was recorded in this treatment. The observation of higher net benefit is in agreement with those of Thakur and Das (1986), Munshi (1996), Vijayakumar *et al.* (1998), Noor Khan *et al.* (2003), Usmani *et al.* (2003) and Chakraborty *et al.* (2005). Overall, highest growth, survival, production and benefits were obtained from the monoculture treatment T_4 and second higher growth, survival, production and highest benefit were recorded in polyculture treatment T_1 which is influenced by a well developed water recirculation method (Chakraborty and Mirza. 2008).

In the present investigation, the amount of supplementary feeds given in different treatments was based on the weight of fish stocked and amount of feed provided per individual was kept at the same level. Hence, the observed low growth, production and lowest benefit in treatment T_2 could be due to higher stocking density, variations in environmental parameters. The results in the present experiment are very similar to those of Saha *et al.* (1988); Kohinoor *et al.* (1998); Hossain (2001) and Chakraborty and Azad (2008).

Conclusion

The survival, growth and production in polyculture and monoculture systems were inversely related to the stocking densities of fingerlings. Monoculture of *A. testudineus* as in treatment T_4 and polyculture of *H. fossilis* and *A. testudineus* as in treatment T_1 can be recommended to fish farmers to produce more protein food for the Bangladesh population. . This may also be helpful towards the protection of shingi from extinction as well as for the conservation of koi.

Acknowledgements

The author thanks private hatchery owners of Fish Farms who helped in conducting the experiments with financial support.

References

Ameen, M., Islam, K.R., Ahmed, K. and Mustafa. G. 1984. Indigenous small fish culture in mini ponds. *Bangladesh J. Zool.* 12: 1-10.

AOAC. 1995. Official methods of analysis of AOAC International, Vol. 2, 16th edition. (ed. W. Howritz), 1018 pp. Association of Analytical Communities, Washington, DC USA.

Boyd, C.E. 1979. Water quality in warm water fish ponds, Alabas Seed Agricultural Experiment Station, Auburn, Alabama, USA. 369 p.

Boyd, C.E. 1982. Water quality management for pond fish culture. Elsevier, The Netherlands. 318 p.

Brett, J.R. 1979. Environmental factors and growth. *In:* Environmental relations and behavior, Fish Physiology Vol. 6 (eds. W.S. Hoar, D.J. Randal and J.R. Brett). pp. 599-677. Academic press, New York.

Brown, M.E. 1957. Experimental studies on growth. *In:* The physiology of fishes. Vol. I, (ed. M.E. Brown). pp. 36-400. Academic Press, New York.

Castell, J.D. and Tiews, K. (eds.). 1980. Report of the EIFAC, IUNS and ICES working group on the standardization of methodology in fish nutrition research, Hamburg, Federal Republic of Germany, 21-23 March, 1979, EIFAC Tech. Paper. 26 p.

Chakraborty, B.K. 2010. Status and position of aquatic biodiversity of four beels and its floodplain level of northern Bangladesh with a good practice of beel nurseries. *In:* Advances in environmental research. (ed. J. A. Daniels) Nova Science Publishers, New York, USA. 8: 121-164.

Chakraborty, B.K. and Azad, S.A. 2008. Rearing and nursing of Singi, *Heteropneustes fossilis* (Bloch) at different densities in earthen nursery ponds of Bangladesh. *Bangladesh J. Life Sci.* 20 (1): 131-141.

Chakraborty, B.K. and Mirza, M.J.A. 2008. Growth and yield performance of threatened Singi, *Heteropneustes fossilis* (Bloch) under semi intensive aquaculture. *J. Fish. Soc., Taiwan.* 35: 117-125.

Chakraborty, B.K., Miah, M.I., Mirza, M.J.A. and Habib, M.A.B. 2005. Growth, yield and returns to *Puntius sarana* (Hamilton) Sarpunti, in Bangladesh under semi intensive aquaculture. *Asian Fish. Sci.* 18: 307-322.

Chiu, Y.N., Das, I. and Ray, A.K. 1989. Growth performance of Indian major carps *Labeo rohita* (Ham.) on duckweed incorporate pelleted feed: a preliminary study. *J. Inland Fish.* 21: 1-6.

Choudhury, H., Rao, N.G.S., Saha, G.N., Rout, M. and Ranaujia, D.R. 1978. Record fish production through intensive fish culture in a farmer's pond. *J. Inland Fish.* 10: 19-27.

Clesceri, L.S., Greenberg, A.E. and Trussell, R.R. 1989. Standard methods of the examination of water and wastewater (17th edition.). American Public Health Association, American Water Works Association and Water Pollution Control Federation, 1015 Washington D. C., USA. pp. 475-481.

Das, I. and Ray, A.K. 1989. Growth performance of Indian major carps *Labeo rohita* (Ham.) on duckweed incorporate pelleted feed: a preliminary study. *J. Inland Fish.* 21: 1-6.

De Silva, S.S., 1989. Reducing feed costs in semi-intensive aquaculture systems in the tropics. NAGA, *ICLARM Quarterly.* 12: 6-7.

De Silva, S.S. and Davy, F.B. 1992. Fish nutrition research system in Asia. *Asian Fish. Sci.* 5: 129-144.

De Silva, S.S. and Gunasekera, R.M. 1991. An evaluation of the growth of Indian and Chinese major carps in relation to the dietary protein content. *Aquaculture.* 92: 237-241.

Duncan, D.B. 1955. Multiple range and multiple *F*-testes. *Biometrics.* 11: 1-42.

Gangadhara, B., Nandeesha, M.C., Varghese, T.J. and Keshavanath, P. 1997. Effect of varying protein and lipid levels on the growth of Rohu, *Labeo rohita. Asian Fish. Sci.* 2: 139-147.

Haque, M.M. and Barua, G. 1989. Rearing of shingi (*Heteropneustes fosilis,* Bloch) fry under laboratory condition-II. Feeding and growth of fry. *Bangladesh J. Fish.* 12: 67-72.

Haque, M.Z., Rahman, M.A. and Hossain, M.M. 1994. Studies on the effect of stocking densities on the growth and survival of mrigal (*Cirrhinus mrigala*) fry in rearing ponds. *Bangladesh J. Zool.* 21: 51-58.

Hossain, Q.Z. 2001. Induced breeding of the fish *Cirrhinus reba* by pituitary gland extract and survival of spawn in nursery ponds. *J. Asiatic Soc.* 27: 205-213.

Houde, E.D. 1975. Effect of stocking density and food on survival, growth and yield of laboratory reared of sea Bream, *Archosargus rhomboidales* (L.) (Sparidae). *J. Fish. Biol.* 7: 115-127.

Hussain, M.G. and Mazid, M.A. 2001. Genetics improvement and conservation of carp species in Bangladesh. Bangladesh Fisheries Research Institute, Mymensingh. 56 p.

Islam, M.S. 2002. Evaluation of supplementary feeds for semi intensive pond culture of Mahaseer, *Tor putitora* (Hamilton). *Aquaculture.* 212: 263-276.

IUCN, Bangladesh. 2000. Red book of threatened fish of Bangladesh. The world conservation Union. Appendix 1, 61 p.

Jhingran, V.G. 1975. Aquaculture lecture, summer intensive fresh water fish culture, Cuttack organized by Central. Inland Fisheries Research Institute, Barracckpore (ICCR), West Bengal, India. 12 p.

Jhingran, V.G. 1991. Fish and fisheries of India, 3rd edition. Hindustan Publishing Corporation Delhi, India. 727 p.

Johnson, W.E. 1965. On mechanism of self regulation of population abundance in *Oncorhynchus nerka. Mitteilung Internationale Vereinigung fuer Theoretische unde Amgewandte.* 13: 66-87.

Kohinoor, A.H.M., Islam, M.L., Wahab, M.A. and Thilsted, S.H. 1998. Effect of mola (*Amblypharyngodon mola,* Hamilton) on the growth and production of carps in polyculture. *Bangladesh J. Fish.* 2: 119-126.

Munshi, J.S.D. 1996. Ecology of *Heteropneustes fossilis*, an air-breathing catfish of South-East Asia. Narendra Publishing House, Delhi. 174 p.

Noor Khan, M., Saiful Islam, A.K.M. and Hussain, M.G. 2003. Marginal analysis of culture of stinging catfish (*Heteropneustes fossilis*, Bloch) effect of different stocking densities in earthen ponds. *Pakistan J. Biol. Sci.* 6: 666-670.

Pechsiri, J. and Yakupitiyage, A. 2005. A comparative study of growth and feed utilization efficiency of sex-reversed diploid and triploid Nile tilapia, *Oreochromis niloticus* L. *Aqua. Res.* 36: 45-51.

Rahman, M.A., Gheyasuddin, H., Rasid, M.H. and Choudhury, M.F.Z. 1982. Proximate composition and native quality of freshwater Zeol fishes of Bangladesh. *Bangladesh J. Fish.* 25: 37-43.

Redddy, S.R. and Katro, S. 1979. Growth and conversion efficiency of air breathing catfish, *Heteropneustes fossilis* in relation to ration size. *Aquaculture.* 18: 35-50.

Ricker, W.E. 1979. Computation and interpretation of biological statistics of fish populations. *Bull. Fish. Res. Board Canada.* 191: 1-382.

Saha, K.C. and Guha, B.C. 1939. Nutritional investigation on Bengal fish. *Indian J. Medical Res.* 26: 921-927.

Saha, S.B., Gupta, M.V., Hossain, M.G. and Shah, M.S. 1988. Growth and survival of rohu (*Labeo rohita*) fry in rearing ponds at different stocking densities. *Bangladesh J. Zool.* 16: 119-126.

Singh Kohli, M.P. and Goswami, U.C. 1989. Studies on age and growth of an air-breathing catfish *Heteropneustes fossilis* (Bloch). *J. Inland Fish. Soc. India.* 21: 17-24.

Thakur, N.K. and P. Das, 1986. Synopsis of biological data on Singhi *Heteropneustes fossilis* (Bloch, 1974). Bull. No. 39, April 1986. Central Inland Fisheries Research Institute, Barrackpore, India, 32 pp.

Tripathi, S.D., Dutta, A., Sen Gupta, K.K. and Pattra, S. 1979. High density rearing of rohu spawns in village ponds. *In:* Symposium of Inland Aquaculture (Abstracts). p.14. February 12-13, 1979. CIFRI. Barrack pore, India.

Usmani, N., Jafrim, A.K. and Khan, M.A. 2003. Nutrient digestibility studies in *Heteropneustes fossilis* (Bloch), *Clarias batrachus* (Linnaeus) and *C. gariepinus* (Burchell). *Aqua. Res.* 34: 1365-2109.

Vijayakumar, C., Sridhar, S. and Haniffa, M.A. 1998. Low cost breeding and hatching techniques for the catfish, *Heteropnuestes fossilius* for small-scale farmers. *NAGA.* 21 (4): 15-17.

Wahab, M.A., Ahmed, Z.F., Haq, M.S. and Begum, M. 1994. Compatibility of silver carp in the poly culture of cyprinid fishes. *Progressive Agriculture.* 5: 221-227.

Zar, J.H. 1984. Biostatistics: Prentice-Hall, Inc. Englewood Cliffs, New Jersey, USA. 718 p.

MORPHOLOGICAL AND MOLECULAR CHARACTERIZATION OF EGGPLANT LINES FOR RESISTANT TO PHOMOPSIS BLIGHT AND FRUIT ROT

M. Ibrahim Khalil[1]*, M. Bahadur Meah[2] and Mirza Mofazzal Islam[3]

Abstract

The F_4 lines of eggplant derived from the crosses of Dohazari G x BAU Begun-1 and Laffa S x BAU Begun-1 were evaluated for resistance to phomopsis blight and fruit rot under confined field conditions. The inoculated plants exhibited differential disease reactions. Among the parents, BAU Begun-1 was resistant whereas Dohazari G and Laffa S were susceptible to *Phomopsis vexans*. All the phenotypes of F_4 progenies showed resistant reaction to the disease. Significant differences were observed among the phenotypes in all the yield components. High genotypic and phenotypic coefficient of variation, heritability and per cent genetic advance were estimated for number of fruits per plant, number of secondary branch per plant, fruit length and fruit breadth. Significant positive correlation was observed between yield contributing characters. Random amplified polymorphic DNA technique was used for assessing genetic variation and relationship among parent cultivars and their F_4 progenies of eggplant. Amplification with five decamer primers generated 69.0% polymorphic bands. Comparatively higher genetic distance was observed between Laffa S vs. green globose (Dohazari G x BAU Begun-1). The dendogram constructed from Nei's genetic distance produced two main clusters, the parent cultivars and six F_4 lines formed cluster 1 and one line in cluster 2. F_4 lines of the tested phenotypes showed similar disease reaction and divided into same sub cluster. The parent cultivars were different in case of disease reaction and finally divided into two groups, susceptible cultivars Laffa S and Dohazari G belonged to group 1 and the resistant parent BAU Begun-1 formed another group.

Keywords: Eggplant, Resistance, Phomopsis, Characterization, RAPD

[1]Plant Pathology Division, Bangladesh Institute of Nuclear Agriculture, Mymensingh-2202, Bangladesh
[2]Department of Plant Pathology, Bangladesh Agricultural University, Mymensingh-2202, Bangladesh
[3]Plant Breeding Division, Bangladesh Institute of Nuclear Agriculture, Mymensingh-2202, Bangladesh

*Corresponding author's email: ibrahim_bina@yahoo.com (M. Ibrahim Khalil)

Introduction

Eggplant also known as Brinjal (*Solanum melongena* L.) is a common fruit vegetable and widely grown in Asia, Africa, and the subtropics, including the southern USA and the Mediterranean region with world production of 42.9 million tons for the year 2009 (FAO, 2010). Asia has the largest eggplant production, which comprises more than 90% of the world production area and 87% of the world production (Choudhary and Gaur, 2009). It is the second most important vegetable crop next to potato in Bangladesh in respect of total acreage and production. It is cultivated on about 15% of total vegetable cultivated land and contributes about 8% to total vegetable production (BBS, 2011).

Among the diseases of eggplant Phomopsis blight and fruit rot caused by *Phomopsis vexans* (Sacc. and Syd.) Harter is very devastating and widespread (Chen *et al.*, 2002). Phomopsis blight ranks second only to bacterial wilt in destructiveness of eggplant and varies in severity depending on area, soil type and weather (Meah *et al.*, 2002). *Phomopsis vexans* is both externally and internally seed borne and remains viable for about 14 months in soil with plant debris and in the seed from infected fruits (Kalda *et al.*, 1977). The pathogen also causes damping off, seedling/stem blight, collar rot, stem canker and leaf spot. Studies showed that about 21% of fruit rot and 7% of seed rot in eggplant is caused by *P. vexans* and gives final yield losses of 15-50% (Das, 1998), equivalent to about US $ 20 million. Farmers have been using expensive pesticides to control the disease, but no satisfactory results could be obtained. Rather, repeated use of pesticides creates environmental pollution and a health hazard. Host plant resistance is the best control method for environmental and financial reasons. BAU Begun-1 is a round shaped variety and only resistant variety in Bangladesh cultivated in north-western region (Meah *et al.*, 2007). This variety was crossed with two commercial and popular varieties, Dohazari G and Laffa S. Almost all the F_1, F_2 and F_3's were found to be the resistant to phomopsis blight and fruit rot (Kabir, 2007; Islam, 2006).

Different scientists (Roychowdhury *et al.*, 2011; Sharmin *et al.*, 2010; Islam and Uddin, 2009; Baswana *et al.*, 2002) studied the genetic diversity, heritability and genetic advance in eggplant genotypes. The extensive variability in growth habit, spiny or non-spiny nature, foliage shape/colour, floral structure, fruit shape/size/colour and yields of eggplant were observed to have influence on phomopsis disease reaction (Singh *et al.*, 1999).

The use of molecular markers to track loci and genome regions in crop plants is now routinely applied in many breeding programs. For an effective breeding program, information concerning the extent and nature of genetic diversity or variation within a crop species is essential. Random amplified polymorphic DNA (RAPD) is a widely applied approach for characterization of DNA from plants and other organisms using PCR. The RAPD technique has provided a relatively simple and inexpensive method for analysis of genetic variation in plants, fungi and bacteria (Fukuan *et al.*, 2003; Bidochka *et al.*, 1994). RAPD have been used to construct genetic maps and for the molecular tagging of various agronomic traits in various crop species (Williams *et al.*, 1990).

RAPD technique has been successfully used for the study of genetic variability analysis in different crops including different species of *Solanum* in Bangladesh and other countries. Laila *et al.* (2012); Sharmin *et al.* (2011); Kabir (2007) and Islam (2006) characterized resistant and susceptible cultivars of eggplant along with their interspecific offspring in Bangladesh.

BAU Begun-1 is yet to be widely cultivated and its popularity is limited to a particular area of the country. So the resistance trait in BAU Begun-1 is needed to be transferred to some widely cultivated varieties covering larger areas of Bangladesh. So BAU Begun-1 was crossed with two popular commercial varieties of eggplant e.g. Dohazari G and Laffa S. The present research work was undertaken to evaluate the advanced lines of the crosses at F_4 generation for the inheritance of the resistance trait against *Phomopsis vexans* and molecular characterization using RAPD markers.

Materials and Methods

Screening of F_4 population of eggplant for resistance to Phomopsis blight and fruit rot

The experiment was conducted in the Field Laboratory of the Department of Plant Pathology, Bangladesh Agricultural University (BAU), Mymensingh, Bangladesh during the winter season of 2006-07.

Source of seed, seedling raising and transplantation

The seeds of the varieties Dohazari G, Laffa S and BAU Begun-1 and advanced lines (F_4 generation) of the intervarietal crosses of eggplant, viz. (1) Dohazari G x BAU Begun-1 (2) Laffa S x BAU Begun-1 were collected from IPM Laboratory, Department of Plant Pathology, BAU. Seedlings were raised in the seedbed following proper management.

The seedlings at 28 days age were transplanted in the field. The experiment was laid out in a Randomized Complete Block Design with four replications. The unit plot size was 12.0 m x 2.0 m with 2 rows in each plot. Fertilizers and manures were applied at recommended doses (Anonymous, 2005). Application of fertilizers and transplantation were done according to Islam (2006). Weeding, irrigation and other intercultural operations were done whenever necessary.

Inoculation of eggplant by Phomopsis vexans

A mixer of cultures of five isolates of *P. vexans* was used for inoculation. The isolates of *P. vexans* were collected from IPM Lab., BAU. The multiplication of *P. vexans* and preparation of spore suspension were done according to the procedure of Islam (2006). All the plants of F_4 generations and parents were inoculated at the flowering and fruiting stage. Spore suspension (5×10^6 spores ml^{-1}) was sprayed @ 70 ml per plant (Islam, 2006). The spraying was done at afternoon and inoculated plants were covered with transparent polythene bag for 48 hours for making favourable condition of infection.

Assessment of Phomopsis blight and fruit rot

After inoculation, symptoms on leaves, flowers and fruits were observed at seven-day interval up to 21 days. Data on leaf infection (%), per cent leaf area diseased (LAD), flower infection (%), fruit infection (%) and per cent fruit area diseased (FAD) were recorded. The disease severity was recorded according to the standard rating scale (1-5) (Islam *et al.*, 1990). The percent disease index (PDI) was calculated according to the formula of Singh (1984).

Harvesting and data recording

The characters responsible for yield of eggplant were studied and data were recorded from the parents and F_4 populations from all the plants. Plant height, number of primary branches per plant, number of secondary branches per plant, number of fruits per plant, fruit length, fruit breadth and individual fruit weight were recorded in the field. The mature fruits were harvested at the edible stage at an interval of seven days. Five fruits per variety/genotype were allowed to ripe and seeds were collected from them for growing plants in the next year.

Data analysis

Data were analyzed to find out the statistical significance. Analysis of variances followed by Duncan's Multiple Range Test (DMRT) was performed to test the differences between the genotypes with the computer based software MSTATC.

Besides, different components of genotypic and phenotypic variance, heritability, genetic advance and correlation were also estimated. Genotypic and phenotypic variances and heritability in broad sense were estimated according to formula given by Johnson *et al.* (1955), genotypic and phenotypic coefficients of variation by the formula of Burton (1952). The expected genetic advance (GA) for different characters under investigation was estimated according to the formula used by Johnson *et al.* (1955) and Allard (1960). Genetic advance in percentage of mean was calculated by the formula used by Comstock and Robinson (1952).

Using the formula of Singh and Chaudhary (1985), genetic and phenotypic covariances were calculated. Genotypic and phenotypic correlation coefficients between different characters in all possible combinations were calculated with the formula given by Miller *et al.* (1958).

Molecular characterization of F_4 population of eggplants through RAPD markers

Plant materials

Three parents (BAU Begun-1, Laffa S and Dohazari G) and seven F_4 lines obtained from their crosses (Dohazari G x BAU Begun-1 and Laffa S x BAU Begun-1) were used for the molecular characterization through random amplified polymorphic DNA (RAPD) technique. The experiment was carried out in the Biotechnology Laboratory of the Bangladesh Institute of Nuclear Agriculture, Mymensingh, Bangladesh.

Genomic DNA Extraction

Fresh leaf samples collected from 14-day old seedling were used for the study. Modified CTAB mini-prep method was followed to extract DNA from leaf samples (Kabir, 2007).

Approximately 25 mg of leaf tissues were cut into small pieces and ground with a pre cooled mortar and pestle. The ground powder was mixed with 670 µl extraction buffer (50 mM Tris-HCl, 25 mM EDTA, 5 M NaCl) and 40 µl SDS (20%). The mixture was then transferred in a 2.0 ml micro centrifuge tube (Eppendorf, Germany) and mixed thoroughly. The samples were vortexes for 20 seconds and incubated at 65°C for 10 minutes in water bath (DSB-1000). The digested mixture was removed from water bath and 100 µl 5 M NaCl was added and mixed by inversion. One hundred micro liters CTAB (BioBasic, Canada) was added and mixed by vortexing. The tubes were again incubated in water bath at same temperature for 10 minutes. Nine hundred micro liter chloroform (chloroform: isoamyl alcohol = 24: 1, v/v) was added to the mixture after cooling in room temperature and mixed gently to separate the DNA from protein. The extracts were centrifuged for 10 minutes at 14000 rpm with a micro centrifuge (SIGMA 1-14) to allow precipitation of the cell debris. The supernatant was carefully transferred to a new 1.5 ml micro centrifuge tubes without disturbing the lower portion. Equal volume of ice cooled Isopropanol was added to the tubes, mixed gently by trapping with finger, and centrifuged for 10 minutes at 14000 rpm. The supernatant was decanted and pellets were air dried for few minutes, washed with 70% ethanol (200 µl), and then centrifuged for 5 minutes at 14000 rpm. Then the liquid was completely removed without disturbing the DNA pellet when air dried for 2 hours at room temperature. Finally, the pellets were suspended in 30.0 µl of TE buffer (10 mM Tris-HCl, 1 mM EDTA) and stored at -20°C.

The quality of the DNA was verified by electrophoreses on a 0.8% agarose gel in TBE (Tris-boric acid-EDTA) buffer. The concentration of DNA samples was determined using a UV Spectrophotometer at 260 nm.

PCR amplification and Electrophoresis

RAPD reactions were maintained following the process of Williams *et al.* (1990) with some modifications. Thirty arbitrary decamer primers (Bengalore Genei, India) against DNA from parental cultivars were screened and five primers producing good scoreable and reproducible bands were selected for subsequent RAPD analysis of eggplant cultivars and phenotypes (Table 5).

PCR reactions were performed on each DNA sample in a 10.0 µl reaction mixture containing 1X PCR buffer (10 mM Tris HCl, pH 8.5; 50 mM KCl and 15 mM $MgCl_2$), 10 mM each dNTPs (Bengalore Genei, India), 5 pmols primer, 2 U of Taq DNA polymerase (Bengalore Genei), 100 ng of genomic DNA and rest amount of sterile deionized water. DNA amplification was carried out in a DNA thermocycler (Biometra, Germany) at the thermal profile: initial denaturation for 3 min at 94°C followed by 41 cycles of 1 min denaturation at 94°C, 1 min annealing at 35°C and extension at 72°C for 2 min. A final extension step at 72°C for 10 min was allowed for complete extension of all amplified fragments (Kabir, 2007). Amplified fragments were separated on a 1.5% agarose (Invitrogen, Canada) gel in 1X TBE buffer along with 20 bp DNA weight marker (Bengalore Genei, India) for 2 hours at 100V. Gel was stained with Ethidium bromide solution (0.1 µg ml^{-1}) for 20 min. Finally fragments were visualized under UV-transilluminator and photographed by Gel Documentation System (Biometra, Germany).

Scoring and Data analysis

The amplified bands were visually scored as present (1) and absent (0) separately for each individual and each primer. The scores obtained were pooled to create a single data matrix. This was used to estimate polymorphic loci, Nei (1972) genetic diversity, genetic distance and a UPGMA (Unweighted Pair Group Method with Arithmetic Means) dendrogram using a computer program, POPGENE (Version 1.31) (Yeh *et al.*, 1999).

Results and Discussion

Screening of F_4 lines of eggplant for resistance to Phomopsis blight and fruit rot

F_4 advanced lines of eggplants and respective parents showed different percentage of leaf infection and severity (leaf area diseased) after inoculation (Table 1). Among the seven F_4 lines and their respective parents, the highest (32.47%) leaf infection was found in cultivar Dohazari G followed by Laffa S (30.57%). The eggplant cultivar BAU Begun-1 did not produce any infection on leaf. In contrast, the F_4 lines (Dohazari G x BAU Begun-1) and (Laffa S x BAU Begun-1) showed leaf infection of 2.87 to 3.12% and 2.83 to 3.16%, respectively. Similarly, the per cent disease index (PDI) was recorded the highest in the cultivars Dohazari G and Laffa S. Less than 1% PDI was recorded for all the F_4 lines of eggplant.

Table 1. Reaction of F_4 lines and parents of eggplant against *P. vexans* in induced condition during 2006-2007 at Mymensingh, Bangldesh

Phenotypes/ varieties	Infected leaves (%)	Per cent disease index (leaf)	Infected fruits (%)	Per cent disease index (fruit)	Disease reaction
(Dohazari G x BAU Begun-1)					
Green round	2.97 c	0.42 c	1.42 bc	0.42 b	Resistant
Green globose	3.12 c	0.43 c	1.83 b	0.42 b	Resistant
Green long	2.87 c	0.47 c	1.79 b	0.39 b	Resistant
(Laffa S x BAU Begun-1)					
Green globose	2.86 c	0.38 c	1.37 bc	0.38 b	Resistant
Green white long	2.83 c	0.45 c	1.38 bc	0.29 b	Resistant
Purple globose	3.12 c	0.41 c	1.61 b	0.30 b	Resistant
Purple long	3.16 c	0.39 c	1.42 bc	0.44 b	Resistant
BAU Begun-1	0.00 d	0.00 d	0.00 c	0.00 b	Resistant
Dohazari G	32.47 a	33.86 a	27.22 a	30.68 a	Susceptible
Laffa S	30.56 b	30.57 b	28.04 a	31.19 a	Susceptible

Means followed by the same letter in a column did not differ significantly at the 1% level by DMRT

The fruit infection of the tested phenotypes ranged from 0.00 to 28.04%. The highest percentage of fruit infection was recorded in parent cultivar Laffa S followed by Dohazari G (Table 1). There was no fruit infection observed in the cultivar BAU Begun-1. All the F_4 lines of eggplants (Laffa S × BAU Begun-1) and (Dohazari G × BAU Begun-1) produced fruit infection less than 2% that were statistically similar. The per cent disease index was also the highest in Laffa S (31.19) followed by Dohazari G (30.68). However, all the F_4 lines had less than 1% PDI, whereas BAU Begun-1 was found free from disease. F_1 and F_2 offspring derived from two crosses, Laffa S x BAU Begun-1 and Dohazari G x BAU Begun-1 were reported free from diseases and graded as resistant to *P. vexans* (Meah, 2007). In F_3 populations, Kabir (2007) also reported similar results.

The tested phenotype were categorized according to Islam *et al.* (1990) and found parent cultivars Dohazari G and Laffa as susceptible to *Phomopsis vexans* and cultivar BAU Begun-1 was resistant. All the F_4 plants were grouped based on colour and shape and graded as resistant though a few plants showed infection (PDI less than 1%). Fruit shape and colour had no effect on the intensity of infection (Table 1). Meah (2007) also reported that Laffa S and Dohazari G were susceptible and BAU Begun-1 (previously known as IPM-31) as resistant.

Performance of the phenotypes

All the yield contributing characters had significant differences amongst the phenotypes (Table 2). This means differences exist between the tested phenotypes even within the same environment where they were grown. Plant height ranged from 58.16 to 70.43 cm. Laffa S was the tallest cultivars among the tested phenotypes which was statistically identical with Dohazari G (Table 2). In contrast, green round line of Dohazari G x BAU Begun-1 was the shortest.

Number of primary branches per plant ranged from 11.68 to 14.58. Green globose of Laffa S x BAU Begun-1 cross produced maximum number of primary branches. Green globose of Dohazari G x BAU Begun-1 cross produced the lowest number of primary branches and it was statistically similar with most of the phenotypes. The highest number (14.76) of secondary branches was recorded in BAU Begun-1 and it was statistically different from all other phenotypes. The lowest number of secondary branches (11.30) was found in green round line (Dohazari G x BAU Begun-1). Number of fruits per plant ranged from 10.52 to 21.13. The parent Laffa S produced the highest number of fruits. All the F_4 phenotypes produced statistically similar number of fruits per plant.

Table 2. Yield attributing characters of F_4 lines and their parents of eggplant affected due to Phomopsis blight and fruit rot in induced condition in 2006-2007 at Mymensingh

Phenotypes/ varieties	Plant height (cm)	Primary branch plant^{-1}	Secondary branch plant^{-1}	Fruits plant^{-1}	Fruit length (cm)	Fruit width (cm)	Individual fruit weight (g)
Dohazari G x BAU Begun-1							
Green round	58.16 e	12.27 de	11.30 e	10.52 d	11.42 d	8.20 b	246.0 e
Green globose	62.24 cd	11.68 e	12.10 de	12.80 d	12.62 c	6.91 ef	259.9 d
Green long	59.97 de	13.18 bcd	13.08 d	13.51 d	15.59 ab	7.45 cde	280.1 bc
Laffa S x BAU Begun-1							
Green globose	60.85 de	14.58 a	13.18 d	12.49 d	12.13 cd	8.09 b	275.0 bc
Green white long	59.04 de	12.58 cde	15.12 c	13.32 d	16.11 a	7.11 def	284.8 ab
Purple globose	61.83 de	11.93 e	11.70 e	12.85 d	11.78 cd	7.79 bc	277.8 bc
Purple long	60.93 de	11.77 e	12.48 de	12.58 d	15.80 ab	6.64 f	266.4 cd
BAU Begun-1	65.45 bc	13.80 ab	17.76 a	17.81 b	11.88 cd	9.43 a	297.8 a
Dohazari G	67.82 ab	11.81 e	16.38 b	20.77 a	14.97 b	9.12 a	272.4 bcd
Laffa S	70.43 a	13.49 bc	16.39 b	21.13 a	15.23 ab	7.59 bcd	283.4 b

Means followed by the same letter in a column did not differ significantly at the 1% level by DMRT

Fruit length varied from 11.42 to 16.11 cm. The highest fruit length was recorded in green white long of F_4 lines (Laffa S x BAU Begun-1). In contrast, the lowest fruit length was observed in the green round of F_4 lines (Dohazari G x BAU Begun-1). Fruit width was recorded as 6.64 to 9.43 cm. The highest fruit width was observed in BAU Begun-1 followed by Dohazari G and both were statistically similar. The lowest fruit diameter was found in purple long (Laffa S x BAU Begun-1). Individual fruit weight ranged from 246.03 to 297.83 gram. The highest fruit weight was recorded in BAU Begun-1 (Figure 5) and the lowest in green round of F_4 lines (Dohazari G x BAU Begun-1).

The significant differences were observed in all the characters studied amongst the phenotypes. Significant difference in plant height, number of branches per plant, number of fruits per plant, fruit length, fruit width and individual fruit weight have also been reported in different eggplant varieties (Islam and Uddin, 2009; Sharmin *et al.*, 2010; Roychowdhury *et al.*, 2011).

Variability, heritability and genetic advance for yield contributing characters

The highest genotypic variance ($\delta^2 g$) and phenotypic variance ($\delta^2 p$) was found in individual fruit weight (195.06 and 243.00) followed by plant height (15.11 and 18.05), number of fruits per plant (13.75 and 14.35). Low magnitude of $\delta^2 g$ and $\delta^2 p$ were observed in number of secondary branches per plant, fruit length, number of primary branches per plant and fruit width (Table 3).

The genotypic coefficient of variation (GCV) and phenotypic coefficient of variation (PCV) ranged from 5.09 to 25.09 and 5.68 to 25.63, respectively. The highest genotypic coefficient of variation and phenotypic coefficient of variation was found in number of fruits per plant followed by number of secondary branch per plant. In contrast, the lowest GCV and PCV were calculated for individual fruit weight. Islam and Uddin (2009); Sharmin *et al.* (2010); Roychowdhury *et al.* (2011) also reported higher phenotypic coefficients of variation than genotypic ones for the yield contributing characters. High GCV indicated high genetic variability and thus results the scope for improvements of the characters.

Table 3. Estimation of genetic parameters for yield contributing characters in seven F_4 lines of eggplant and their parents

Characters	Mean range	Genotypic variance ($\sigma^2 g$)	Phenotypic variance ($\sigma^2 p$)	Genotypic coefficient of variation (%)	Phenotypic coefficient of variation (%)	Heritability (%)	Genetic advance	Genetic advance (%)
Plant height (cm)	58.16-70.43	15.11	18.05	6.20	6.78	83.71	7.33	11.69
Number of primary branches plant^{-1}	11.68-14.58	0.94	1.22	7.63	8.69	77.05	1.75	13.79
Number of secondary branches plant^{-1}	11.30-17.76	5.11	5.44	16.20	16.72	93.93	4.51	32.35
Number of fruits plant^{-1}	10.52-21.13	13.75	14.35	25.09	25.63	95.82	7.48	50.59
Fruit length (cm)	11.42-16.11	3.67	3.91	13.93	14.38	93.86	3.82	27.81
Fruit width (cm)	6.64-9.43	0.80	0.89	11.42	12.05	89.89	1.75	22.31
Individual fruit weight (g)	246.03-297.83	195.06	243.00	5.09	5.68	80.27	25.78	9.40

Heritability of the phenotypes ranged from 77.05 to 95.82%. Almost all the characters showed more than 80% heritability. The highest heritability was calculated for number of fruits per plant followed by number of secondary branches per plant (93.93) and fruit length (93.86). High heritability for most of the characters of different eggplants was also reported (Islam and Uddin, 2009, Sharmin *et al.*, 2010; Roychowdhury *et al.*, 2011). The genetic advance and per cent genetic advance varied from 1.75 to 25.78 and 9.40 to 50.59, respectively. The highest genetic advance was recorded for individual fruit weight; whereas genetic advance in percentage was found in number of fruits per plant, which means there is possibility for improving the characters. Similar results were also reported by Islam and Uddin, (2009); Sharmin *et al.* (2010) and Roychowdhury *et al.* (2011). Sharmin *et al.* (2010) and Roychowdhury *et al.* (2011) showed high heritability for number of fruits per plant with high genetic advance as percentage of mean, which is in agreement with the present study.

Relationship between different yield contributing characters

Different yield contributing characters and their relationships were studied through analysis of correlation coefficients between characters (Table 4).

Plant height was positively correlated with number of primary branches per plant, fruit length, fruit width and individual fruit weight. But the relationship of plant height was significant with number of secondary branch per plant and number of fruits per plant. Positive correlation between plant height and eggplant yield also reported by Muniappan *et al.* (2010).

The relationship of number of primary branches per plant with number of secondary branches per plant, number of fruits per plant, fruit width and individual fruit weight was positive. The negative significant correlation was observed for primary branches per plant with fruit length. The number of secondary branches per plant was positively related with other characters. Significant relationships were calculated with number of fruits per plant and individual fruit weight. Insignificant relationship was observed with fruit length and fruit width.

The number of fruits per plant had positive correlation with fruit length, fruit width and individual fruit weight. Fruit length had negative correlation with fruit width. There was positive insignificant correlation between fruit length with individual fruit weight. Number of fruits has positive correlation with fruit yield also reported by Muniappan *et al.* (2010). Individual fruit weight was positively correlated with plant height, number of primary branches per plant, number of secondary branches per plant, number of fruits per plant, but insignificant with fruit length and fruit width.

Table 4. Correlation coefficients between different yield contributing characters in eggplant cultivars and their F_4 lines

Characters	Plant height (cm)	No. of primary branches plant^{-1}	No. of secondary branches plant^{-1}	No. of fruits plant^{-1}	Fruit length (cm)	Fruit width (cm)	Individual fruit weight (g)
Plant height (cm)		0.140	0.720**	0.951***	0.159	0.407	0.418
No. of primary branches plant^{-1}			0.392	0.170	- 0.154	0.336	0.531
No. of secondary branches plant^{-1}				0.854***	0.291	0.564	0.742**
Number of fruits plant^{-1}					0.315	0.491	0.524
Fruit length (cm)						- 0.430	0.240
Fruit width (cm)							0.304
Individual fruit weight (g)							

*, ** and *** indicate significant at 5%, 1% and 0.1% level of probability, respectively.

Molecular characterization of F_4 lines of eggplants through RAPD markers

RAPD profiles and analysis

The genome DNA from three eggplant cultivars and their F_4 lines when amplified with five different RAPD primers, various banding patterns were observed. Among the RAPD profiles, only two are presented in Photograph 1 and 2. The number of band ranged from 4 to 10. The primer 67AB10G7, OPB10 and OPC05 produced the highest numbers of bands as well as polymorphic bands whereas the other two primers (OPA03 and OPB09) produced the lowest number of bands. So first three primers showed a higher level of polymorphism (Table 5). The band size ranged from 100 to 800bp. RAPD analysis of BAU Begun-1, Laffa S, Dohazari G and their F_4 lines revealed that the selected five primers generated 37 bands of which 26 (69.0%) were polymorphic and 11 (31.0%) were monomorphic.

Table 5. RAPD primers with corresponding bands scored and their size range together with polymorphic bands observed in three parents (Dohazari G, Laffa S and BAU Begun-1) and their seven F_4 lines of eggplant

Primer code	Primer sequences (5´-3´)	Band size (bp)	Total bands scored	Polymorphic bands	Polymorphic loci (%)
67AB10G7	TTG GCA CGG G	100-600	10	8	80.00
OPA03	AGT CAG CCA C	100-300	4	3	75.00
OPB09	TGG GGG ACT C	140-300	5	3	60.00
OPB10	CTG CTG GGA C	160-800	10	8	80.00
OPC05	GAT GAC CGC C	110-500	8	4	50.00
Total			37	26	345.00
Average			7.4	5.2	69.00

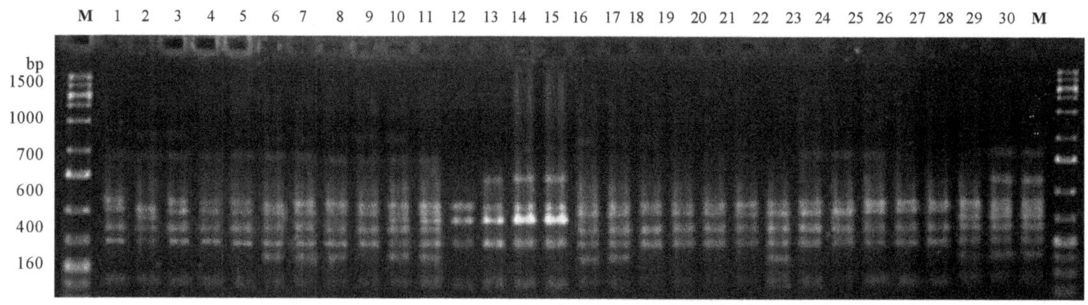

Photograph 1. RAPD profiles of 3 parents and their 7 F_4 progenies using primer OPB10 (M: 20bp ladder, Lane 1-3: BAU Begun-1, Lane 4-6: Dohazari G, Lane 7-9: Laffa S, Lane 10-12: Green round (Dohazari G x BAU Begun-1), Lane 13-15: Green globose (Dohazari G x BAU Begun-1), Lane 16-18: Green long (Dohazari G x BAU Begun-1), Lane 19-21: Green globose (Laffa S x BAU Begun-1), Lane 22-24: Green white long (Laffa S x BAU Begun-1), Lane 25-27: Purple globose (Laffa S x BAU Begun-1) and Lane 28-30: Purple long (Laffa S x BAU Begun-1).

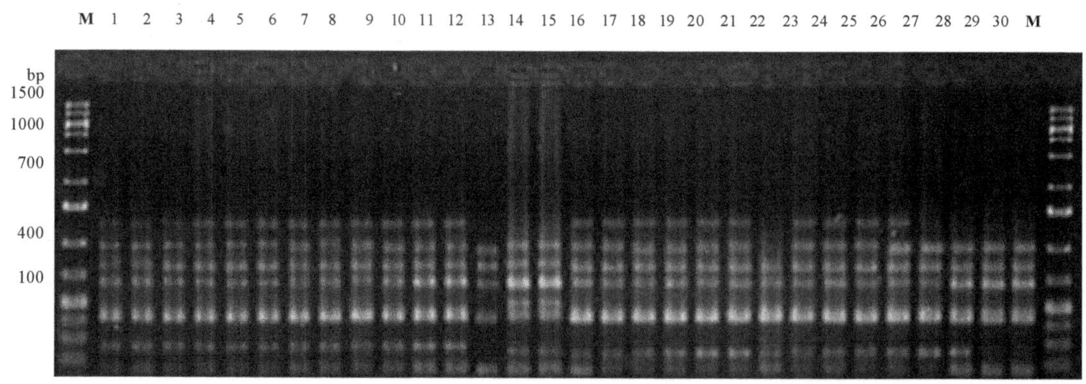

Photograph 2. RAPD profiles of 3 parents and their 7 F_4 progenies using primer OPC05 (M: 20bp ladder, Lane 1-3: BAU Begun-1, Lane 4-6: Dohazari G, Lane 7-9: Laffa S, Lane 10-12: Green round (Dohazari G x BAU Begun-1), Lane 13-15: Green globose (Dohazari G x BAU Begun-1), Lane 16-18: Green long (Dohazari G x BAU Begun-1), Lane 19-21: Green globose (Laffa S x BAU Begun-1), Lane 22-24: Green white long (Laffa S x BAU Begun-1), Lane 25-27: Purple globose (Laffa S x BAU Begun-1) and Lane 28-30: Purple long (Laffa S x BAU Begun-1).

The number of band and polymorphic band per primer obtained 7.4 and 5.2, respectively (Table 5). The number of polymorphic band per primer have been reported 4.0, 5.3 and 5.9 in eggplant by Laila *et al.* (2012), Sharmin *et al.* (2011) and Karihaloo *et al.* (1995), respectively. The difference might be due to the use of different primers and eggplant varieties. The percentage of polymorphic loci was 69.00 for Laffa S, Dohazari G, BAU Begun-1 and their F_4 lines. Similar level of polymorphism expressed by arbitrary primers in eggplant also available (Laila *et al.*, 2012; Sharmin *et al.*, 2011).

Frequency of polymorphic loci

The number and proportion of polymorphic loci was found to be the highest in green white long line of Laffa S x BAU Begun-1 (27.03%) and the lowest on green long (F_4 line of Dohazari G × BAU Begun-1) and green globose (F_4 line of Laffa S x BAU Begun-1) (Table 6). The highest proportion of Nei's gene diversity value and Shannon's Information index were found in green white long (F_4 line of Laffa S x BAU Begun-1) which was 0.1166 and 0.1675, respectively. On the other hand, the lowest proportion of polymorphic loci and Nei's gene diversity value were found in purple globose (F_4 line of Laffa S × IPM-31) (Table 6).

Table 6. Estimates of genetic variation, number and proportion of Polymorphic loci, gene diversity and Shannon's Information index obtained in 3 cultivars and 7 F_4 lines of eggplant

Phenotypes/ varieties	No. of polymorphic loci	Proportion of polymorphic loci (%)	Gene diversity (h)	Shannon's Information index (I)
BAU Begun-1	7	18.92	0.0720	0.1068
Laffa S	6	16.22	0.0639	0.0939
Dohazari G	5	13.51	0.0558	0.0810
Green round (D)*	8	21.62	0.1004	0.1417
Green globose (D)	4	10.81	0.0477	0.0681
Green long (D)	1	2.70	0.0132	0.0184
Green globose (L)**	1	2.70	0.0132	0.0184
Green white long (L)	10	27.03	0.1166	0.1675
Purple globose (L)	3	8.11	0.0294	0.0442
Purple long (L)	4	10.81	0.0426	0.0626

*Green round (D) = Green round progeny of Dohazari G x BAU Begun-1
**Green globose (L) = Green globose progeny of Laffa S x BAU Begun-1

The DNA polymorphisms are detected by band presence versus absence and may be caused by failure to prime a site in some individuals due to nucleotide sequence difference or by insertions or deletions between priming sites (Clark and Langigan, 1993).

Genetic identity and genetic distance

Genetic identity between varieties was found for the 5 primers, ranged from 0.5661 to 0.9883 (Table 7). Comparatively the higher genetic identity was found in green globose (Laffa S x BAU Begun-1) vs. purple globose (Laffa S x BAU Begun-1) followed by purple globose (Laffa S x BAU Begun-1) vs. green white long (Laffa S x BAU Begun-1). The lowest genetic identity was observed between green globose (Laffa S x BAU Begun-1) vs. Laffa S.

The values of pair-wise comparisons of Nei's (1972) genetic distance between varieties were computed from combined data for the 5 primers, ranged from 0.0177 to 0.5689 (Table 7). Comparatively, higher genetic distance was observed between Laffa S vs. green globose (Dohazari G x BAU Begun-1), BAU Begun-1 vs. green globose (Dohazari G x BAU Begun-1) and Dohazari G vs. green globose (Dohazari G x BAU Begun-1) phenotype pairs than other phenotype combinations. The lowest genetic distance was found in green globose (Laffa S x BAU Begun-1) vs. purple globose (Laffa S x BAU Begun-1) phenotype pair. Considering the genetic distance values, the varieties were genetically different from each other.

Table 7. Summary of Nei's (1972) genetic identity (above diagonal) and genetic distance (below diagonal) values for ten phenotype pairs of eggplant

Phenotypes/ varieties	BAU Begun-1	Laffa S	Dohazari G	Green round (D)	Green globose (D)	Green long (D)	Green globose (L)	Green white long (L)	Purple globose (L)	Purple long (L)
BAU Begun-1	****	0.9148	0.9326	0.8616	0.6029	0.8608	0.8934	0.8943	0.8964	0.7665
Laffa S	0.0891	****	0.9322	0.8404	0.5661	0.8147	0.8562	0.8626	0.8590	0.7025
Dohazari G	0.0698	0.0703	****	0.8663	0.6262	0.8500	0.8723	0.8593	0.8752	0.7435
Green round (D)	0.1490	0.1738	0.1435	****	0.7770	0.8563	0.8929	0.8674	0.8897	0.8525
Green globose (D)	0.5060	0.5689	0.4681	0.2523	****	0.6507	0.7005	0.6760	0.7217	0.7379
Green long (D)	0.1499	0.2050	0.1626	0.1552	0.4298	****	0.9586	0.8904	0.9389	0.8326
Green globose (L)	0.1127	0.1552	0.1366	0.1133	0.3559	0.0423	****	0.9438	0.9883	0.8326
Green white long (L	0.1117	0.1478	0.1516	0.1422	0.3915	0.1161	0.0578	****	0.9604	0.8445
Purple globose (L)	0.1094	0.1520	0.1333	0.1169	0.3262	0.0630	0.0117	0.0404	****	0.8482
Purple long (L)	0.2659	0.3532	0.2964	0.1596	0.3039	0.1832	0.1832	0.1690	0.1646	****

*Green round (D) = Green round progeny of Dohazari G x BAU Begun-1
**Green globose (L) = Green globose progeny of Laffa S x BAU Begun-1

Dendrogram of F_4 lines and their parents

Dendrogram based on Nei's (1972) genetic distance using Unweighted Pair Group Method of Arithmetic Means (UPGMA) indicated segregation of the phenotypes of eggplant into two main clusters: nine phenotypes grouped into cluster I and only green globose (F_4 line of Dohazari G x BAU Begun-1) grouped in cluster 2 (Fig. 1).

In cluster 1, BAU Begun-1, Dohazari G, Laffa S, green round and green long (F_4 line of Dohazari G x BAU Begun-1), green globose, purple globose and green white long (F_4 lines of Laffa S x BAU Begun-1) formed sub cluster 1 while purple long (F_4 line of Laffa S x BAU Begun-1) was in sub cluster 2. Among the phenotypes of sub cluster 1, the three parent cultivars BAU Begun-1, Dohazari G and Laffa S formed sub sub cluster 1 and F_4 lines of eggplant belonged to sub sub cluster 2. F_4 lines of the tested phenotypes showed similar disease reaction and also produced similar number of fruits and formed same sub cluster. The parent cultivars are produce similar number of fruits, but they are different in case of disease reaction and finally divided into two groups, susceptible cultivars Laffa S and Dohazari G belonged to group 1 and the resistant parent BAU Begun-1 formed another group. On the other hand sub sub cluster 2 were divided into two groups, green round of Dohazari G x BAU Begun-1 formed sub group 1 and other four phenotypes were in sub group 2.

Eggplant germplasms of the Indian subcontinent are very diverse. Wide variation in the desirable phenotypes in different regions substantiates the high level of genetic variability (Karihaloo *et al.*, 1995). Thus, the present study showed different types of reaction and grouping.

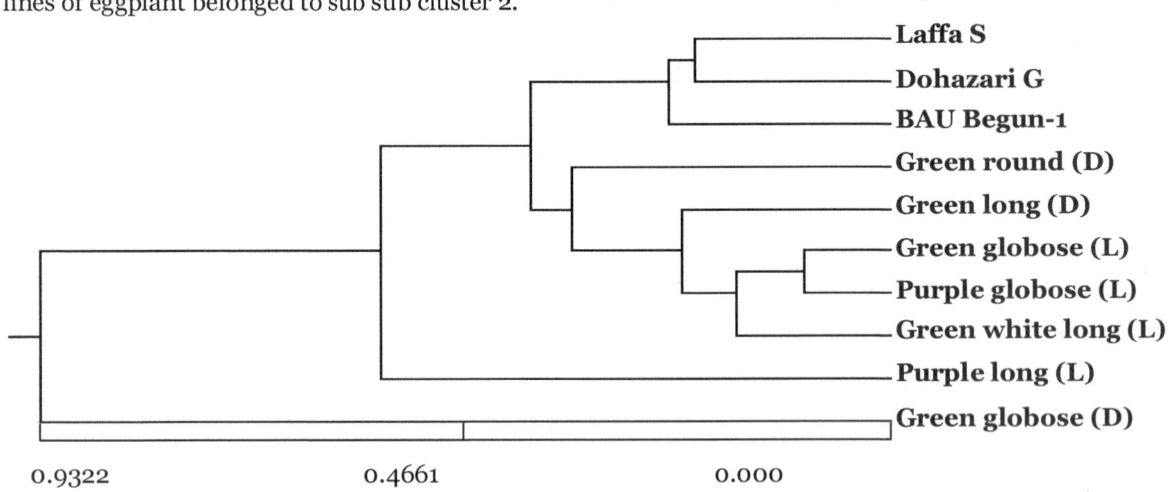

Fig. 1. UPGMA dendrogram based on Nei's (1972) genetic distance summarizing the data on differentiation between 3 cultivars and 7 F_4 lines of eggplant, according to RAPD analysis [Green round (D) = Green round progeny of Dohazari G x BAU Begun-1, Green globose (L) = Green globose progeny of Laffa S x BAU Begun-1].

Among the three parent and their 7 F_4 lines studied, the cultivar BAU Begun-1 and F_4 lines showed resistant reaction and rest parents showed susceptible reaction to phomopsis blight and fruit rot. According the RAPD analysis, three cultivars and 6 F_4 lines grouped in same cluster and finally resistant parent formed separate sub sub cluster. Therefore, the result of the present study indicates that the variability and diversity present among the phenotypes, which will be useful for development of new eggplant cultivars.

References

Allard, R.W. 1960. Principles of Plant Breeding. John Willey and Sons. Inc. New York. p. 36.

Anonymous. 2005. Fertilizer Recommendation Guide, Bangladesh Agricultural Research Council, Farmgate, Dhaka. p. 108.

Baswana, K.S., Bhatia, M.K., Dharamveer, D. and Duhan, D. 2002. Genetic variability and heritabilty studies in rainy season brinjal (*Solanum melongena* L.). *Harayana J. Hort. Sci.* 31: 143-145.

BBS (Bangladesh Bureau of Statistics). 2011. Statistical Yearbook of Bangladesh. Bangladesh Bureau of Statistics. Ministry of Planning, Government of the Peoples' Republic of Bangladesh. 139 p.

Bidochka, M.J., McDonald, M.A., Leger R.J. and Roberts. D.W. 1994. Differentiation of species and strains of entomopathogenic fungi by random amplification of polymorphic DNA (RAPD). *Current Genet.* 25: 107–113.

Burton, G.M. 1952. Quantitative inheritance in grasses. Proceedings of 6th International Grassland Congress. pp. 277-283.

Chen, N.C., Kalb, T., Talekar, N.S., Wang, J.F. and Ma, C.H. 2002. AVRDC Training Guide: Suggested Cultural Practices for Eggplant. p. 4.

Choudhary, B. and Gaur, K. 2009. The development and regulation of Bt brinjal in India (Eggplant/Aubergine). International Service for the Acquisition of Agri-biotech Applications (ISAAA) Brief No. 38. ISAAA, Ithaca, NY, USA. 15 p.

Clark, A.G. and Langigan, C.M.S. 1993. Prospects for estimating nucleotide divergence with RAPDs. *Mol. Evolution.* 10: 1096-1111.

Comstock, R.E. and Robinson, H.F. 1952. Genetic parameters, their estimation and significance. Proc. 6th International Congress. pp. 284-291.

Das, B.H. 1998. Studies on Phomopsis fruit rot of brinjal. M.S. thesis. Department of Plant Pathology, Bangladesh Agricultural University, Mymensingh. 64 p.

FAO. 2010. Food and Agriculture Organization (FAO)database.http://www.faostat.fao.org/site/567/DesktopDefault.aspx.

Fukuan, Z., Gao, X.H., Cheng, G.H., Fran, S.X. and Ying, Y. 2003. RAPD analysis and identification of chilling resistance of cellular variant of the eggplant. *Acta Agric. Boreali Sinica.* 18: 17-19.

Islam, M.R. 2006. An integrated approach for management of Phomopsis blight and fruit rot of eggplant. Ph.D. thesis. Department of Plant Pathology, Bangladesh Agricultural University, Mymensingh. 133 p.

Islam, M.S. and Uddin, M.S. 2009. Genetic variation and trait relationship in the exotic and local eggplant germplasm. *Bangladesh J. Agril. Res.* 34 (1): 91-96.

Islam, S.K., Sintansu, P. and Pan, S. 1990. Efficacy of humidity and temperature on Phomopsis fruit rot of brinjal. *Environ. Ecology.* 8: 1309-1310.

Johnson, H.W., Robinson, H.F. and Comstock, R.E. 1955. Estimation of genetic and environmental variability in Soybean. *Agron. J.* 47: 314-318.

Kabir, M.M. 2007. Molecular characterization of F_3 offspring of eggplant crosses for resistance to Phomosis blight and fruit rot. M.S. thesis. Department of Plant Pathology, Bangladesh Agricultural University, Mymensingh. 102 p.

Kalda, T.S., Swarup, V. and Choudhury, B. 1977. Resistance to Phomopsis blight in eggplant. *Veg. Sci.* 4: 90-101.

Karihaloo J.I., Brauner, S. and Gottlieb, L.D. 1995. Random amplified polymorphic DNA variation in the eggplant, *Solanum melongena* L (*Solanaceae*). *Theor. Appl. Genet.* 90: 767-770.

Laila, R., Siddiqua, M.K., Khalil, I., Robin, A.H.K. and Meah. B. 2012. Molecular characterization of *Solanum melongena* using RAPD marker for collar rot resistance. *Int. Res. J. Appl. Life Sci.* 1 (4): 38-65.

Meah, M.B. 2007. Development of an integrated approach for management of Phomopsis blight and fruit rot of eggplant in Bangladesh. Final research report (2001-2006). Dept. of Plant Pathology, Bangladesh Agricultural University, Mymensingh, Bangladesh. pp. 67-75.

Meah, M.B., Ahmed, S.A., Islam, M.R., Islam, M.M. and Tusher, M.K. 2007. Eggplant of Bangladesh (2nd edition). IPM Laboratory, Department of Plant Pathology, Bangladesh Agricultural University, Mymensingh, Bangladesh. 27 p.

Meah, M.B., Hossain, M.D. and Islam, M.R. 2002. Development of an integrated approach for management of Phomopsis blight and fruit rot of eggplant in

Bangladesh. Annual research report (2001-2002). Department of Plant Pathology, Bangladesh Agricultural University, Mymensingh, Bangladesh. pp. 8-10.

Miller, P.A., Williams, J.C., Robinson, H.F. and Comstock, R.E. 1958. Estimates of genetic and environmental variance and covariance and their implication in selection. *Agron. J.* 50: 126-131.

Muniappan, S., Saravanan, K. and Ramya, B. 2010. Studies on genetic divergence and variability for certain economic characters in eggplant (*Solanum melongena* L.). *Electronic J. Plant Breed.* 1 (4): 462-465.

Nei, M. 1972. Genetic distance between populations. *American Naturalist.* 106: 283-292.

Roychowdhury, R., Roy, S. and Tah, J. 2011. Estimation of heritable components of variation and character selection in eggplant (*Solanum melongena* L.) for mutation breeding programme. *Continental J. Biol. Sci.* 4 (2): 31-36.

Sharmin, D., Khalil, M.I., Begum, S.N. and Meah, M.B. 2011. Molecular characterization of eggplant crosses by using RAPD analysis. *Int. J. Sust. Crop Prod.* 6 (1): 22-28.

Sharmin, D., Meah, M.B. and Moniruzzaman, M. 2010. Inheritance of resistance to phomopsis blight and fruit rot in brinjal. *J. Agroforestry Environ.* 3 (2): 135-140.

Singh, N., Gyanendra, S. and Kalda, T.S. 1999. Genetic diversity in eggplant. *IPGRI-Newsletter for Asia, the Pacific and Oceana.* 29: 22.

Singh, P.K. and Chaudhary, B.D. 1985. Biometrical Methods in Quantitaive Genetic Analysis. Kalyani Publishers, New Delhi, India. 318 p.

Singh, R.S. 1984. An Introduction to Principles of Plant Pathology. 2nd edn. IBH Publishing Co. Delhi, India. 608 p.

Williams, J.G.K., Kubclik, A.R., Livak, K.J., Rafalski, J.A. and Tingey, S.V. 1990. DNA polymorphisms amplified by arbitrary primers are useful as genetic markers. *Nucleic Acids Res.* 18: 6531-6535.

Yeh, F.C., Yang, R.C., Boyle, T.B.J., Ye, Z.H. and Mao, J.X. 1999. POPGENE. The user-friendly software for population genetics analysis. Molecular Biology and Biotechnology Centre, university of Albetra, Canada. 29 p.

10

DEVELOPMENT AND EVALUATION OF A DRUM SEEDER FOR ONION

M.A. Hoque[1] and M.A. Wohab[2]

Abstract

To achieve optimum yield of onion by placing the seeds at a certain depth in lines, a manually operated drum seeder for onion has been designed, fabricated and tested in Farm Machinery and Postharvest Process (FMP) Engineering Division of Bangladesh Agricultural Research Institute (BARI). The seeder is suitable for the small farmers who have practice of direct seeding of onion under dry land cultivation. The seeder consists of a frame, a PVC drum, furrow openers, furrow closer, wheels and handle. In 2009-2010, the seeder was fabricated in FMP Engineering divisional workshop and tested in experimental plot, BARI, Gazipur. Seed rate of the seeder was 4 kg ha^{-1} and field capacity was 0.08 ha hr^{-1}. Weight and price of the seeder were 5.5 kg and Tk 2500 per unit, respectively. In 2010-2011, a drug chain was added in the seeder and field performance evaluation of the seeder was repeated comparing with manual line sowing and traditional broadcasting methods in experimental plot of BARI, Gazipur. There was no significant variation between average yield of onion by using the seeder (13.26 ton ha^{-1}) and in that of manual line sowing (12.90 ton ha^{-1}) but there was significantly higher yield of onion by using the seeder than that in traditional (10.25 ton ha^{-1}). But sowing time (6 h ha^{-1}) was significantly lower for seeder use than manual line sowing (48 h ha^{-1}) which contribute to reduce the cost of sowing for direct seeded onion.

Keywords: Onion, Drum Seeder, Field Capacity, Yield and Sowing Time

[1]Senior Scientific Officer, F M P Engineering Division, BARI, Gazipur, Bangladesh
[2]Principal Scientific Officer, F M P Engineering Division, BARI, Gazipur, Bangladesh

*Corresponding author's email: arshadulbari@yahoo.com (M.A. Hoque)

Introduction

Onion is the most important spices crops grown in Bangladesh. It is cultivated almost throughout Bangladesh and amongst all spices ranks first in production and second in acreage (Asaduzzaman *et al.*, 2012). The area under onion cultivation is about 36,842 ha with annual production of about 8,49,000 mt (BBS, 2007). The national average yield of onion is very low as 6.94 ton ha^{-1} while the world average yield is about 20.64 ton ha^{-1} (FAO, 2006). To meet up the demand of increased population, onion production should be increased with lower production cost.

Generally, the onion seeds are sown in nursery and transplanted. Macias *et al.* (2004) observed that the system of direct seed obtained better yield (18%), bulb weight (21%) and precocity (11 days) in comparing to the transplant methods. In case of rain fed condition, the onion seeds are broadcasted to save the labour and the yield of around 4.5 tones ha^{-1} is obtained. The poor yield is recorded due to non-uniform spacing of seeds that affects the size of the onion bulbs. Considering the fact, an experiment was undertaken on development of a drum seeder for onion for reducing production cost with higher yield in collaboration with Spices Research Centre, Bogra. The objectives of the study were:

a) To develop a manually operated onion seeder by using locally available materials.
b) To test the seeder in laboratory and field conditions.

Materials and Methods

A drum seeder for onion was designed and fabricated in Farm Machinery and Postharvest Process Engineering Division of Bangladesh Agricultural Research Institute (BARI), Joydebpur, Gazipur during 2009-2010. The design of the drum seeder was drawn by AutoCAD software 2004 and then constructed according to drawing dimensions in the FMP Engineering divisional workshop. The seeder is made of locally available materials like M S flat bar, M S sheet, M S pipe, M S hollow square bar, M S rod, PVC pipe and M S nut bolts etc. The seeder consists of a frame, a drum, furrow openers, wheels and handle. Functional views of the drum seeder for onion are shown in Fig. 1.

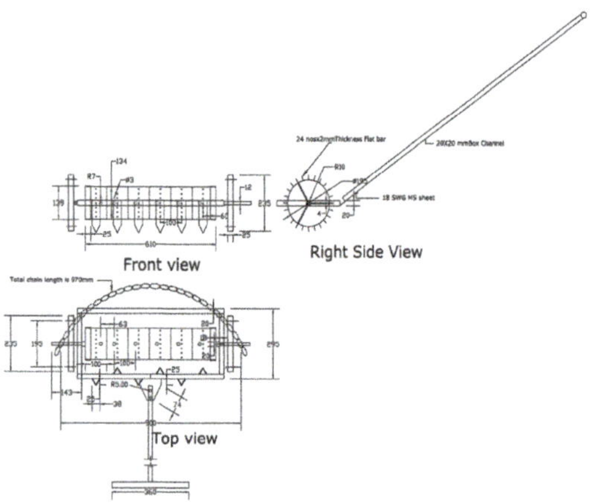

Fig. 1. Functional views of the drum seeder

The functions of the parts of the drum seeder are described briefly as follows.

1) **Frame:** It is made of MS hollow square bar of 18 mm size. Length and width of frame is 650 mm and 300 mm, respectively. Other parts of seeder are fitted with the frame.

2) **Wheel**: MS flat bar of 18 mm size was used to make the wheel. There are lugs welded at the out side of round flat bar. Wheel diameter is 195 mm. It is fitted in both side of the drum by a shaft.

3) **Drum:** PVC pipe of 139 mm diameter and 610 mm length was used in the drum. It has holes of 3 mm diameter size, in 6 rows with spacing of 100 mm (line to line spacing). To achieve 20 mm spacing between seeds, the drum has holes along the circumference at a spacing of 13.5 mm. Five partitions are made inside the seeder drum between the rows to avoid accumulation of seeds to one side of the seeder drum. The entire assembly is attached in the middle of the frame. In order to rotate the drum, the seeder is pulled in forward direction.

4) **Furrow opener:** Single pointed shovel type six furrow openers are fitted to the frame in front of the seeding drum. It is made of MS angle bar.

5) **Furrow closer:** A 700 mm long drug chain was used behind the frame to close the furrows.

6) **Handle:** The seeder has a handle of 1500 mm length for easy pulling. It is made of MS hollow square bar of 18 mm size.

A photographic view of the drum seeder for onion and onion field sown by the seeder are shown in Fig. 2 and Fig.3 respectively.

Laboratory test

To ascertain various design parameters of the seeder, laboratory test was carried out in the FMP Engineering division, BARI, Gazipur. The drum seeder was tested in lab bed to determine the performance parameters. The seeder was calibrated to appraise seed rate and seed depth.

Machine operation

Before putting the seeds inside the drum, seeds were soaked for 24 hours for sprouting. Then small amount of seeds (100 g) were put inside the drum in each chamber. Seeding operation was started from a corner of the land. After placement the seeder in a corner, the seeder was pulled forward. When the seeder moved forward, wheel as well as the drum rotated and seeds dropped in the furrows made by furrow openers. Then the seeds were covered by drug chain and leveled the land.

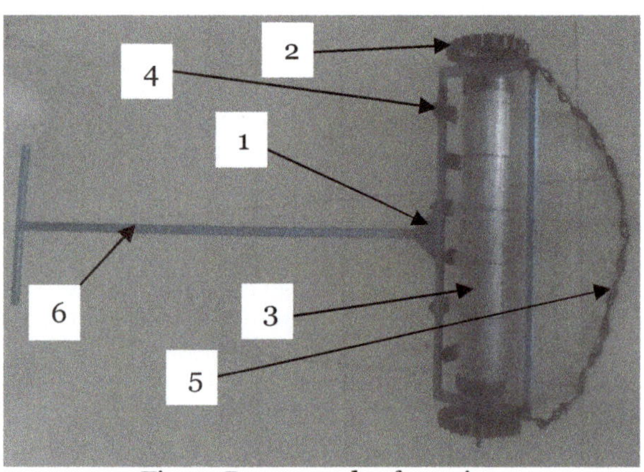

Fig. 2. Drum seeder for onion

Fig. 3. Onion field seeded by the seeder, Gazipur

Field test

In 2009-2010, the seeder was tested after fabrication in experimental plot, BARI, Gazipur. In 2010, a drug chain was incorporated to improve the seeder performance in divisional workshop and tested in experimental plot, BARI, Gazipur. The seeder was tested with 3 treatments. Treatments were: T_1- Seeding by seeder; T_2- Hand seeding in line; T_3-Broadcast. The seeder was tested in well prepared land. After germination, irrigation and weeding operations were done as and when necessary. At the age of 45 days of plants, excess plants were thinned out and transplanted in other plots.

Results and Discussion

The test results of the drum seeder in 2009-2010 are shown in Table 1. Width of the seeder was within the reach of an operator to pull it at 1.5 cm sowing depth. Diameter of the holes in the seeder was 3 mm to release onion seed easily. It depends on the variety of the onion seed. Average geometric mean diameter of the onion seed was observed 1.98mm (Chhina and Sharma, 2011). Seed rate of the seeder was 4 kg ha^{-1}. But 45 days after sowing, plants were thinned out keeping optimum plant population. The field capacity of the seeder was 0.08 ha hr^{-1}. Weight of the seeder was 5.5 kg, which was comfortable for the operator to turn at head of the land. The price of the seeder (Tk 2500) was also reasonable for an onion growing farmer.

Table 1. Test results of onion drum seeder in 2009-2010

Sl. No.	Particulars	Results
1	Seeder drum width, mm	600
2	Depth of seed sowing, mm	10-15
3	Number of rows	6
4	Row spacing, mm	100
5	Seed rate, kg ha^{-1}	4.0
6	Field capacity, ha h^{-1}	0.08
7	Sowing time, man-day ha^{-1}	2
8	Weight, kg	5.5
10	Yield of onion, kg ha^{-1}	10660

Comparative study of drum seeder for onion cultivation with conventional practices is shown in Table 2. There is no significant difference on yield between T_1 and T_2. Nevertheless, significantly lowest yield was found in T_3 due to less plant population and 100-bulb weight. In T_3, plant population was less due to uneven distribution of seed and uneven seed depth. However, there was no significant variation of sowing time between treatment T_1 and T_3, but sowing time of onion was significantly highest in T_2 than T_1 and T_3.

Table 2. Comparative study of drum seeder for onion cultivation with conventional practices in Gazipur

Treatment	Plant population (Plant/m^2)	Plant hight (cm)	Bulb dia. (cm)	100 bulb weight (kg)	Yield (ton ha^{-1})	Sowing time (hr ha^{-1})
T_1	42a	40.36	4.41	3.56a	13.26a	6a
T_2	40a	39.16	4.43	3.41a	12.90a	48b
T_3	34b	37.86	4.40	2.85b	10.25b	4a

Conclusion and Recommendations

The seeder was capable to use for onion seed sowing in particular depth (10-15 mm). The field capacity of the seeder for onion was 0.008 ha h^{-1}. Onion yield of seeder sown field and that of manual line sowing was statistically identical though sowing time was significantly varied. The seeder should be further developed for multi crop use like cucumber, amaranth, carrot, radish etc.

References

Asaduzzaman, M., Hasan, M.M., Hasan, M.M. and Moniruzzaman, M. 2012. Quality seed production of onion (*Allium cepa* L.): an integrated approach of bulb size and plant spacing. *J. Agril. Res.* 50 (1): 119-128.

BBS. 2007. Statistical Bulletin of Bangladesh. Bangladesh Bureau of Statistics, Ministry of Planning, Government of the Peoples Republic of Bangladesh.

Chhina, R.S. and Sharma, A.D. 2011. Studies on physical properties of onion seeds (*Allium cepa* L.). *Prog. Hort.* 43 (1): 76-80.

FAO. 2006. Bulletin of Statistics. Food and Agriculture Organization of the United Nations, Rome.

Macias, R., Grijalva, R.L., Valenzuela, M.J. and Robles, F. 2004. Yield and quality of onion varieties under direct seed and transplant seedling. *Hort. Sci.* 39 (4): 801.

MARKER-ASSISTED BACKCROSSING FOR IDENTIFICATION OF SALT TOLERANT RICE LINES

M. Moniruzzaman[1], M.S. Alam[2*], J.A. Rashid[3], S.N. Begum[4] and M.M. Islam[5]

Abstract

SSR or microsatellite markers are proved to be ideal for making genetic maps, assisting selection and studying genetic diversity in germplasm. SSR markers are playing important role to identify gene for salt tolerance that can be helpful for plant breeders to develop new cultivars. The experiment was conducted during the period from July 2009 to November 2010 in the experimental field and Biotechnology Laboratory of Plant Breeding Division, Bangladesh Institute of Nuclear Agriculture (BINA), Mymensingh to identify salt tolerant rice line of BC_1F_1 progenies of Binadhan-5 x FL-478 using SSR markers. Salt tolerant genotype, FL-478 was crossed with high yielding variety, Binadhan-5. Randomly selected 40 BC_1F_1 progenies along with their two parents (Binadhan-5, FL-478 and F1) were genotyped with microsatellite or SSR markers for identification of salt tolerant rice lines. Parental polymorphism survey was assayed by 10 SSR markers and three polymorphic SSR markers viz., RM 336, RM 510, and RM 585 were selected to evaluate BC_1F_1 rice lines for salt tolerance. In respect of Primer RM 336, 11 lines were found as salt tolerant and 25 lines were heterozygous and 3 lines were susceptible. Primer RM 510 identified two tolerant, 14 heterozygous and 22 susceptible lines. And primer RM 585 identified 4 lines as tolerant and 35 lines as susceptible. Thus, these markers could be efficiently used in tagging salt tolerant genes, in marker-assisted selection and quantitative trait loci (QTL) mapping. The selected BC_1F_1 could be used for developing BC_2F_1 and BC_2F_2 and mapping genes for salinity tolerance.

Keywords: Salinity, Marker-assisted backcrossing (MABC), and SSR marker

[1]Biotechnologist (Field Research), Bangladesh Jute Research Institute, Manikgonj, Dhaka, Bangladesh
[2]Scientific Officer (Plant Breeding), Regional Agricultural Research Station, BARI, Akbarpur, Moulvibazar
[3]Executive, R&D (Biotechnology/ Agriculture/ Microbiology/ Enzyme), MATEX Bangladesh Ltd. Dhaka
[4]Senior Scientific Officer, Plant Breeding Division, Bangladesh Institute of Nuclear Agriculture, Mymensingh.
[5]Principal Scientific Officer & Head, Biotechnology Division, Bangladesh Institute of Nuclear Agriculture, BAU campus, Mymensingh, Bangladesh

*Corresponding author's email: asarowar04bau@gmail.com (M.S. Alam)

Introduction

Rice (*Oryza sativa* L.) is the most important food crop of the world. Eighty five per cent of it is devoted for human consumption (IRRI, 1997). The Green Revolution of the 1970s resulted in remarkable increases in rice production. Since then, the rate of production in most rice-growing countries has slowed down and has now reached a plateau (Maclean *et al.*, 2002).

Salt stress is a major problem in coastal areas because of seawater intrusion during high tide and the rising shallow saline groundwater, particularly during the dry season. Saline soil covers our earth's surface, estimated to be from 400 to 950 million ha (Lin *et al.*, 1998). More than 54 million ha of rice land in Asia are now affected by salinity. About 9.5 million ha of saline soils can be managed by large-scale irrigation and drainage schemes and by chemical treatment of soil, but the scale of the problem renders these solutions too costly (Gregorio *et al.*, 2002). Salinity is one of the major obstacles in increasing production in rice growing areas worldwide, which is an ever-present threat to crop yield. Accumulation of salt in the soil has deleterious effects and leads to a reduction in crop production, including rice. Use of salt-tolerant variety considered the most economical and most effective way of increasing crop production on saline soils. Therefore, development of salt tolerant varieties considered as one of the strategies to increase rice production in saline prone coastal areas.

Microsatellite markers based on SSRs have been developed in many crop species, including rice, tomato, soybean and others. These markers detect simple sequence length polymorphism

(SSLP) and are rapidly replacing RFLPs or others for genetic studies, due to their technical simplicity, the small amount of starting DNA required, the relatively low cost for the user, the level of allelic diversity and high power of resolution (Panaud *et al.*, 1995; McCouch *et al.*, 1997). Secondly, PCR based assay makes them easily detectable in gel electrophoresis. SSR or microsatellite markers are proved to be ideal for making genetic maps (Islam, 2004; Niones, 2004), assisting selection and studying genetic diversity in germplasm. Microsatellite marker analysis is promising to identify major gene locus for salt tolerance that can be helpful for plant breeders to develop new cultivars. The objective of this study was to identify salt tolerant rice line of BC_1F_1 progenies of Binadhan-5 x FL-478 using SSR markers.

Materials and Methods

Location and time

For the marker assisted backcrossing (MAB) the experiment was conducted during the period from July 2009 to November 2010 in the experimental field and Biotechnology Laboratory of Plant Breeding Division, Bangladesh Institute of Nuclear Agriculture (BINA), Mymensingh.

Plant materials

Forty BC_1F_1 progenies were selected for this study from the backcross made between rice of F_1 (Binadhan-5 x FL-478) and Binadhan-5. Binadhan-5 is a salt susceptible but one of the popular HYV of irrigated rice ecosystem in Bangladesh. On the other hand, FL-478 is one of the salt tolerant lines of IRRI. Binadhan-5 was the recurrent parent and FL-478 was the non-recurrent donor line. F_1 seeds were collected and grown for developing BC_1F_1 seedlings during Aman season 2010. F_1 plants were crossed with Binadhan-5 following of backcrossing and 133 BC_1F_1 seeds were developed (Fig. 1). Among these, 40 seedlings were selected to collect the leaf to make foreground selection genotypically at the molecular level (by using SSR markers).

Development of F1 population (Binadhan-5 x FL 478) of rice

On the basis of research works based on phenotypic evaluation conducted by Islam *et al.* (2007), Siddika *et al.* (2007), one rice genotype (FL-478) was selected as parent for transferring salt tolerant genes to high yielding rice genotype (Binadhan-5). Crossing programe was conducted during Boro season in 2009, where Binadhan-5 was the recurrent parent and FL-478 was the non-recurrent donor line. Binadhan-5 and FL-478 were selected for development of F1 based on phenotypic performance of fifteen rice germplasm at reproductive stage.

F_1 seeds were collected and grown for BC_1F_1 seedlings during Aman season in 2010. Eight seeds of selected F_1 lines were grown in the pot. Then F_1 plants were crossed with Binadhan-5 following of þackcrossing and 133 BC_1F_1 seeds were developed. Among these, 40 seedlings were selected to collect the leaf to make foreground selection genotypically.

Field experiment

For easy handling and carrying, the experiment was conducted on the plastic pot (10L bucket). All kinds of recommended doses of fertilizers were applied properly. Cultural practices were followed as and when necessary. The pots were labeled with pot labeler.

Production of BC_1F_1 plants:

Three sets of recurrent parental line (Binadhan-5) and one set of F_1 (Binadhan-5/FL-478) were seeded at an interval of 10 days starting from 11 June 2010. All the intercultural operations, diseases and pest management were done properly. Backcrossing was done with Binadhan-5 by applying synchronization of flowerings. 133 F_1 seeds were developed through successful backcrossing. Seeds were dried into seed dryer at 65°C for seven days. Then the well-dried seeds were soaked by clear water at 24 hours into a Petri dish by covering a tissue paper. After that, water was removed and kept for sprouting within two days. Then the well-sprouted BC_1F_1 seeds were placed into seed box for the development of BC_1F_1 seedling. After 24 to 26 DAS, seedlings were transplanted into the previously prepared pots in the field. 85 seedlings were successfully survived for the next backcrossing activities.

Collection of leaf sample (Binadhan-5, FL-478, BC_1F_1 Lines) for DNA extraction

The seedlings of two Parents (Binadhan-5, FL-478), F_1 and BC_1F_1 population rice germplasm were raised in pots at the glasshouse of BINA at the different growing season. Young vigorously growing fresh leaf samples from these seedlings were collected from 21-day-old seedlings to extract genomic DNA. Initially, healthy portion of the youngest leaves of the tiller were cut 3" apart with sterilized scissors and washed in distilled water and ethanol (70%), and dried on fresh tissue paper to remove spore of microorganisms and any other source of foreign DNA. The collected leaf samples were then kept in polythene bags for avoiding any damage of the leaf tissues. The bags were placed in an icebox to carry it in Laboratory and finally, the samples were stored in -80°C freezer with marking.

Genomic DNA extraction

Extraction of DNA was done using the mini preparation Cetyl Trimethyl Ammonium Bromide (CTAB) method from 3" young leaves tissues of 21-day old seedlings. Ten random primers viz. RM21, RM80, RM152, RM310, RM336, RM510, RM585, RM10720, RM28102, and RM28502 were used for surveying and among them; three primers (RM336, RM510 and RM585) showed polymorphisms during parental survey. These three primers were then utilized for amplification of the DNA sequences of segregating BC_1F_1 lines. The sequence and size of the microsatellite markers (SSRs) are mentioned in Table 1.

Table 1. The sequence and size of the microsatellite markers (SSRs) used for BC_1F_1 survey

Primer Name	Expected PCR product size (bp)		Primer Sequence	Annealing Temp. (°C)
RM 510	122	For.	AACCGGATTAGTTTCTCGCC	55
		Rev.	TGAGGACGACGAGCAGATTC	
RM 585	233	For.	CAGTCTTGCTCCGTTTGTTG	55
		Rev.	CTGTGACTGACTTGGTCATAGG	
RM 336	154	For.	CTTACAGAGAAACGGCATCG	55
		Rev.	GCTGGTTTGTTTCAGGTTCG	

Amplification of SSR markers

PCR amplification of simple sequence repeats (SSR) was performed with 1.5 µl 10X PCR buffer, 0.75 µl dNTPs, 1.0 µl forward, 1.0 µl reverse primers, 0.5 µl Taq polymerase, 2.0 µl of each template DNA and 8.25 µl ddH_2O using DNA thermal cycler. The PCR reactions were: initial denaturation at 94°C for 5 minutes, then final denaturation at 94°C for 1 minute and annealing at 55°C for one minute. Polymerization was carried out at 72°C for 2 minutes to complete a cycle and cycle was repeated for 34 times. The final extension was at 72°C for 7 minutes. After PCR, products were mixed with 3 µl of 2X gel loading dye. Polymorphisms in the PCR products were detected by ethidium bromide staining after electrophoresis on 1.5% agarose gel using UV transilluminator.

Scoring of bands

The banding pattern of BC_1F_1 lines was compared with their parents Binadhan-5 (susceptible), FL-478 (tolerant) and F_1. The BC_1F_1 population having similar banding pattern to Binadhan-5 were considered as salt susceptible, similar to FL-378 were considered as salt tolerant and having one allele from Binadhan-5 and one allele from FL-378 considered as heterozygous. In each marker, allelic bands were designated as **T** for tolerant, **S** for susceptible and **H** for heterozygous.

Results and Discussion

DNA concentration check

The extracted DNA was checked through agarose gel to observe the presence of DNA. After electrophoresis, the gel was placed on the UV trans-illuminator in the dark chamber of the large Documentation system. Isolated DNA of selected population showed sickle shape or nearly sickle shape bands that indicated good quality of DNA.

Table 2. Genotypic performance of parents, F_1 and Backcross population (BC_1F_1) against 3 SSR markers

*SL No.		Sample no.	Name of the marker		
			RM510	RM585	RM336
1	Parents	Binadhan-5	S	S	S
2	Parents	FL-478	T	T	T
3	Parents	F1 line (Binadhan-5× FL-478)	H	S	T
BC_1F_1 lines					
4		5/478-5-3	S	S	T
5		5/478-5-5	S	S	H
6		5/478-5-6	T	T	T
7		5/478-5-7	S	S	H
8		5/478-5-8	S	S	H
9		5/478-5-10	S	S	H
10		5/478-5-11	S	T	T
11		5/478-5-12	H	S	T
12		5/478-5-17	S	S	H
13		5/478-5-18	S	S	T
14		5/478-5-19	H	S	H
15		5/478-5-20	S	S	S
16		5/478-5-22	T	S	T
17		5/478-5-23	S	S	H
18		5/478-5-25	S	S	H
19		5/478-5-26	H	S	H
20		5/478-5-28	S	S	H
21		5/478-5-29	S	S	H
22		5/478-5-30	H	S	N
23		5/478-5-34	H	S	H
24		5/478-5-35	S	S	T
25		5/478-5-36	A	S	T
26		5/478-5-44	H	S	H
27		5/478-5-46	N	S	H
28		5/478-5-53	H	S	H
29		5/478-5-54	N	S	H
30		5/478-5-55	H	S	H
31		5/478-5-56	S	S	H
32		5/478-5-57	S	S	H
33		5/478-5-63	S	S	S
34		5/478-5-64	S	S	H
35		5/478-5-66	S	S	T
36		5/478-5-67	H	S	T
37		5/478-5-69	H	T	T
38		5/478-5-70	H	S	T
39		5/478-5-71	H	S	H
40		5/478-5-72	H	S	H
41		5/478-5-73	H	S	H
42		5/478-5-74	S	N	H
43		5/478-5-75	S	S	H

NB: S= Susceptible, T= Tolerance, H= Heterozygous, and N= Not recognizable
(*)= This sequence was used in the banding pattern of salt tolerance of 40 BC_1F_1 lines using RM510, RM585 and RM336.

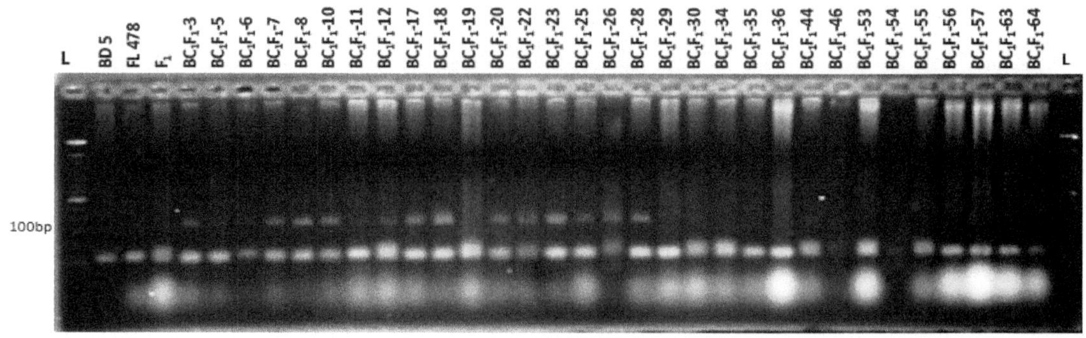

Fig. 1(a). Banding pattern of BC_1F_1 population of Binadhan-5/FL-478 using SSR marker RM510 from Sl. No. 1 to 34 (Table 2)

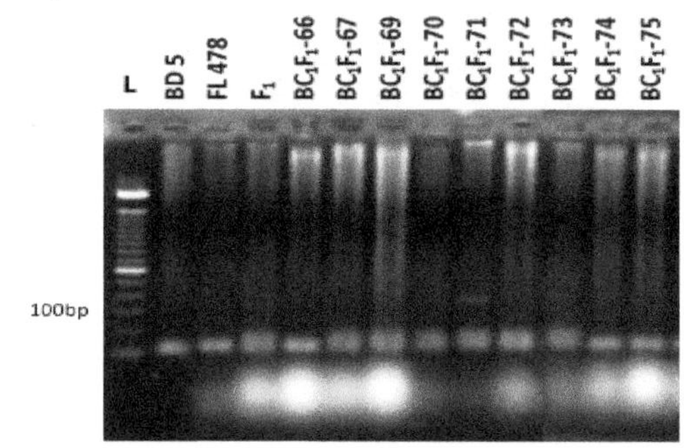

Fig. 1(b). Banding pattern of BC_1F_1 population of Binadhan-5/FL-478 using SSR marker RM510 from Sl. No. 1 to 3 and 35 to 43 (Table 2)

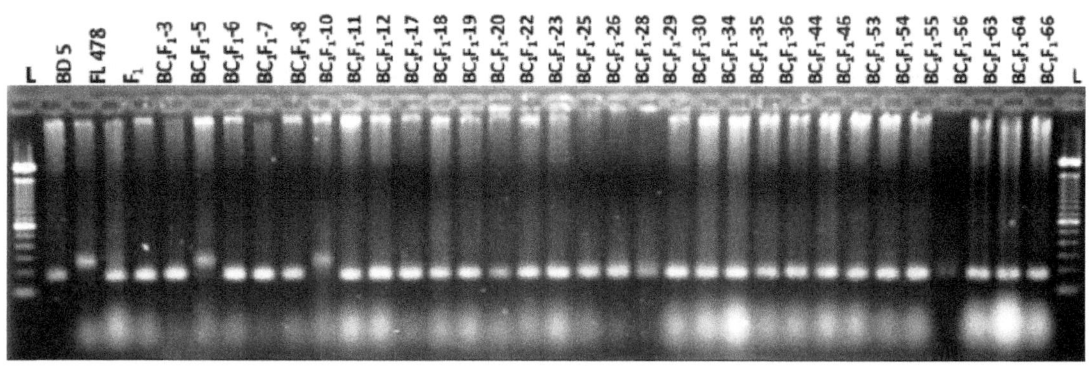

Fig. 2(a). Banding pattern of BC_1F_1 population of Binadhan-5/FL-478 using SSR marker RM585 from Sl. No. 1 to 34 (Table 2)

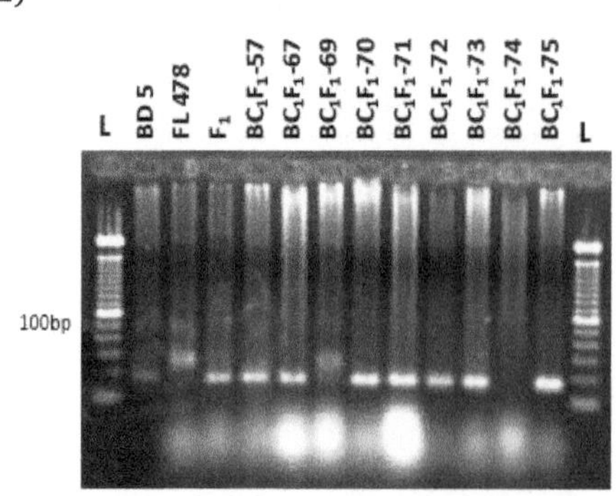

Fig. 2(b). Banding pattern of BC_1F_1 population of Binadhan-5/FL-478 using SSR marker RM585 from Sl. No. 1 to 3 and 35 to 43 (Table 2)

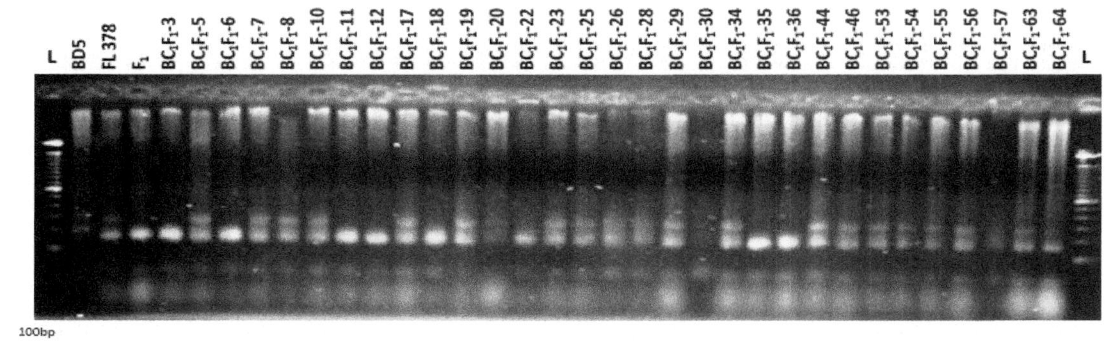

Fig. 3 (a). Banding pattern of BC_1F_1 population of Binadhan-5/FL-478 using SSR marker RM336 from Sl. No. 1 to 34 (Table 2)

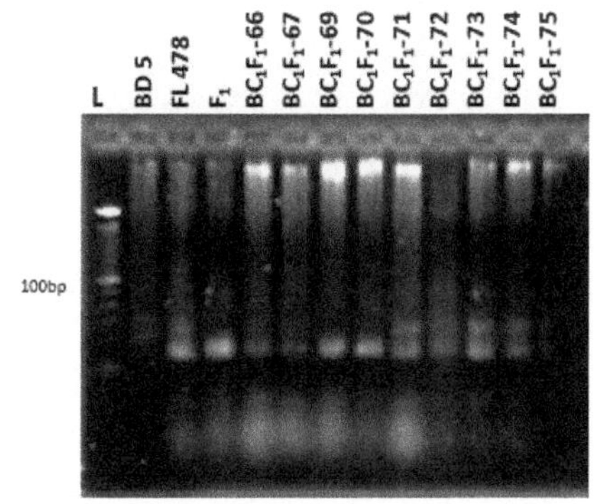

Fig. 3 (b). Banding pattern of BC_1F_1 population of Binadhan-5/FL-478 using SSR marker RM336 from Sl. No. 1 to 3 and 35 to 43 (Table 2)

Parental polymorphism survey

Polymorphism of 10 microsatellite primers was examined in the 3 rice samples; two parents, and F_1. Of 10 SSR primers, 3 showed marker polymorphisms to the parents. Of the remainder, some showed no amplification and hazy bands; some gave amplified products, but these were monomorphic. These three primers (RM336, RM510, and RM585) showed clear bands among the 40 BC_1F_1 rice samples. Out of 85 BC_1F_1 rice leaf samples, randomly 40 leaf samples were used for genotyping performance.

Genotypic performance of parents, F_1 and BC_1F_1 against SSR markers

The banding patterns were scored with reference to those of FL-478 and Binadhan-5, since FL-478 is salt tolerant and Binadhan-5 is salt susceptible. In each marker, allelic bands were scored based on the parental bands and designated as T for tolerant, S for susceptible and H for heterozygous (Table 2).

Overall performance of BC_1F_1 rice lines against the RM510, RM585 and RM336

BC_1F_1 rice lines showed wide variation with three primers; RM510, RM585 and RM336. In respect of primer RM510, a total of 16 lines (2 lines were similar to donor parent type FL-478 and 14 were heterozygous) were identified [Fig. 1 (a, b)] having salt tolerant gene derived from backcrossing between F_1 (Binadhan-5×FL478) and Binadhan-5.

With respect to the primer RM585, total 4 lines (lines were similar to parent FL-478) were identified [Fig. 2 (a, b)] having salt tolerant gene derived from backcrossing between F_1 (Binadhan-5×FL478) and Binadhan-5.

With respect to the primer RM336, total 34 lines (29 lines were recurrent parent type Binadhan-5 and 5 were heterozygous) were identified [Fig. 3 (a, b)] having salt tolerant gene derived from backcrossing between F_1 (Binadhan-5×FL478) and Binadhan-5.

Shaon (2009) conducted a study to identify salt tolerant rice lines of segregating F_2 progenies of Mut-1-1×Charnock (DA6) using SSR markers. According to his study, 12 microsatellite (SSR) markers (RM9, RM155, RM336, RM32, RM169, RM80, RM310, RM18, RM594, RM585, RM7075 and OSR17 primers) were used for parental polymorphism survey of Mut-1-1 and Charnock (DA6). Among the tested primers, three SSR primers (RM9, RM336 and RM585) showed polymorphisms between the parents [Mut-1-1 and Charnock (DA6)]. In the present study similar type of result were found for parental polymorphism of Binadhan-5 and FL-478. Among the used primers RM336, RM510 and RM585 showed polymorphisms between the parents. Thus in this study 25 BC_1F_1 heterozygous were identified through RM336 having salt tolerant gene derived from backcrossing between F_1 (Binadhan-5×FL478) and Binadhan-5. With respect to the primer RM585, 4 lines were identified having salt tolerant gene derived from FL-478 (donor parent).

Three primers (RM336, RM510 and RM585) showed polymorphisms in the studied parents, their F_1 and their BC_1F_1 population and they showed different banding patterns and discriminate tolerant lines from susceptible with relation to their tolerant (FL-478) and susceptible (Binadhan-5) parents. Those markers were also reported as highly polymorphic in IR29 × Pokkali for tagging salt tolerant genes (Islam, 2004; Niones, 2004). El-Refaee *et al.* (2006) also reported that 80 % of the all tested SSR primers showed polymorphic pattern in rice while they studied 272 SSR primers on nine rice genotypes.

The tested markers RM510, RM585, and RM336 could be efficiently used to identify salt tolerant lines in rice and can also be used in marker-assisted selection (MAS) for breeding, quantitative trait loci (QTL) mapping and gene pyramiding in rice salinity breeding. The selected salt tolerant rice lines would be further tested in saline areas to observe yield potentiality for developing high yielding salt tolerant varieties.

Further investigation will be needed to the selected lines of BC_1F_1 population i.e., recombination selection and background selection. But due to some constraints like limitation of stipulation of the study, unavailable time table, limitation of our resources and man power, recombinant and background selections were not possible to carry out. After the recombination selection and background selection have been completed, the selected lines of BC_1F_1 population will be further backcrossed for the development of BC_2F_1. After that BC_2F_1 population will be transformed into BC_2F_2 population through self-pollination for increasing the number of the population. And that population will be undertaken for the QTL mapping for salinity tolerance.

The identification of a major gene locus for salt tolerance near a microsatellite marker can be used by plant breeders to select more efficiently to better understand salt tolerance at the reproductive stage. Other microsatellite markers could be used to trace the flow of genes or quantitative trait loci of interest in rice and to make predictions about the outcome of the crossing and selection programs that will help in improving salt tolerant rice varieties.

Conclusions

Salt tolerant rice lines were identified genotypically (identification with SSR markers) in BC_1F_1 generation. The tested primers RM336, RM510 and RM585 in the BC_1F_1 rice lines could be efficiently used to identify salt tolerant lines and could also be used in marker assisted breeding, quantitative trait loci (QTL) mapping, genetic diversity study and gene pyramiding in rice salinity breeding. Thus, these markers could be efficient for tagging salt tolerant genes, in marker assisted selection (MAS) and quantitative trait loci (QTL) mapping.

References

El-Refaee, Y.Z., Atallah, M. and Porceddu, E. 2006. Fine QTLs mapping for salinity tolerance in rice. Proceedings of the 50th Italian society of Agricultural Genetics Annual Congress. Ischia, Italy. (ISBN 88-900622-7-4).

Gregorio, G.B., Senadhira, D., Mendoza, R.D., Manigbas, R.D., Roxas, N.L. and Guerta, C.Q. 2002. Progress in breeding for salinity and associated abiotic stresses in rice. *Field Crops Res.* 76: 91-101.

IRRI (International Rice Research Institute). 1997. Rice Almanac. IRRI-WARDA-CIAT, Los Baños, Laguna, Philippines. 181 p.

Islam, M.M. 2004. Mapping salinity tolerance genes in rice (*Oryza sativa* L.) at reproductive stage. Ph.D. dissertation, University of the Philippines Los Baños College, Laguna, Philippines. pp. 1-149.

Islam. M.M., Mondol, M.N.H., Emon, R.M., Begum, S.N., Bhowmik, S.K. and Hasan, A.K. 2007. Screening of salt tolerant rice genotypes using SSR markers at seedling stage. *Bangladesh J. Prog. Sci. Tech.* 5 (1): 45-48.

Lin, H.X., Yanagihara, S., Zhuang, J.Y., Senboku, T., Zheng, K.L. and Yashima, S. 1998. Identification of QTL for salt tolerance in rice via molecular markers. *Chinese J. Rice Sci.* 12 (2): 72-78.

Maclean, J.L., Dawe, D.C., Hardy, B. and Hettel, G.P. 2002. Rice Almanac. Los Baños (Philippines): International Rice Research Institute, Bouake (Cote d'Ivorie): West Africa Rice Development Association, Cali (Colombia): International Center for Tropical Agriculture, Rome (Italy): Food and Agriculture Organization. 253 p.

McCouch, S.R., Chen, X., Panaud, O., Temnykh, S., Xu, Y., Chao, Y.G., Huang, N., Ishii, T. and Blair, M. 1997. Microsatellite marker development, mapping and applications in rice genetics and breeding. *Plant Mol. Biol.* 35: 89-99.

Niones, J.M. 2004. Fine mapping of the salinity tolerance gene on chromosome 1 of rice (*Oryza sativa* L.) using near-isogenic lines. MSc. Dissertation, University of the Philippines Los Baños College, Laguna, Philippines. pp. 76-78.

Panaud, O., Chen, X. and McCouch, S.R. 1995. Frequency of microsatellite sequences in rice (*Oryza sativa* L.). *Genome*. 36: 170-176.

Shaon, S.A. 2009. Identification of salt tolerant rice lines of segregating F_2 progenies of Mut-1-1× Charnock (DA 6) using SSR markers. MS Thesis, Department of Biotechjnology, Bangladesh Agricultural University, Mymensingh. pp. 35-56.

Siddika, A., Islam, M.M., Begum, S.N., Bhowmik, S.K. and Nasiruddin, K.M. 2007. Screening of Rice Germplasms under salt stress and Identification of Salt Tolerant Genes via SSR Markers. *Bangladesh J. Environ. Sci.* 13 (1): 65-70.

12

IN-VITRO STUDIES OF SOME SELECTED BOTANICALS AND BAU-BIOFUNGICIDE ON MYCELIAL GROWTH AND CONIDIAL GERMINATION OF *Cercospora arachidicola* AND *Cercosporidium personatum*

M.H. Hossain[1*] and I. Hossain[2]

Abstract

Efforts have been made to assess some plant extracts namely, *Lycopersicon esculentum, Tagetus patula, Achras sapota, Azadirachta indica, Datura metel, Cymbopogon citrates, Polyalthia longifolia, Allium sativum* and *Allium cepa in vitro* for the management of leaf spot (tikka) disease of groundnut cultivar Dhaka-1 caused by *Cercospora arachidicola* and *Cercosporidium personatum*. Results indicated that all the tested plant extracts and BAU-Biofungicide suppressed the growth of mycelium and inhibition of conidial germination of *C. arachidicola* and *C. personatum*. Among the treatments, the leaf extracts of *L. esculentum* showed the most effective followed by leaf extract of *D. metel, A. indica* and BAU-Biofungicide in case of mycelial growth and conidial germination. Other plant extracts also had inhibitory effects. In case of conidial germination and germination inhibition, the least effective plant extract was *C. citrates*. Leaf extract of *A. sapota* was the least effective in case of mycelial growth.

Keywords: *In-vitro*, Botanicals, BAU-Biofungicide, Mycelial growth and Conidia germination

[1]Senior Scientific Officer, Plant Pathology Division, Regional Agricultural Research Station, BARI, Comilla, Bangladesh
[2] Professor, Department of Plant Pathology, Bangladesh Agricultural University, Mymensingh, Bangladesh

*Corresponding author's email: hdrhssn8@gmail.com (M.H. Hossain)

Introduction

Groundnut (*Arachis hypogaea* L.) is an important annual legume crop belonging to the family Fabaceae growing in many tropical and subtropical countries of the world (Wudiri and Fatoba, 1992). It is a multipurpose and highly nutritious crop containing oil, food and its foliage or haulm provides a valuable fodder for livestock. More than 40 fungal diseases attack groundnut (Jackson and Bell, 1969), but in Bangladesh, the crop is subjected to attack by twenty one diseases (Talukder, 1974; Ahmed and Hossain, 1985). Among the diseases, early leaf spot caused by *Cercospora arachidicola* and late leaf spot caused by *Cercosporidium personatum* are the most devastating and economically important foliar fungal disease and major yield reducing factor of groundnut worldwide (Backman and Crawford, 1984; Khaleque, 1985). The early and late leaf spots of groundnut although caused by two fungal species, *C. arachidicola* and *C. personatum*, they are commonly referred together as Tikka disease. This disease results in early defoliation thereby affecting the pod formation (Worthington and Smith, 1973). Loss in pod yield due to the diseases was recorded as 70% in groundnut (Subrahmanyam *et al.,* 1980). The yield loss was calculated in the groundnut variety Dhaka-1 due to early and late leaf spot (Tikka) by over 30-48% in Bangladesh (Hossain *et al.,* 2005). In addition to direct yield loss, they hamper seed quality by reducing seed size and seed weight (Souta, 1912; Arthur, 1929) and oil content (Gupta *et al.,* 1988). Severe attack of the disease resulting in heavy defoliations of groundnut leaves (Harrison, 1969). The disease can be controlled by developing resistant varieties, seed treatment with non-conventional chemicals (Maiti *et al.,* 2005), spraying fungicides, influence by sowing times (Naidu and Vasanthi, 1995), by using indigenous medicinal plants and biological control means (Kishore and Pande, 2005). The most acceptable method for controlling this disease is cultivation of resistant variety but there was no absolute resistant variety in the world (Wynne *et al.,* 1991). Increasing concerns about environment hazards caused by excessive usage of chemical fungicide, necessitates the development of more economical and eco-friendly alternative components of disease management. Biological control represents a less cost, environmental friendly natural and ecological approach for controlling diseases that reduces chemical inputs and their effect on environment. The main purpose of the study was to evaluate selected nine plant extracts for their *in vitro* antifungal activity against *C. arachidicola* and *C. personatum*.

Materials and Methods

Laboratory experiment was carried out at Disease Resistant Laboratory, Department of Plant Pathology, Bangladesh Agricultural University, Mymensingh following Completely Randomized Design (CRD) with four replications for bioassay of nine botanical extracts, BAU-Biofungicide (*Trichoderma* based preparation) and Bavistin (Carbendazim) against mycelial growth of *C. arachidicola* and *C. personatum*. The treatments were also tested in the present experiment following cavity slide technique to find out their effect on conidial germination of *C. arachidicola* and *C. personatum*.

Preparation of plant extracts

Plant extracts were prepared from fresh leaves of tomato (*Lycopersicon esculentum*), marigold (*Tagetus patula*), sapota (*Achras sapota*), neem (*Azadirachta indica*), datura (black) (*Datura metel*), lemongrass (*Cymbopogon citrates*) and debdaru (*Polyalthia longifolia*) and bulb of garlic (*Allium sativum*) and onion (*Allium cepa*). The collected fresh leaf samples of different plants were washed in running tap water to make free from dust. The plant materials were cut into small pieces and extracts were prepared by grinding in a mortar and pestle followed by crashing plant parts in an electric blender with sterilized distilled water at different doses. The mixtures were filtered through linoleum cloth. The extracts were kept in conical flasks separately before use. Concentration (weight/volume) of tomato leaf, datura leaf (black) and sapota leaf was 20%, neem leaf, marigold leaf, garlic clove and onion bulb was 25%. Lemongrass, BAU-Biofungicide and Bavistin were use tested at 1:2, 2.5% and 0.1% concentration, respectively.

Bioassay of botanicals, BAU-Biofungicide and Bavistin on C. arachidicola and C. personatum

Carrot leaf extract agar medium was prepared and poured into 90 mm Petri plates at 20 ml/plate. After solidification, three 5 mm discs of the medium were scooped from three places maintaining equal distance of 4 cm from the centre using a sterilized disc cutter. One millimeter of each of plant extracts and suspensions of BAU-Biofungicide and Bavistin was put into each hole and the plates were stored overnight in refrigerator for diffusion of the test materials into the medium surrounding the hole. Next day, the plates were inoculated at the center with 6 mm blocks of 15 days old culture of *C. arachidicola* and *C. personatum*. Three plates (replications) were maintained for each material. Control plates did not receive any material. To prevent contamination, the plates were covered with the Para film and the plates were incubated at 24 ± 1°C.

The isolates of *C. arachidicola* and *C. personatum* were grown on carrot leaf extract in Petri plates for 12 days at room temperature (24 ± 1°C). The culture plates were kept under NUV light for 3 days for maximum production of conidia. Conidia were collected from the plates by scraping with a sterilized glass slide and conidial suspension was prepared in sterilized distilled water. The suspension was filtered through two-ply cheese cloth to remove mycelial fragments and lumps of agar. The concentration of conidia in suspension was adjusted to 2×10^4 per milliliter using a hemocytometer (Krishna *et al.*, 2001).

Observation was made regularly to record the mycelial growth. After inoculation periods of 10, 13 and 15 days the linear growth of mycelial of *C. arachidicola* and *C. personatum* was measured (McKeen *et al.*, 1986; Nene and Thaplial, 1993; Islam *et al.*, 2001) and percent inhibition of growth was calculated using the following formula as suggested by Sundar *et al.* (1995):

$$\text{Inhibition (\%)} = \frac{X - Y}{X} \times 100$$

Where,
X = Mean mycelial growth (radial) of pathogen in control plate
Y = Mean mycelial growth (radial) of pathogen in treatment

Growth inhibition of C. arachidicola and C. personatum.

To assay the antifungal activity of the botanical extracts, Bavistin and BAU-Biofungicide, 50µl of conidial suspension and equal volume of the suspension of test materials transferred to the well of each cavity slide and mixed thoroughly. Three single cavity slides (replications) were used for each treatment. All slides were kept in a humid chamber prepared by lining 90 mm diameter Petri dishes with wet tissue paper and incubated in the dark at 24± 1°C. The slides were directly observed under light microscope for conidial germination at 24, 48 and 72 hours after incubation. Immediately after inoculation, a drop of lacto phenol-cotton blue was added to each well to prevent further germination of conidia. Fifty conidia in a well were observed under a compound microscope and number of germinated and non-germinated conidia was counted. Percentage inhibition of conidial germination in each treatment was calculated from the formula:

Percentage inhibition = (number of conidia germination in control–number of conidia germination in treatment/ number of conidia germination in control ×100 (Krishna *et al.*, 2001).

Results and Discussion

Effect of nine different selected botanicals, Bavistin and BAU-Biofungicide on mycelial growth of C. arachidicola and C. personatum

The efficacy of different nine selected botanicals, Bavistin and BAU-Biofungicide were evaluated against mean mycelial growth and growth inhibition of *C. arachidicola* and *C. personatum* at 10, 13, 15 days after inoculation (DAI). The results are presented in Table 1.

Table 1. *In-vitro* evaluation of nine different selected botanicals, Bavistin and BAU-Biofungicide on mycelial growth of *C. arachidicola* and *C. personatum*

Treatments	10 DAI		13 DAI		15 DAI	
	Mycelium growth (cm)	Growth inhibition (%)	Mycelium growth (cm)	Growth inhibition (%)	Mycelium growth (cm)	Growth inhibition (%)
Tagetus patula	3.9	41.9	4.5	50.0	4.7	47.8
Datura metel	1.3	80.8	1.5	83.3	1.6	81.9
Azadirachta indica	2.9	57.9	2.2	75.6	2.3	74.4
Lycopersicon esculentum	0.9	86.3	1.1	87.4	1.2	86.7
Allium sativum	4.2	38.0	4.6	49.3	4.6	48.6
Achras sapota	5.6	17.3	6.4	28.6	8.9	14.4
Polyalthia longifolia	2.5	62.6	3.9	56.7	4.4	51.4
Cymbopogon citrates	5.2	23.5	8.5	5.6	8.9	0.8
Allium cepa	5.3	21.3	5.9	34.1	6.0	33.8
BAU-Biofungicide	1.9	71.8	2.2	75.9	2.4	73.7
Bavistin	1.0	85.2	1.4	84.8	1.4	84.8
Control (untreated)	6.8	-	9.0	-	9.0	-
LSD($P \geq 0.001$)	1.04	-	3.24	-	3.42	-

Data represent the means of three replications

All treatments with botanicals, Bavistin and BAU-Biofungicide reduced mycelial growth on culture plates significantly over control. At 10 days after inoculation (DAI), the mean mycelial growth varied from 0.9 to 6.8 cm, where maximum and minimum mycelial growths were recorded under control (plain water) and leaf extract of *Lycopersicon esculentum*, respectively. Radial colony diameter under Bavistin, leaf extracts of *Datura metel* and BAU-Biofungicide was statistically similar to that under leaf extract of *Lycopersicon esculentum*. Maximum of 86.3% growth inhibition over control was obtained with leaf extract of *Lycopersicon esculentum* followed by Bavistin, *Datura metel*, BAU-Biofungicide, leaf extract of *Polyalthia longifolia* and *Azadirachta indica*. Minimum of 17.3% growth inhibition was recorded under the treatment with leaf extract of *Achras sapota* leaf. At 13 and 15 DAI, trends in mean mycelial growth and growth inhibition were identical as recorded at 10 DAI (Table 1). The findings of this trial are in agreement with the results of Natarajan *et al.* (2005), Kishore and Pande (2005), Patni *et al.* (2005), Buckinghum (1993), Srivastava and Lal (1997).

Effect of nine different selected botanicals, Bavistin and BAU-Biofungicide on conidial germination of C. arachidicola and C. personatum

The comparative effects of different nine selected botanicals, Bavistin and BAU-Biofungicide on conidial germination of *C. arachidicola* and *C. personatum* are presented in Table 2. All the botanicals, Bavistin and BAU-Biofungicide significantly reduce germination of conidia of the pathogens in the well of cavity slides compared to control (only water) at 24, 48 and 72 hours after incubation (Table 2).

At 24 hours of incubation, conidial germination ranged 0.0 to 68.0%, where the lowest conidial germination was observed in leaf extract of *Lycopersicon esculentum*, which was followed by leaf extract of *Datura metel*, Bavistin and leaf extract of *Azadirachta indica*. The highest conidial germination was recorded in untreated control. The extracts of onion bulb, *Tagetus patula*, *Achras sapota* and *Cymbopogon citrates* showed statistically similar effect on conidial germination after 24 hours of incubation. Among the treatments, leaf extract of *Lycopersicon esculentum* showed maximum (100%) inhibition of conidial germination. All the botanical extracts and BAU-Biofungicide inhibited conidial germination over 80% except leaf extract of *Tagetus patula*, *Cymbopogon citrates* and bulb of *Allium cepa*. BAU-Biofungicide, Bavistin, leaf extract of *Datura metel* and *Azadirachta indica* showed statistically similar effect on the inhibition of conidial germination (Table 2).

Table 2. Effect of nine different selected botanicals, Bavistin and BAU-Biofungicide on conidial germination of *C. arachidicola* and *C. personatum* following cavity slide method

Treatments	24 hour		48 hour		72 hour	
	Conidial germination (%)	Germination inhibition (%)	Conidial germination (%)	Germination inhibition (%)	Conidial germination (%)	Germination inhibition (%)
Tagetus patula	30.7	69.3	46.7	53.3	52.0	48.0
Datura metel	1.3	98.7	16.0	84.0	25.3	74.7
Azadirachta indica	9.3	90.7	24.0	76.0	29.3	70.7
Lycopersicon esculentum	0.0	100.0	1.3	98.7	8.0	92.0
Allium sativum	18.7	81.3	26.7	73.3	39.3	60.7
Achras sapota	29.3	70.7	25.3	74.7	36.0	64.0
Polyalthia longifolia	21.3	78.7	30.7	69.3	26.7	73.3
Cymbopogon citrates	36.0	64.0	56.0	44.0	57.3	42.7
Allium cepa	32.0	68.0	36.0	64.0	37.3	62.7
BAU-Biofungicide	14.7	85.3	28.0	72.0	26.7	73.3
Bavistin	9.3	90.7	16.0	84.0	17.3	82.7
Control (untreated)	68.0	32.0	100.0	0.0	100.0	0.0
LSD(P ≥0.001)	20.54	20.54	22.17	22.17	21.37	21.37

Data represent the means of three replications

At 48 hours of incubation, significantly the highest conidial germination (100%) was recorded from control, whereas the lowest germination was found in leaf extract of *Lycopersicon esculentum* followed by *Datura metel* and Bavistin. The germination of conidia was 24% in leaf extract of *Azadirachta indica*. Conidial germination in the suspension of BAU–Biofungicide, extracts of *Allium sativum*, *Polyalthia longifolia*, *Achras sapota* and *Allium cepa* was statistically similar. In case of inhibition of conidial germination, the highest per cent of inhibition was observed in leaf extract of *Lycopersicon esculentum*, whereas the lowest was found in plain water (control).

At 72 hours of incubation, conidial germination was reduced significantly over control under all the treatments with plant extracts, Bavistin and BAU-Biofungicide. Conidial germination ranged 8.0 to 100.0%, while the lowest conidial germination was recorded in leaf extract of *Lycopersicon esculentum* followed by Bavistin, leaf extracts of *Datura metel*, BAU-Biofungicide, leaf extracts of *Azadirachta indica* and *Polyalthia longifolia*. The highest (100.0%) conidial germination was observed in plain water (control). The conidial germination inhibition ranged 0.0 to 92.0%, where maximum and minimum conidial germination inhibition was recorded from the wells containing leaf extract of *Lycopersicon esculentum* and water (control), respectively (Table 2). The findings of the present research work are in consonance with the findings of Natarajan *et al.* (2005), Kishore and Pande (2005), Abdulrahman and Alkhail (2005), Aage *et al.* (2003).

Among the treatments, the most effective was the leaf extracts of *Lycopersicon esculentum* followed by *Datura metel*, BAU-Biofungicide, leaf extract of *Azadirachta indica* in case of mycelial growth and conidial germination. Other plant extracts also had inhibitory effects but not as much as the leaf extracts of *Lycopersicon esculentum*, *Datura metel*, BAU-Biofungicide, leaf extract of *Azadirachta indica*. Leaf extract of *Achras sapota* was the least effective against mycelial growth of *C. arachidicola* and *C. personatum*. In case of conidial germination and germination inhibition the least effective plant extract was *Cymbopogon citrates*. The potentials of these plant extracts for pathogen control have not been fully realized largely because the experiment was performed *in vitro*. However, their effectiveness in field condition could be of a potential advantage as it will help to determine the *in vivo* inhibitory effect of the botanicals and BAU-Biofungicide.

References

Aage, V.E., Gaikwad, S.J., Behere, G.T. and Tajame, V.S. 2003. Efficacy of extracts of certain indigenous medicinal plant against *Cercospora* leaf spot of groundnut. *J. Soils & Crops*. 13 (1): 140-144.

Abdulrahman, A. and Alkhail, A. 2005. Antifungal activity of some extracts against some plant pathogenic fungi. *Pakistan J. Biol. Sci.* 8 (3): 413-417.

Ahmed, H.U. and Hossain, M.M. 1985. Crop disease survey and establishment of a herbarium at BARI. Final Report, Plant Pathology Division, BARI. Joydebpur, Gazipur. p. 107.

Arthur, J.C. 1929. The plant rusts. John Wiley, New York, USA. Castellani, E. 1959. La ruggine dell arachide. *Olearia*. 13: 261-270.

Backman, P.A. and Crawford, M.A. 1984. Relationships between yield loss and severity of early and late leaf spot diseases of peanut. *Phytopathol.* 74: 1101-1103.

Buckingham, J. 1993. Dictionary of natural products. Chapman and Hall, London.

Gupta, S.K., Gupta, P.P. and Yadava, T.P. 1988. Estimation of oil content by Wide Line NMR. *Oil Crops Newsl.* 5: 15-16.

Harrison, A.L. 1969. Control of rust and *Cercospora* leaf spot of peapods in south Texas, USA. *Phytopathol.* 62: 803.

Hossain, M.S., Khalequzzaman, K.M., Hossain, M.A., Hossain, M. and Mollah, M.R.A. 2005. Screening of groundnut genotypes for leaf spots and rust resistance. *J. Subtropical Agril. Res. & Dev.* 3 (1): 57-60.

Islam, M.R., Akhter N., Chowdhury S.M., Ali, M. and Ahmed, K.U. 2001. Evaluation of fungicides against *Alternaria porri* causing purple blotch of onion. *J. Agric. Sci. Tech.* 2 (1): 27-30.

Jackson, C.R. and Bell, D.K. 1969. Diseases of peanut (groundnut) caused by fungi rust *Puccinia arachidis* Speg. Research Bulletin no. 56. Georgia Agricultural Experiment Station, Athens, Georgia, USA. pp. 47-51.

Khaleque, M.A. 1985. Manual of oil crop cultivation in Bangladesh. Department of Agricultural Extension, Ministry of Agriculture, Government of Peoples' Republic of Bangladesh. pp. 33-43.

Kishore, G.R. and Pande, S. 2005. Integrated management of late leaf spot (*C. personata*) and rust diseases of groundnut (*Arachis hypogaea*) with *Prosopis* Juliflora leaf extract. *Int. J. Pest Management.* 51 (4): 327-334.

Krishna, K.G., Pande, S. and Narayana Rao, J. 2001. Control of late leaf spot of groundnut by extracts from non-host plant species. *Plant Path. J.* 17 (5): 264-270.

Maiti, M.K., Raj, S.K. and Dey, S. 2005. Management of leaf spot (*Cercospora arachidicola* and *C. personata*) of groundnut by seed treatment with non-conventional chemicals. *Indian J. Agric. Sci.* 75 (7): 452-453.

McKeen, C.D., Reilly, C.C. and Pusey, P.L. 1986. Production and partial characterization of antifungal substances antagonistic to *Monilia fructicola* from *Bacillus subtilis*. *Phytopathol.* 76: 136-139.

Naidu, P.H. and Vasanthi, R.P. 1995. Influence of sowing dates on Tikka late leaf spot in rabi and summer groundnut (*Arachis hypogaea* L). *Legume Res.* 25 (4): 275-281.

Natarajan, D., Srinivasan, K., Mohanasundari, C., Perumal, G., Dheen, M.A.N., Ganapathi, G.A. and Rajarajan, T. 2005. Antifungal properties of three medicinal plant extracts against *Cercospora arachidicola*. *Advances in Plant Sci.* 18 (1): 45-47.

Nene, Y.L. and Thaplial, P.N. 1993. Fungicides in plant disease control. Oxford and IBH publishing Co. PVT. Ltd. New Delhi. pp. 531-534.

Patni, C.S., Kolte, S.J. and Awasthi, R.P. 2005. Efficacy of botanicals against *Alternaria* blight (*Alternaria brassicae*) of mustard. *Indian Phytopathol.* 58 (4): 426-430.

Souta, F.W. 1912. Report on the prevalence of some pests and diseases in the West Indies for 1910 & 1911. *West Indian Bulletin.* 12: 432.

Srivastava, A.K. and Lal, B. 1997. Studies on bio-fungicidal properties of leaf extract of some plants. *Indian Phytopathol.* 50 (3): 408-411.

Subrahmanyam, P., Mehan, V.K., Nevill, D.J. and MacDonald, P. 1980. Research on Fungal diseases of groundnut (ICRISAT). pp. 193-198. *In:* Proc. International Workshop on Groundnut (ICRISAT). Patancheru A.P. India.

Sundar, A.R., Das, N.D. and Krishnaveni, D. 1995. *In-vitro* antagonism of *Trichoderma spp.* against two fungal pathogens of Castor. *Indian J. Plant Prot.* 23 (2): 152-155.

Talukder, M.J. 1974. Plant diseases of Bangladesh. *Bangladesh J. Agril. Res.* 1: 61-86.

Worthington, R.E. and Smith, D.H. 1973. Effect of different foliar fungicides on the fatty acid composition and stability of peanut oil. *J. Agric. Food Chem.* 21: 619-621.

Wudiri, B.B. and Ftoba, I.O. 1992. Cereals in the food economy of Nigeria. pp: 13-32. *In:* Proc. Workshop on Recent Development in Cereal Production in Nigeria. Kaduna, 2-4 Sept. 1991. Organized by IITA Ibadan, Nigeria.

Wynne, J.C., Beute, M.K. and Nigam, S.N. 1991. Breeding for disease resistance in pea nut (*Arachis hypogaea* L.,). Annual Review of Phytopathol.29: 279-303.

13

GENETIC ANALYSIS OF SOME AGRONOMIC TRAITS IN GROUNDNUT (*Arachis hypogaea* L.)

M.K. Alam[1*] , U.K. Nath[2] , M.A.K. Azad[3], M.A. Alam[4] and A.A. Khan[5]

Abstract

A 10×10 half diallel experiment was conducted on groundnut (*Arachis hypogaea* L.) to ascertain the gene action and genetic parameters of ten traits including 50% flowering, no. of pods per plant, plant height, harvest index, pod index, 100 pod weight, 100 kernel weight, pod size, diseases infection and yield per plot. The experiments were carried out in the Department of Genetics and Plant Breeding, Bangladesh Agricultural University (BAU), Mymensingh during the cropping season of 2010-2011. The estimates of gene effects indicated that significance of both additive and non-additive variance for pod size, 100 pod weight and diseases infection among the traits and presence of over dominance satisfying assumptions of diallel except dormancy. However, both the additive and non-additive gene affects together importance to control of most quantitative traits in the groundnut. The average degree of dominance (H_1/D) 1/2 (H_1 = dominance variance, D = additive variance) was higher than one, indicating over dominance for all the traits. The narrow-sense heritability was high for 50% flowering (38%), harvest index (35%), pod size (52%), 100 pod weight (35%) and yield per plot (41%) indicating that great genetic gain could be achieved for them.

Keywords: Additive Variance, Dominance Variance, Diallel, Groundnut

[1]Scientific Officer, Plant Breeding Division, ORC, Regional Agricultural Research Station, BARI, Ishurdi, Pabna, Bangladesh
[2]Professor, Department of Genetics and Plant Breeding, Bangladesh Agricultural University, Mymensingh, Bangladesh
[3]Principal Scientific Officer, Genetics and Plant Breeding Division, Bangladesh Institute of Nuclear Agriculture, Mymensingh, Bangladesh
[4]Scientific Officer, Plant Breeding Division, Regional Agricultural Research Station, BARI, Ishurdi, Pabna, Bangladesh
[5]Scientific Officer, Plant Breeding Division, WRC, Regional Agricultural Research Station, BARI, Ishurdi, Pabna, Bangladesh

*Corresponding author's email: khorshed_bari79@yahoo.com (M.K. Alam)

Introduction

Groundnut (*Arachis hypogoea* L.) is the second most important oilseed crop in Bangladesh after mustard/rapeseed, with an annual production of 143,000 metric tons (AIS, 2012). Its seeds cointain 45-56% high quality edible oil, 25-30% protein and 20% carbohydrate together with vitamins E and B. Being a multipurpose crop. Breeding for improved and high yielding groundnut cultivars has been hampered by the lack of information on the genetics of yield and yield components (Hammons, 1973; Norden, 1973; Gibori *et al.*, 1978). Wynne and Coffelt (1973) pointed out that despite the availability of methods for characterizing the genetic variability in self-fertilizing species, little information has been obtained on the various types of gene action and their relative importance in the inheritance of important traits in groundnut. Plant breeders are primarily concerned with the improvement of those traits which are directly or indirectly related to the economic values. Such traits are generally quantitative in nature and governed by several numbers of genes each having small effect and acting in a cumulative manner called polygenes. Among the various biometrical tools, diallel analyses furnish useful information and identification of superior parents and crosses with their gene effects. It is also provides information on the nature and magnitude of genetic variance on which success of plant breeding.

Materials and Methods

The experiments were carried out in the Department of Genetics and Plant Breeding, Bangladesh Agricultural University, Mymensingh. The parental material consisted of ten groundnut genotypes; P_1 (Dacca-1), P_2 (Zhinghabadam), P_3 (BARI Chinabadam-7), P_4 (J×57015-SL-1), P_5 (ICGV-95063), P_6 (Binachinabadam-1), P_7 (Binachinabadam-2), P_8 (Binachinabadam-3), P_9 (Binachinabadam-4) and P_{10} (ICGV-90227). The genotypes were crossed in half diallel fashion during February to March, 2010. The F_1 seeds of all crosses along with their parents were planted in the field during 2010-2011 for evaluation in a randomized complete block design with three replications. A unit plot size 1.0 m x 0.8 m. One seed was sown per hole at a depth of 2.5-3.0 cm. Seeds were sown at 15 cm distances within rows of 30 cm apart from lines. All cultural practices (soil preparation, sowing, fertilizer and cultivation) followed by recommended procedures. The data were recorded on 50% flowering, no. pods/plant, plant height, harvest index, pod size, 100pod weight, 100kernel weight, pod index, diseases infection, yield per plot. The experiment was used to study additive and dominant gene effect was involved in the inheritance of this character. The genetic components were analysis by Hayman's approach (1954b). Genetic components were also shown in Morley Jones (1965) modification of Hayman's approach.

Components of genetic variance and genetic parameters

After testing the validity of hypothesis that epistasis is absent, determination of genetic variance components along with allied genetic parameters, which were derived by Hayman (1954b). The genetic components were calculated as follows:

$D = V_{OLO} - E$ = Variance component due to additive effects of the genes.

$H_1 = V_{OLO} + 4\ V_{OL1} - 4W_{OLO1} - 4W_{OLO1} - (3n - 2)\ E/n$ = Variance component due to dominance deviation

$H_2 = 4\ V_{1L1} - 4\ V_{1L1} - 2E$ = proportion of positive and negative genes in the parents

$h_2 = 4\ (M_{L1} - M_{LO})^2 - 4\ (n - 1)\ E/n^2$ = Algebraic sum of dominance effects over all loci in heterozygous phase in all crosses

$F = 2\ V_{OLO} - 4\ W_{OLO1} - 2\ (n - 2)\ E/n$ = Mean of the covariance of additive and dominance effects over all the arrays

$E = \left[\dfrac{\text{ErrorSS} + \text{ReplicationSS}}{df}\right]$/number of replications = The expected environmental components of variation

Where,

V_{OLO} = Variance of parents

V_{OL1} = Variance of the means of arrays

V_{1L1} = Mean of all the array variances

W_{OLO1} = Mean of all the covariance values

$(M_{L1} - M_{LO})^2$ = Dominance relationship

In order to test the significance of each component: D, F, H_1, H_2, h_2, E, the SE is calculated for each of them by the formula:

$SE = (Cs^2 / n^5)^{0.5}$

Where, $s^2 = \dfrac{1}{2}\,[\text{var}\ (Wr - Vr)]$

C= a multiplier specific to each component and was calculated as follows:

For, $D = (n^5 + n^4) / n^5$

$F = (4n^5 + 20n^4 - 16n^3 + 16n^2) / n^5$

$H_1 = (n^5 + 41n^4 - 12n^3 + 4n^2) / n^5$

$H^2 = 36\ n^4 / n^5$

$E = n^4 / n^5$

The significance of each component was tested by t test at (n – 2) df. The calculated value of 't' for each component was obtained dividing each component by their respective standard errors.

The allied genetic parameters were as follows:

$\sqrt{(H_1 / D)}$ = Mean degree of dominance

$H_2 / 4H_1$ = Proportion of dominant genes with positive and negative effect

$\dfrac{\sqrt{4DH_1} + F}{\sqrt{4DH_1} - F}$ = Proportion of dominant and recessive genes

h^2 / H^2 = Number of gene blocks exhibiting dominance

h^2_{ns} = Heritability in narrow sense

$= \dfrac{D/4}{D/4 + H_1/4 + E - F/4}$

Morley Jones modification for diallel without reciprocal

The analysis of variance for the complete diallel table was given by Hayman (1954b). Assuming the absence of reciprocal differences. Morley-Jones (1965) brought some modification of Hayman's approach. In this modification as Hayman, determination of the sum of squares corresonding to additive effects (a), and on the assumption of no epistasis to mean dominance (b_1), to additional dominance effects that can be accounted for genes having one allele present in only one line (b_2) and to residual dominance effects (b_3), is in essence a straight forward application of fitting constants by least squares.

Morley Jones modified analysis for diallel without reciprocals is as follows:

Item	df	Sum of squares	Mean squares
a	n-1	$\frac{1}{n+2}$ dev²ur	Ma
b	n(n-1)/2	b_1 ss + b_2 ss + b_3 ss	Mb
b_1	1	$\frac{1}{n(n^2+1)}$ [2X. -(n+1)X.]²	Mb_1
b_2	n-1	$\frac{1}{n^2+4}$ dev²tr	Mb_2
b_3	n(n-3)/2	Total SS- (a ss + b_1 ss + b_2 ss)	Mb_3
Error	(r-1) (t-1)	ESS	Me

Where,

a = Additive effects

b_1 = Mean dominance

b_2 = Additional dominance effects that can be accounted for genes having one allele present in only one line the remaining n-1 lines being assumed to carry the same alternative allele (dominance deviation due to arrays).

b_3 = Residual dominance effects

dev² = Sum of square of deviations from the mean

Vr = $X_i + X_{ii}$

Wr = $2(X_{i.} + X_{ii}) - (n+2) X_{ii}$

Results and Discussion

Different components of variance were estimated using the approaches of Hayman (1954b). The application of this approach is based upon various assumptions viz. diploid segregation, homozygous parents, no multiple allelism, no reciprocal differences and independent action and distribution of non-allelic genes. Failure of any one of the assumptions invalidates drawn. For testing the validity of assumptions, 't^2' test and Wr, Vr regression coefficient (b) tests were employed. The hypothesis of these tests in that if the assumptions are satisfied, the 't^2' value should be non-significant and regression coefficient (b) should be significantly different from zero but not from unity (Jinks and Hayman, 1953). If the assumptions are not satisfied, the Wr-Vr is plotted against Y (Mean of array) to arrive at a decision of either using transformation of scale or dropping some of the interacting lines or crosses for satisfying the assumptions (Hayman, 1954b). In the present study the assumptions were satisfied for pod size, 100 pod weight and diseases infection. A consequence of these assumptions to that when satisfied the homogeneity of difference between Wr and Vr is constant. Homogeneity of difference (Wr-Vr) while always implied by the validity of assumptions may also be attained in certain cases of balanced failure of these assumptions (Jinks, 1954; Hayman, 1954a). In the present study, homogeneity of difference (Wr-Vr) was constant as all the parents were situated along the regression line.

The components of additive effect (D) and dominance effects (H_1 and H_2) were significant for pod size, 100 pod weight and diseases infection, indicating importance of both additive and non-additive components in the inheritance of this traits. The magnitude of dominance (H_1) was significantly higher than additive components (D) for all the traits, indicated the presence of over-dominance for the above traits. The F component was non-significant for 50% flowering, no. of pods per plant, plant height, harvest index, 100 kernel weight, pod index and yield per plot, indicating presence of equal frequency of dominant and recessive genes for these characters. However, for pod size, 100 pod weight and diseases infection the significant F component indicated unequal frequency of dominant and recessive genes. Similar genetic control was exhibited by the order of dominance (Wr-Vr) and array per se performance (Yr) relationship which indicated that 50% flowering, no. of pods per plant, plant height, harvest index, 100 kernel weight, pod index and yield per plot were determined by equal proportion of dominant and recessive genes. Again exhibited that pod size, 100 pod weight and diseases infection was determined by dominant genes. From the above discussion for gene actions it is clear that both additive and non-additive components were involved in the expression at the above mentioned traits. The magnitude of dominance component was higher where they did not differ significantly. The significant values of h^2 indicated that there was significant difference between the parents and F_1's for all the traits. $\sqrt{H_1 / D}$ value was higher than unit suggesting the presence of over dominance in expressing all the traits. The value of $H_2/4H_1$ was nearly 0.25, indicating symmetrical distribution of positive and negative alleles for no. of pods per plant, plant height, harvest index, 100 kernel weight, pod index and yield per plot while unequal value for 50% flowering, pod size, 100 pod weight and diseases infection as indicated asymmetrically distributed of positive and negative alleles.

The ratio of[$\sqrt{4DH_1 + F}$] / [$\sqrt{4DH_1 - F}$] was less than unity suggesting an equal distribution of dominance and recessive genes in the parents for plant height, harvest index and yield per plot. Since the parameter h^2/H_2 had a value greater than unity, therefore the character was controlled by more than two genes or group of genes for all the traits. Heritability estimates in narrow sense. The heritability in narrow sense was high for 50% flowering (38%), harvest index (35%), 100 pod weight (35%), pod size (52%) and yield per plot (41%), and low heritability viz. no. of pods per plant (27%), plant height (24%), 100 kernel weight (15%), pod index (19%), diseases infection (17%). Presence of significant additive and non-additive gene effects was also reported by Makne and Bhale (1989), Verman *et al.* (1990), Verman and Paramasivam (1992), Makne (1992).

From the result and discussion for gene action, obviously additive and non- additive components were involved in the expression at the above mentioned traits. The magnitude of dominance component was higher except dormancy where they did not differ significantly.

References

AIS. 2012. Krishi Diary (Agricultural Diary). Agricultural Information Service, Khamarbari, Farmgate, Dhaka 1215. p. 14 (In Bangla).

Gibori, A., Hillel, J., Cahaner, A. and Ashri, A. 1978. A 9x9 diallel analysis of peanut (*Arachis hypogaea* L.) flowering times, tops weight, pod yield per plant and pod weight. *Theor. Appl. Genet.* 53: 169-179.

Hammons, R.C. 1973. Artificial cross-pollination of *Arachis hypogaea* L. with bee collected pollen. *Crop Sci.* 3: 562-563.

Hayman, B.I. 1954a. The theory and analysis of diallel crosses. *Genetics*. 39: 789-809.

Hayman, B.I. 1954b. The analysis of variance of diallel tables. *Biometrics*. 10: 235-245.

Jinks, J.L. 1954. The analysis of continuous variation in a diallel crosses of *Nicotiana rustica*. *Genetics*. 39: 567-788.

Jinks, J.L. and Hayman, B.I. 1953. The analysis of diallel crosses. *Maize Genetics Cooperative Newsl.* 27: 48-54.

Makne, V.G. 1992. Diallel analysis for studying the inheritance of branches, developed pods, and harvest index in groundnut. *J. Maharashtra Agril. Uni.* 17: 153-154.

Makne, V.G. and Bhale , N.L. 1989. Inheritance of pod yield and its components in groundnut. *J. Maharashtra Agril. Uni.* 14: 30-33.

Morley-Jones, R. 1965. The analysis of variance of the half diallel table. *Heredity*. 20: 117-121.

Norden, A.J. 1973. Breeding of the cultivated peanut, *Arachis hypogaea* L. (C.T. Wilson, ed, Peanut: Culture and Uses). Amer. Peanut Res. Educ. Assoc., Stillwater, Oklahoma. pp. 175-208.

Verman, V.P. and Paramasivam. 1992. Genetic architecture of yield and quality characters in groundnut. *Madras Agril. J.* 79 (12): 688-694.

Verman, V.P. Manoharan, V. and Rathinaswami, R. 1990. Studies on gene action in groundnut. *Madras Agril. J.* 77 (9): 571-573.

Wynne, J.C. and Coffelt, T.A. 1973. Genetics of *Arachis hypogaea* L. (H.E. Pattee and C.T. Wilson, eds, Peanut Science and Technology). Amer. Peanut Res. Educ. Soc., Yoakum, Texas. pp. 50-94.

Table 1. Components of variance and genetic parameters for different quantitative traits of 10×10 half diallel cross of groundnut

Components	50% Flowering	No. of pod splant^{-1}	Plant height	Harvest index	Pod size	100Pod weight	100Kernel weight	Pod index	Diseases infection	Yieldplot^{-1}
D	8.25 ±4.03	1.69 ±7.05	10.04 ±59.84	25.42 ±23.46	0.46*** ±0.05	281.75** ±67.38	7.19 ±4.65	26.23 ±15.29	182.76*** ±28.05	2846.48 ±1939.04
F	13.90 ±9.29	0.43 ±16.26	-28.93 ±138.06	-21.81 ±55.05	0.48*** ±0.11	404.22* ±155.46	8.66 ±10.74	5.77 ±35.27	307.83** ±64.72	-20.14 ±4473.95
H_1	46.58** ±8.57	61.89** ±15.0	509.81** ±127.37	216.78** ±50.78	0.85*** ±0.10	717.79** ±143.42	85.78*** ±9.90	128.71** ±32.54	398.10*** ±59.71	17854.78** ±4127.43
H_2	30.73** ±7.28	48.41** ±12.75	466.34** ±108.25	194.93** ±43.16	0.53*** ±0.09	448.70** ±121.89	76.87*** ±8.42	123.52** ±27.66	244.64** ±59.71	14956.05** ±3507.86
h^2	3506.82** ±4.88	423.41** ±8.33	1646.79** ±72.46	3421.17** ±28.89	8*** ±0.06	6322.56** ±81.59	443.55*** ±5.63	4270.00*** ±18.51	1080.90*** ±64.72	61629.40** ±2348.02
E	0.50 ±1.21	7.21** ±2.12	10.35 ±18.04	15.22 ±7.19	0.004 ±0.01	23.02 ±20.13	2.24 ±1.40	21.14** ±4.61	5.41 ±8.46	302.57 ±584.64
$\sqrt{H_1/D}$	2.38	6.44	7.23	2.93	1.36	1.6	3.5	2.22	1.48	2.5
$H_2/4H_1$	0.16	0.2	0.23	0.22	0.16	0.16	0.22	0.24	0.15	0.21
$\frac{\sqrt{4DH_1}+F}{\sqrt{4DH_1}-F}$	2.1	1.01	0.65	0.74	2.24	2.63	1.41	1.1	3.66	1
h^2/H_2	114.69	8.82	3.54	17.59	15.08	14.11	5.81	34.73	4.43	4.12
h^2ns	0.381	0.27	0.244	0.35	0.52	0.351	0.15	0.19	0.17	0.41

*and ** indicating significance at 0.05 and 0.01 level of probability, respectively

Table 2. Hayman analysis of variances (MS) following Morley jones modification for different morpho-physiological agronomic traits in 10×10 half diallel cross of groundnut

Source	df	50% flowering	No. of pods plant^{-1}	Plant height	Harvest index	Pod size	100 Pod Weight	100 kernel weight	Pod index	Diseases infection	Yieldplot^{-1}
a	9	64.78**	110.85**	602.12**	559.56**	2.57**	1232.03**	59.68**	236.42**	313.82**	44166.83**
b	45	29.99**	55.83**	378.74**	189.20**	0.52**	454.39**	76.08**	143.97**	259.26**	12435.09**
b_1	1	353.05**	66.28^{ns}	603.68**	1163.13**	0.18 ^{ns}	1143.66**	1325.34**	825.57**	2384.15**	18.88
b_2	9	47.41**	59.19*	155.30**	106.37*	1.18**	898.48**	34.25**	82.75 ^{ns}	506.54**	98.87.92**
b_3	35	16.29**	54.67***	429.77**	182.67**	0.36**	320.50**	51.14**	140.27**	134.96**	13444.83**
Error	108	2.07	28.02	29.23	45.79	0.07	69.62	6.10	70.71	14.81	925.33

*and ** indicating significance at 0.05 and 0.01 level of probability, respectively
df→degree of dominance

CONTENT ANALYSIS OF AGRICULTURAL TRAINING ADVERTISEMENTS IN NIGERIAN NEWSPAPERS

A. Oloruntoba[1*], J.O. Oladeji[2] and E.F. Odedele[3]

Abstract

It is obvious that the use of newspapers is one of the ways through which literate farmers could access agriculture-related information and are acquainted with innovations in agriculture. This study investigated the content of agricultural training advertisements in three Nigerian newspapers: The Guardian, Nigerian Tribune and Daily Times. Using multistage sampling techniques, 240 editions of these newspapers containing 609 advertisements for five years (2001–2005) were selected. In terms of agricultural sub-sectoral coverage, 30.3% were on veterinary services while forestry services recorded the least advertisement (4.0 %). The Nigerian Tribune newspaper has the highest coverage of agricultural training advertisements 40.7% compared to the 34.7% and 12.2% for Daily Times and Guardian newspapers, respectively. The result also showed that majority of agricultural training advertisements (81.10%) were placed on the non-prominent pages of the selected newspapers. Chi-Square analysis of association between the categories of agricultural training advertisements in the focal newspapers showed that there is significant relationship in the rate at which the sampled newspapers advertise different category of agricultural training advertisements in the newspapers ($p<0.05$). Chi square analysis also indicated that there is significant association between the placements of agricultural training advertisements and type of newspapers ($p<0.05$). This implies that advertisement placement on prominent pages of newspapers is determined by the policy of print media organization which also varies with cost of advertisement. It is therefore recommended that agricultural news items should be given more prominence as a panacea for increased information source to new entrants, especially the literate farmers to encourage farming.

Keywords: Agricultural Training Advertisement, Nigerian Newspapers, Literate Farmers

[1]Department of Agricultural Extension, Rural Development & Gender Studies, University for Development Studies, Tamale, Ghana
[2]Department of Agricultural Extension and Rural Development, University of Ibadan, Ibadan, Nigeria
[3]Department of Agricultural Extension and Rural Development, University of Agriculture, Abeokuta, Nigeria

*Corresponding author's email: yomitoba@hotmail.com (A. Oloruntoba)

Introduction

Dissemination of different agricultural information are priorities for technological transfer and acquisition of new skills in agriculture especially now that literate but less technically sound individuals are engaging in agriculture. Oladeji (1999) posited that the movements of the new entrant (literate farmers) who usually have higher education than the traditional farmers, into the farming enterprise have brought about an upswing in the literacy level of the farming audience. This group of people, mostly urban based, may rely on the news media especially the print media for the much needed agricultural information. Reilly and Clark (1990) indicate that training is a process through which knowledge, skill and attitude for a purpose of a specific job, task or vocation is developed through instruction and practice. Newspapers serve as the major advertising medium for the smallest retailers as well as some of the largest corporations. Therefore, there are agricultural content in Nigerian daily news; contents that must be content analyzed for efficiency. Content analysis according to Oladeji and Fawole (1999) is a systematic way of reporting social and economic life that is dominated or influenced by casual observations about the content of communication.

A broad channel of communication in extension is the mass media. Most mass media have a wide coverage and a far-reaching effect on agricultural information dissemination. Communication is a vital link in agricultural and rural development. However, agricultural news items have suffered limited coverage in Nigerian Newspaper (Olowu 1990; Salawu, 1983. While advertisements are predominant in most Nigerian dailies, most of the advertisements are non-agricultural (Oladeji and Olowu, 2001). Studies have been undertaken on content analysis of advertisements in

Guardian newspaper (Oloruntoba and Olowu, 2001) on agricultural news and stories, visuals and information clues in agricultural advertisements (Oladeji, 1999) but none has focused on agricultural training advertisements in Nigerian newspapers. It is against this background that this study will provide answers to the following questions:

- What are the content categories of agricultural training advertised in the selected Nigerian newspapers?
- What is the nature of placement of agricultural training advertisements on pages of newspapers?

Methodology

There are 28 daily newspapers in Nigeria (Media World, 1996) and these newspapers constitute the study population. Sampling of newspapers was based on purposive selection and stratification based on circulation figures, location of publishing house and ownership. A five year period (July 1st 2001 to June 30th 2005) was chosen to determine the coverage given to agricultural training advertisement in three Nigerian newspapers. The period was chosen because it marked the emergence of many agricultural programmes and most especially poverty reduction programmes like Fadama II, Local Empowerment and Environmental Management Programme (LEEMP), Special Programme on Food Security (SPFS) and Community Based Poverty Reduction Project (CPRP) in the nation. Newspapers that were in circulation throughout the five year period of study were purposively considered for sampling. Consequently, The Guardian, Nigeria Tribune and Daily Times Newspapers that have wide coverage both in space and in content were selected.

Using simple random sampling technique in selecting the months in order to determine the number of editions to be examined, 50% of the months in the five year period were selected for a total of 30 months (6 x 5), 50% of the weeks in the months were selected for a total of 60 weeks (30 x 2), while three days were chosen from Monday to Friday as week days and a day from Saturday and Sunday, that is, weekends. This exercise made a total of two hundred and forty days (60 x 4) for the study. Hence, a total of 240 editions of newspapers were selected, used and data were analyzed with both descriptive and inferential statistics. The content of all agricultural training advertisements were analyzed based on their themes. The specific content areas prominently reported are those that have immediate functional relevance for the target audience. However, to facilitate the content analysis, six content categories that were mutually exhaustive and exclusive were developed based on the training category as follows:

- Veterinary Services (VS) training category: This includes training advertisement on livestock related studies: Veterinary nursing and livestock management, veterinary physiology.
- Agricultural Services (AS) training category. This includes training advertisements on fish farming, animal feeds and nutrition, livestock production, pig husbandry, poultry keeping, crop protection, crop production, commercial horticulture, glass house and protected cropping, nursery practice, fruit production, vegetable, flower production, agricultural, food marketing, farm business administration and management, landscape.
- Forestry Services (FS) training category. This includes training adverts on Wood Science, Agro-forestry and forestry related activities.
- Agricultural Technology (AT) training category. This includes training adverts on agricultural mechanization, meat technology, food processing machine and any other agricultural related machine.
- Food Related (FR) training category. This include training advertisements on meat inspection, food and drink inspection, food marketing, food sciences and fisheries.
- Other Agricultural Related Training (OAR) training category. This includes training advertisements on other aspects of agriculture not mentioned.

Position of advertisement in selected newspapers and frequency are the two variables of prominence used in analyzing the editions. All agricultural training advertisements contained in the front page, middle page, back page or centre/special page of the sampled newspaper editions were considered prominent while those on other pages were rated as not prominent. The number of times agricultural training advertisements appears in front and back pages were counted and recorded. Hence, frequency is the number of times agricultural training advertisements appear on the prominent pages of the newspapers.

Reliability of Coding

To ensure reliability of data, code and re-code method of reliability was employed to cross-check accuracy of coding using the Spearman Rank Correlation (r). Here 20 newspaper editions out of the total sample were selected. The proportion of agreement was 87 per cent.

Results and Discussion

Content category of agricultural training advertisement

Findings showed that all the categories of Agricultural Training Advertisements were contained in the selected newspapers, except in the Guardian Newspaper that does not contain Veterinary Services and Agricultural Technology Advertisements. Olowu (1990) posited that virtually all aspects of agriculture production activities are newsworthy because they are found in newspapers. Table 1 presents the distribution of the categories of agricultural training advertisements. Findings further revealed that the Nigerian Tribune has relatively the highest coverage with 40.7% compared with the 34.7% and 12.2% for Daily Times and The Guardian respectively.

Table 1. Categories of agricultural training advertisements

Advertisement Category	Daily Times		The Guardian		Nigerian Tribune		Overall	
	Freq	%	Freq	%	Freq	%	Freq	%
Veterinary Services (VS)	32	39.0	0	00	29	27.6	61	30.3
Agricultural Services (AS)	12	14.6	6	42.9	15	14.3	33	16.4
Forestry Services (FS)	4	4.9	1	7.1	3	2.9	8	4.0
Agricultural Tech. (AT)	10	12.2	0	0.0	25	23.8	35	17.4
Food Related (FR)	15	18.3	3	21.4	15	14.3	33	16.4
Other Agriculture Related (OAR)	9	11.0	4	28.6	18	17.1	31	15.4
Sub Total	82	100	14	100	105	100	201	100
Non-Agricultural Advertisements (NAA)	154	100	101	100	153	100	408	100

The dominance of almost a third (30.0%) of the Veterinary Advertisements training categories may not be unconnected to the outbreak of epidemics such as the avian Influenza (Bird flu), Swine Fever and Swine flu in the livestock sub-sector within the period in the country. Again, livestock and fish farming are now in vogue among new entrant farmers than other aspects of agriculture. This brings to the fore the specific training needs advertisements for this category of farmers for optimum service delivery. According to FAO (1997), training is useful only when it is designed to meet the need of trainee, and is offered to people who will benefit from it.

Placement of agricultural training advertisements

Findings on Table 2 showed that the majority of agricultural training advertisements were placed on the non-prominent pages of the selected newspapers (81.10%) while few (18.90%) were displayed on prominent pages. This report is in agreement with previous findings that agricultural news or activities are mostly placed on non prominent newspaper pages (Agumagu, 1988; Fawole, 1995). Placement of agricultural advertisements on prominent pages of newspapers is a function of request for the pages by manufacturers, marketers or advertising agencies of agricultural products or services as well as the ability to pay the advertisement rates of prominent pages as advertisement rates differ for prominent and non-prominent pages of Nigerian Newspapers (Oladeji, 1999; Oloruntoba and Olowu, 2001).

Table 2. Prominence of agricultural training advertisements in selected Nigerian newspapers

Position of Advertisement	Daily Times		The Guardian		Nigerian Tribune		Overall	
Agricultural Advertisement	Freq.	%	Freq.	%	Freq.	%	Freq.	%
Front Page	2	2.4	0	0.0	4	3.8	6	3.0
Back Page	7	8.5	1	7.1	10	9.5	18	9.0
Centre Page	2	2.4	0	0.0	3	2.9	5	2.5
Special Page	4	4.9	1	7.1	4	3.8	9	4.5
Other Page	67	81.7	12	85.7	84	80.0	163	81.1
Total	82	100	14	100	105	100	201	100
Position of Advertisements	Daily Times		The Guardian		Nigerian Tribune		Overall	
Non-agricultural Advertisements	Freq.	%	Freq.	%	Freq.	%	Freq.	%
Front Page	19	12.3	24	23.8	34	22.2	77	18.9
Back Page	39	25.3	48	47.5	36	23.5	123	30.1
Centre Page	20	13.0	5	5.0	44	28.8	69	16.9
Special Page	21	13.6	3	3.0	16	10.5	40	9.8
Other Pages	55	35.7	21	20.7	23	15.0	99	24.3
Total	154	100	101	100	153	100	408	100

Table 3 presents the result of chi square analysis which shows that there is significant association between the placements of agricultural training advertisements and type of newspapers (χ^2 = 60.53, p<0.05). This implies that advertisement placement on prominent pages of newspapers is determined by the policy of print media organization which also varies with cost of advertisement. The Cramer's V test of strength of association shows that the association between the categories of advertisements and the focal newspapers is strong (Cramer's V = 0.19, p<0.05).

Table 3. Chi square analysis of relationship between categories of agricultural training advertisements in the focal newspapers (n=201)

Variables	Tool	df	value	*p*	Cramer V	Decision
Categories of Advertisements * Newspapers	χ^2	12	45.11	0.00	0.19	Significant

Significant at p<0.05

Conclusion and Recommendations

The study concluded that the advertisements of agricultural training placed in Nigerian newspapers were not given adequate coverage and prominence. The study has revealed that while the coverage of agricultural items in the papers is all-inclusive, such news items are not given positions of prominence in the newspapers. This is a serious cause for concern when viewed against the need to provide exposure to new entrants literate farmers in the agricultural sector. Indeed, agriculture should be accorded its dominance role in the Nigerian economy as it was before the discovery of crude oil, which is now the main stay of the economy. It is therefore recommended that agricultural news items should be given more prominence and sponsors of agricultural programmes should be encouraged to help in reducing the cost of advertisement as agricultural training advertisements can be a panacea for increased information source to new entrants, especially the urban farmers on agricultural training programmes.

References

Agumagu, A.C. 1988. Agricultural reporting in Nigerian news paper: The case of Daily Times and the New Nigeria, 1985-1986. *Nigerian Agril. J.* 23 (2): 71-80.

FAO. 1997. Issues and Opportunities for Agricultural Education and Training in the 1990s and beyond. Rome. pp. 49-50.

Fawole, P.O. 1995. Coverage of Women's Agricultural activities in Nigerian Daily Newspapers. PhD Thesis. Department of Agricultural Extension Services, University of Ibadan, Ibadan, Nigeria. 217 p.

Media World. 1996. Newspapers and Magazines list in Nigeria. http://newslink.org.

Oladeji, J.O. 1999. Content analysis of Visuals and Information Cues of Agricultural Advertisements in Nigerian Daily Newspapers. PhD Thesis. Department of Agricultural Extension Services, University of Ibadan, Nigeria. 213 p.

Oladeji, J.O. and Fawole, O.P. 1999. Content Analysis. pp. 23-29. Oladele. O.I., Akinbile, L.A. and Adekoya, A.E. (eds). *In:* Social-Science Research: Approaches, Techniques, and Reporting. Ijebu-Ode: Shanu Books Nigeria Limited.

Oladeji, J.O. and Olowu, T.A. 2001. Agricultural Advertisements in Nigerian Newspapers - A Content Analysis. *J. Soc. & Agril. Res.* 1 (1): 34-37.

Oloruntoba, A. and Olowu, T.A. 2001. Content Analysis of Training Programme Advertisements in The Guardian Newspaper. *Nigerian J. Agril. Dev. Stud.* 2 (1): 90-96.

Olowu, T.A. 1990. Reportage of Agricultural News in Nigerian Newspapers. *Journalism Quart.* 67 (1): 195-200.

Reilly, W. and Clark, R. 1990. Training for Public Management. A Handbook for Management Development; Human Resources Development Group, London. pp. 23-25.

Salawu, S. 1983. The Nigerian press and the neglect of the rural majority. A paper presented at the Inaugural Conference of the Nigerian Mass- Communication Association, Lagos. p. 7.

15

VALUE-CHAIN ANALYSIS OF FRESHWATER APPLE SNAIL (*Pila globosa*) USED FOR ON-FARM FEEDS IN THE FRESHWATER PRAWN FARMING SECTOR IN BANGLADESH

S.A.A. Nahid*[1,3], P.J.G. Henriksson[2] and M.A. Wahab[3]

Abstract

Growth of the freshwater prawn (*Macrobrachium rosenbergii*) sector in Bangladesh since 1970s has been supported by natural availability of freshwater apple snail (*Pila globosa*), used for on-farm prawn feeds. The present study identified the current configuration of the value-chain benefits and constraints of freshwater apple snail in south-western Bangladesh in August 2011, based upon Rapid Market Appraisal (RMA) approach. The site of snail collection was Chanda Beel in Gopalganj district, while trading, processing and final consumption was represented by Rayer Mahal Bazar in Khulna district. There were seven different nodes recognized throughout the value chain. Snail marketing was identified as a seasonal business and took place during June to November each year. Between 1995 and 2011 the price of whole snail, meat and shell has increased by 800%, 325% and 315%, respectively. The abundance of snail had been reduced and its demand has increased due to the expansion of the prawn farming industry. Prawn farmers preferred snail meat due to its' low cost (US$ 0.21 kg^{-1}) as a source of protein compared to commercial prawn feed (US$ 0.41 kg^{-1}). Snail harvesting and processing were considered as additional livelihood options for the poor, where 60% of the labour involved in snail harvesting were women, and 95% the de-shelling workforce. Induced breeding in captivity and sustainable management in nature as well as development of commercial production of apple snails might reduce the pressure on ecosystems and positively contributed to the continued expansion of freshwater prawn farming in Bangladesh.

Keywords: Bangladesh, Prawn Farming, Snail, Value Chain

[1]Faculty of Fisheries, Chittagong Veterinary and Animal Sciences University, Bangladesh
[2]CML, Institute of Environmental Sciences, Leiden University, The Netherlands
[3]Faculty of Fisheries, Bangladesh Agricultural University, Mymensingh 2202, Bangladesh

*Corresponding author's email: nahid83bau@gmail.com (S.A.A. Nahid)

Introduction

There are about 450 species of snails in Bangladesh (Gain, 1998), of which freshwater apple snail, *Pila globosa* (Swainson) is one of the most abundant and commercially valued molluscs. It is widely distributed in all types of temporary and permanent water bodies, such as ponds, canals, ditches, *beels* (a large surface static water body that accumulates surface runoff water through internal drainage channel), *haors* (a wetland ecosystem which physically is a bowl or saucer shaped shallow depression, also known as a *back swamp*), and *baors* (closed water body equivalent to an ox-bow lake, up to several hundred hectares) (Nath *et al.*, 2008). Snail plays an important role in the ecosystem in maintaining healthy aquatic environments by acting as a biofilter, a pre-requisite for conserving biodiversity (Jahan *et al.*, 2001). According to a study of the *P. globosa* in Chanda Beel, Bangladesh, the snail flesh contain 85.5±0.02% moisture, 2.599±0.02% ash, 8.272±0.05% protein, 0.725±0.03% fat, 2.902±0.03% carbohydrate and 0.0256±0.01% crude fibres, with each g of shell containing 3.04 mg calcium, 1.33 mg phosphorus, 1.00 mg iron, 0.43 mg sodium, and 0.32 mg potassium (Baby *et al.*, 2010). The energy content of snail is between 4.745 and 5.594 Kcal g^{-1} dry weight and also enriched with an essential amino acid index (EAAI) of 0.84, is a useful alternative source of protein (Sing, 1991; Bombeo-Tuburan *et al.*, 1995). It can easily be cultured in fertilised fish ponds and rice fields (Xoing *et al.*, 1987).

Snails are not consumed by humans within the Muslim community of Bangladesh. Saha (1998), however, identified 29 groups of tribal people that consume snail flesh. Snail meat is instead more extensively used in freshwater prawn (*Macrobrachium rosenbergii*) farming in the south-western part of the country (Baby *et al.*,

2010), which (prawn) is one of the major exports earning sectors in Bangladesh (DoF, 2011). The average application of snail meat to prawn ponds is 66.5 kg $ha^{-1}day^{-1}$ during June to October (Ahmed *et al.*, 2008). Snail meat is also utilized as supplementary feed in indigenous catfish (*Clarius batrachus*) farms and domestic ducks throughout most of the year (Banglapedia, 2006). Snail shell is used to produce lime and animal feed additive due to its rich $CaCO_3$ content (Nath *et al.*, 2008).

However, recently, as a result of excessive fishing pressure and lack of regulations, drastic reductions of wild snail populations have occurred in Bangladesh (Nath *et al.*, 2008). Regulating the collection of snails is difficult as it is an essential livelihood for the poor (Jahan *et al.*, 2001). After collection, snails are distributed to prawn farmers in the south-western regions through a line of middlemen. Although Ahmed (2001) described the value chain of snails from Chanda Beel of Gopalganj to Bagerhat district in 1999 and Nath *et al.* (2008) within Khulna in 2006, the chain between Chanda Beel and Khulna district has, to date, remained unreported.

The aim of this study was to obtain a more detailed understanding of the actors, activities, inputs, costs, opportunities and constrains related to the flow of snails throughout the value chain, starting with harvesters and ending with the targeted buyers and/or consumers. The outcomes will provide increased understanding of the designing, implementation, evaluation, and scaling the relevant enterprises. It will also provide an up-to-date description of the snail value chain to aid researchers and policy makers in improving future regulations within this sector.

Materials and Methods

Study locations were identified through a preliminary interview of 10 DoF (Department of Fisheries) personnel in Khulna and fieldwork was conducted in August 2011. The site of snail collection was Chanda Beel in Gopalganj district, where a major portion of all snails originate from. Trading and processing is represented by Rayer Mahal Bazar in Khulna district and the final utilization is also representative of practices in Khulna where a high density of prawn farms are located (Fig. 1). Data was collected using a Rapid Market Appraisal (RMA) approach, as described by Holtzman (2002). This method relies on a combination of secondary data, a review of trends and primary data collected by interviews. After identifying chain actors, respondents were opportunistically approached for each market stage. Data for the specific inventories were collected directly from the stakeholders by personal interview (total 35 interviews, 5 for each chain actors) and focus group discussion (total 3 FGD for harvesters, small retailers and farmers), using a questionnaire. Secondary information was collected from published reports.

Results and Discussion

Actors and stages of the snail marketing chain

A cradle-to-farm gate approach ensured the inclusion of all lifecycle stages, from snail capture in wetlands to application in prawn farms. Seven different chain actors were identified, namely harvester, trader-1, trader-2, commission agent, small retailer, prawn farmer, and supplier of shell to lime and/or animal feed factory. The stages of the snail marketing chain are shown in Fig. 2.

Snail harvesting from wetland

Chanda Beel is the most important wetland of the Madhumati river floodplain ecosystem in the south central Bangladesh (Baby *et al.*, 2010). It is also one of the largest freshwater wetlands in the country, covering an area of 10,870 ha (Khan, 1998). Seven species of snails are found in Chanda Beel, amongst them *P. globosa* is of greatest commercial importance (Khan *et al.*, 2005). Snail collection first started in 1992 in this area and is seasonally restricted to six months of the year, from *Jaishtha* (May-June) to *Kartik* (October-November) (Ahmed, 2001). The peak season for snail collection lasts for three months, from *Ashar* (June-July) to *Vadro* (Aug-Sep). *Ashar* is the rainy season when snails are commonly found in high abundances in natural wetlands. In the winter season, harvesters are unable to collect snails as they dig down in the mud and demand for snail meat declines.

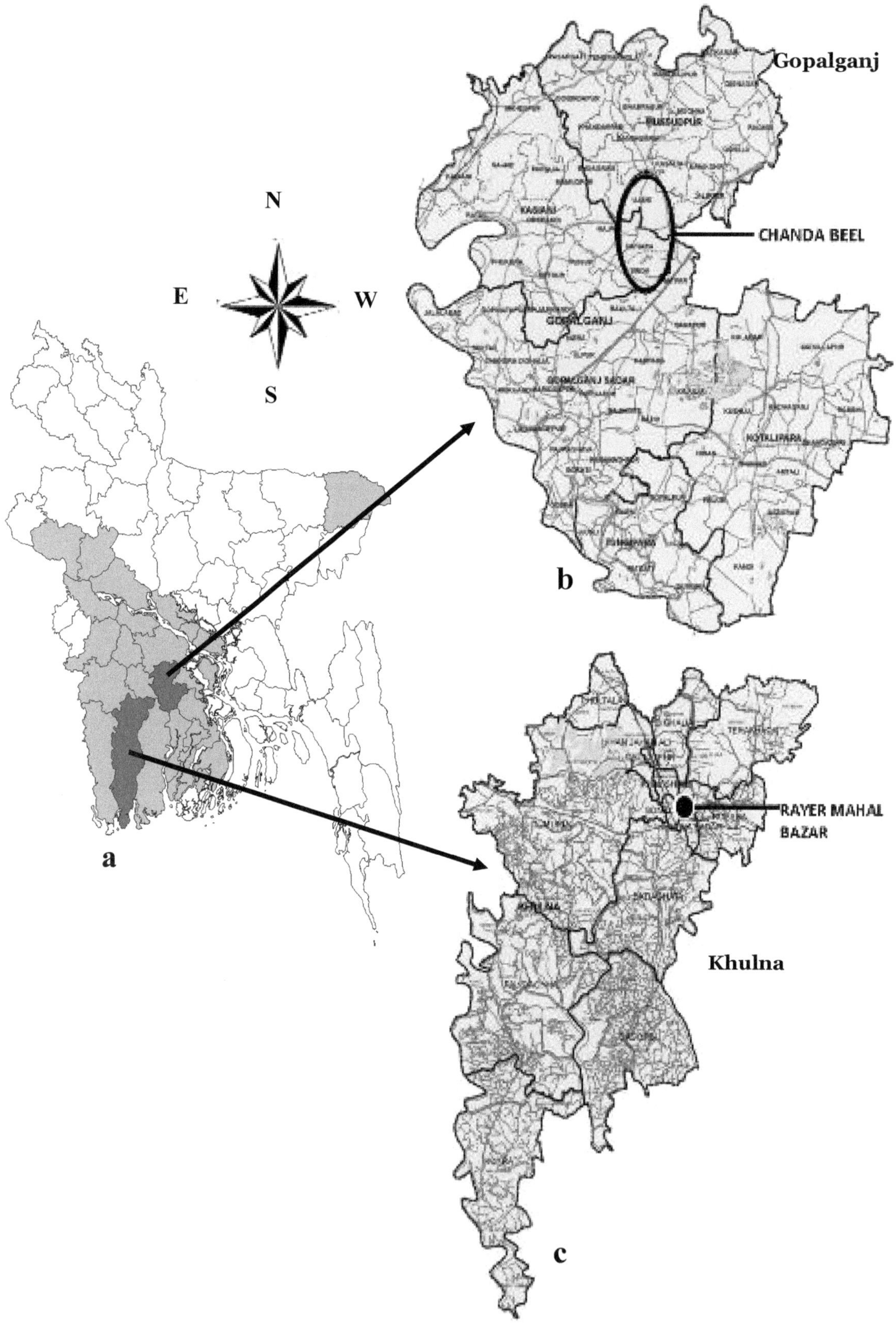

Fig. 1. a) The shadow areas show the major snail harvesting locations in Bangladesh, b) Gopalganj district showing Chanda Beel, c) Khulna district showing Rayer Mahal Bazar.

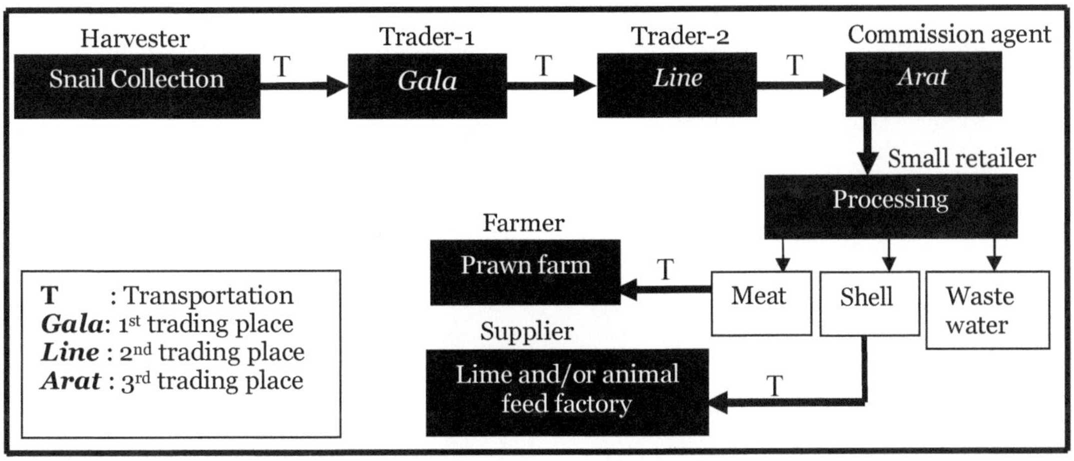

Fig. 2. Flow chart of snail marketing chain from Chanda Beel of Gopalganj to Khulna district

Transportation to 1st trading place (Gala)

At the day of capture, harvesters go 5 km to the floating market place in the *beel* area (locally called as "*Gala*") to sell the recently collected snails. The same boat as used for harvesting is used to transport the snails from the site of collection to the *Gala*.

1st trading place (Gala)

There are between 150 and 200 boats selling snails in each *Gala* every day in season. Traders (here referred as trader-1) come here to buy the captured snails from the *Khetoyas*.

Transportation to 2nd trading place (Line)

Trader-1 then transports the snails to another market, about 20 km away, (locally called "*Line*") using a motor boat. Each boat can carry about 500 kg snails and travels two hours to get from *Gala* to *Line*.

2nd trading place (Line)

In *Line*, other traders (here referred as trader-2) buy snail from trader-1. He packs the snails into jute bags, each containing 55 kg of snails. The activities run at the afternoon to the evening.

Transportation to 3rd trading place (Arat)

After packaging the snails, trader-2 transports them 120 km overnight by truck to the market (Locally called *Arat*) at Rayer Mahal Bazar (studied location), one of the most important snail markets in Khulna. Different types of trucks of different carrying capacity are used for this transport.

3rd trading place (Arat)

The *Arat* is the third trading place for snails. It opens at four in the morning and goes on for only two hours. At the *Arat* there are several commission agents. The commission agents are the middlemen between seller (trader-2) and buyer (small retailer and/or farmer). They take the risks of selling the snails, collecting the money, and forwarding the payments to the trader-2. The commission for this is US$ 0.27 bag^{-1} of snail (1 bag contain 55 kg snail) for which they sometimes take the risk of accepting credit.

Processing or de-shelling by the small retailers

Processing locations are near to the *Arat*. In Bengali the small retailers are called "*Khuchra-Pyker*" in the snail market. They buy the snails from trader-2 at wholesale price after which they de-shell and sell the snail meat to prawn farmers at retail price. Small groups consist of 5-7 people and large groups of 20-30, most of which are women. The women members also have household duties on the side and therefore only break snail's part of the day, with the occasional help of their men. The children also help when not in school.

One large group buy 30-35 bags day^{-1} during peak season on the credit given by the commission agent. An efficient member can break up to 11 kg of snails per hr. Two hand-made small devices (locally called *Futoni* and *Kholani*) made of wood and iron pin are used for this purpose. Large farm owners buy the snail directly from the *Arat* in bulk amount and transport it to their farms for processing at farm-gate. Small farmers may instead group and buy snails together, after which they transport them to their farms using different hired vehicles. At the farm, they pay local snail breaking groups US$ 0.41 bag^{-1} of snail for de-shelling.

Selling of snail meat and shell

De-shelling of one kg snail yields 0.51 kg meat, 0.33 kg shell, and 0.16 kg waste water. Meat is subsequently spread into the prawn farms on the day of de-shelling. The shells on the other hand are stored before they are supplied to a lime and/or an animal feed factory.

Amount of snail harvested in different years

The study revealed that amount of available snails has decreased by 50% between 1995 and 2011 in the Chanda Beel (Fig. 3). Overharvesting is believed to be the reason for the rapid disappearance of *P. globosa* from the wetland ecosystem of Chanda Beel (Baby *et al.*, 2010). Chowdhury (1999) reported a total catch of 365, 849 mt *P. globosa* in Bangladesh in 1999. However, Sultana *et al.* (2002) identified a 48% reduction in snail catches in between 1997 to 2000 in Goakhola-Hatiara Beel of Narail district, whereas Nath *et al.* (2008) showed a 90% reduction in the abundance of snail in the Khulna district between 1997 and 2006.

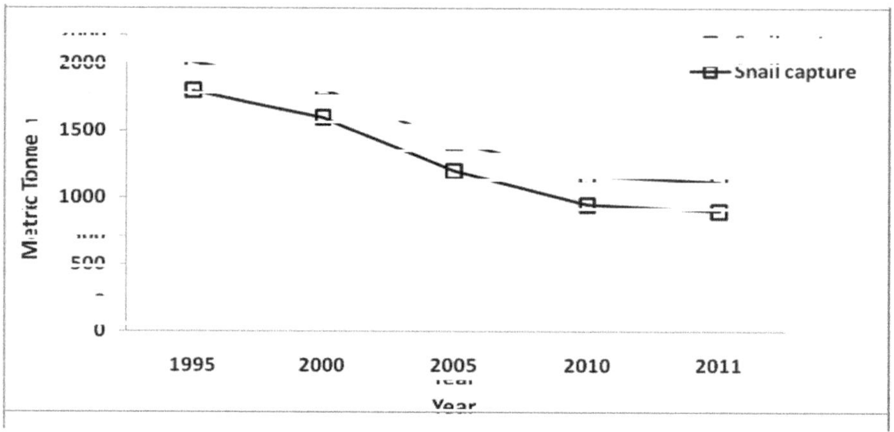

Fig. 3. Amount of *P. globosa* capture in Chanda Beel in different years

Seasonality, supply and demand of snail

Demand, availability and price of snails remain high from mid-June to mid-October. Between June and August, the rainy season offers a suitable breeding environment for the snails (Andrews, 1964) and abundances increase. Prawn farmers also utilize more snail meat during this period as the prawns are of 4-5 months age (stocked in April) and need quick growth before harvest.

In the Chanda Beel, the flood water starts to ebb at the end of October and at the end of December the entire area becomes dry with exception for some deep pockets. During this period, the snails aestivate in the sediments as their habitat dries up (Andrews, 1964). Similarly does the demand of snail meat decreased during this period of year as it is the beginning of the prawn harvest. The price of snails varies over the year depending upon supply and demand (Table 1).

Price trend of snail meat and shell

The price of snail meat and shells were recorded to have increased by 800%, 325% and 315% in 2011 compared with 1995 (Increased rate are based on BDT (to understand the actual growth rate) and figure is based on US$. Exchange rate of one US$ to BDT varied in different years which were 40 in 1995; 50 in 2000; 60 in 2005; 70 in 2010; and 74 in 2011 (gocurrency.com, 2011) (Fig. 4 and Fig. 5). The reason for this rapid increase is due to a decreasing supply alongside an increasing demand for snail meat. Availability of snail is decreasing in an alarming rate in different water bodies, whereas prawn farms have been reached to 50,000 ha (DoF, 2010), compared with only 3,500 ha in the mid 1980s (Muir, 2003).

During the early 90s, snail shells were of little demand and were mainly dumped in nearby canals after removing the meat. However, in the mid-90s, the demand for snail shells increased as a result of lime production and in 2000 it started being used as a cheap calcium source (US$ 0.01 kg^{-1}) in animal feed factories. The price has therefore steadily been increasing and today shell suppliers even give advance payments to the shell breaking groups to ensure smooth supply.

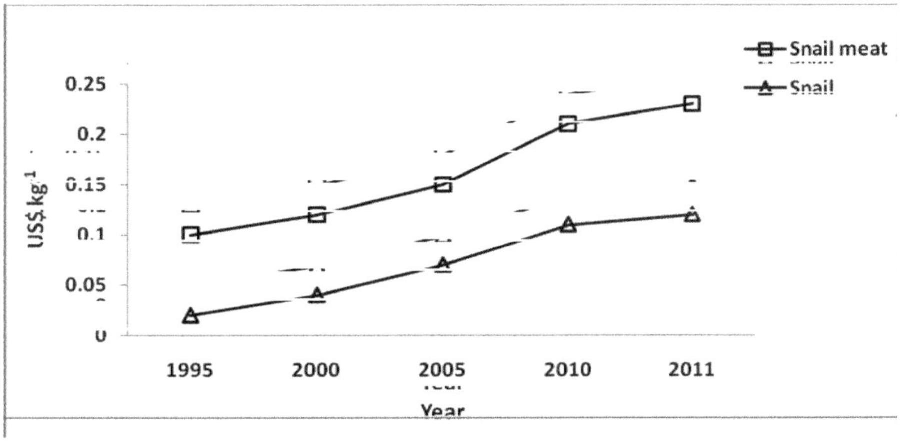

Fig. 4. Price of snail meat and snail in different years at farm level

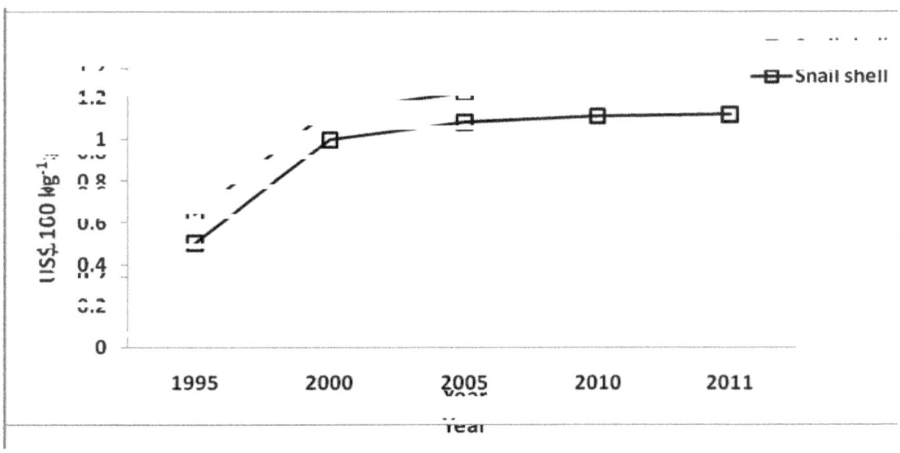

Fig. 5. Price of snail shell in different years (Source: Survey outcomes)

Snail price at different stages of the production chain

At each stage of the market chain, the cost of the snails increases, often with some form of value addition (Table 2). For example, trader-1 pays to the harvester US\$ 0.04 kg^{-1} snail, while trader-2, in turn, pays US\$ 0.06 kg^{-1} snail and sells the snails for US\$ 0.10 to the retailer. After breaking the snails the retailers charges US\$ 0.11 for 0.51 kg of meat and US\$ 0.003 for 0.33 kg of shells.

Income of different actors of snail marketing chain

Snail harvesting has become an important additional income for the poor people of Chanda Beel without permanent positions. The additional income is of importance for these casual labourers, despite that they make US\$ 0.03 less per hour compared to their normal labuor wage (Labour wage US\$ 2.03 day^{-1} for 10 hours of work in Gopalganj in 2011), as they invest little physical effort and is early in the morning before other duties. Moreover, the harvesting of snails is physically very easy and only requires a boat (common for monsoon transportation) and bamboo handled scoop net. Additional earning is offered to other actors in the snail marketing chain, including traders, transporters and day labourers (Table 2 and Fig. 6). Trader-1 and trader-2 have a brokerage rate of 26.59% and 18.45%, respectively, in proportion to their investment. The small retailers, in turn, need to invest US\$ 0.003 in labour wage to break the snails.

Table 1. Seasonality of snail marketing

Bengali month and season	English month	Demand	Supply	Price (US$ per 55 kg snail) at *Arat*	Remarks
Jaishtha (Summer)	May-June	Few	Very few	4.73	Supply depends on rain Demand so-so to feed over wintered big size stocked prawns Current year's stocked prawns (stocked in April) are not too big to take snail meat
Ashar (Rainy)	June-July	Few	High	3.78-4.00	Due to few demand and high supply price goes down
Shraban (Rainy)	July-Aug	High	Remain same	4.73-5.41	Due to high demand price goes up
Vadro (Autumn)	Aug-Sep	Increase	Remain same	5.41-5.68	Increased demand led to high price
Asshin (Autumn)	Sep-Oct	Remain same	Remain same	5.41-5.68	Price remained same
Kartik (Late autumn)	Oct-Nov	Decrease	Decrease	5.41	Prawn harvesting season
Agrahaon (Late autumn)	Nov-Dec	Few	Very few	4.73	Prawn harvesting season
Poush & *Magh* (Winter); *Falgun* & *Chaitra* (Spring)	Dec-April	No	No	-	It is not the prawn culture season Snails are not captured in winter
Baishakh (Summer)	April-May	No	No	-	Prawn post larvae stocking season

Note: 1 US$ = 74 BDT (in 2011)

However, like harvesters the retailers also consider the business as an additional income besides other duties. Retailers also receive credits from the commission agents, which allows them to avoid initial invest. For the prawn farmers, one kg of snail cost US$ 0.19 less compared to commercial pellet feeds.

Table 2. Pricing of one bag (55 kg) snail (provides 28 kg meat and 18 kg shell) at different stages of the marketing chain

Stage of chain	Place	Purpose	Cost (US$)	Final price (US$)
Snail harvester	Wetland	Capture of snail (Physical labour) (PL):	2.68[a]	
		Total cost	2.68	
	Gala	Sell to Trader-1	-	2.23
Trader-1	*Gala*	Buy snail	2.23	
	Gala to *Line*	Transportation (T)	0.37	
	Line	Miscellaneous	0.07	
		Total cost	2.67	
	Line	Sell to Trader-2	-	3.38
Trader-2	*Line*	Buy snail	3.38	
	Line	One jute made bag to pack 55 kg snail with labour cost	0.34	
	Line to *Arat*	T: Gopalganj-Khulna	0.34	
	Arat	Labour cost to off-load bag from truck	0.068	
	Arat	Commission to the Agent	0.27	
	Arat	Tax to the snail market	0.014	
		Total cost	4.412	
	Arat	Sell to small retailer	-	5.41
Small retailer (processor)	*Arat*	Buy snail	5.41	
	Beside *Arat*	Processing	PL: 1.01[b]	
		Total cost	6.42	
	Beside *Arat*	Sell to factory (18 kg snail shell obtained from 55 kg snail)	-	0.20 (18 kg shell)
	Beside *Arat*	Sell to farmer (28 kg snail meat obtained from 55 kg snail)	-	6.05 (28 kg meat)
Farmer	Beside *Arat*	Buy snail meat (28 kg meat)	6.05	
	Arat to farm	T (28 kg meat)[c]	0.39	
		Total cost	6.44	
	Farm	Feed to prawn	-	6.44 (28 kg meat)

Note: 1 US$ = 74 BDT (in 2011)

[a] Calculation was done in context of labour wage (US$0.203 hr^{-1}) at Gopalganj in 2011 and to capture 55 kg snail a harvester takes 13.2 hours

[b] Calculation was done in relation to labour wage (US$0.203 hr^{-1}) at Khulna in 2011 and to break 55 kg snail a person takes 5 hours

[c] Transportation cost varies depends upon distance of farm from *Arat* and types of vehicle used.

Gender involvement at different stages of snail marketing channel

The contributions from women have been found dominant among harvesters (60%) and small retailers (95%), with no women working as traders or commission agents (Fig. 7). Women also represented a negligible percentage (0.50% only) of the farm owners. Children (below age 15) were also found amongst the harvesters and snail breaking (small retailer) groups at 15% and 25%, respectively.

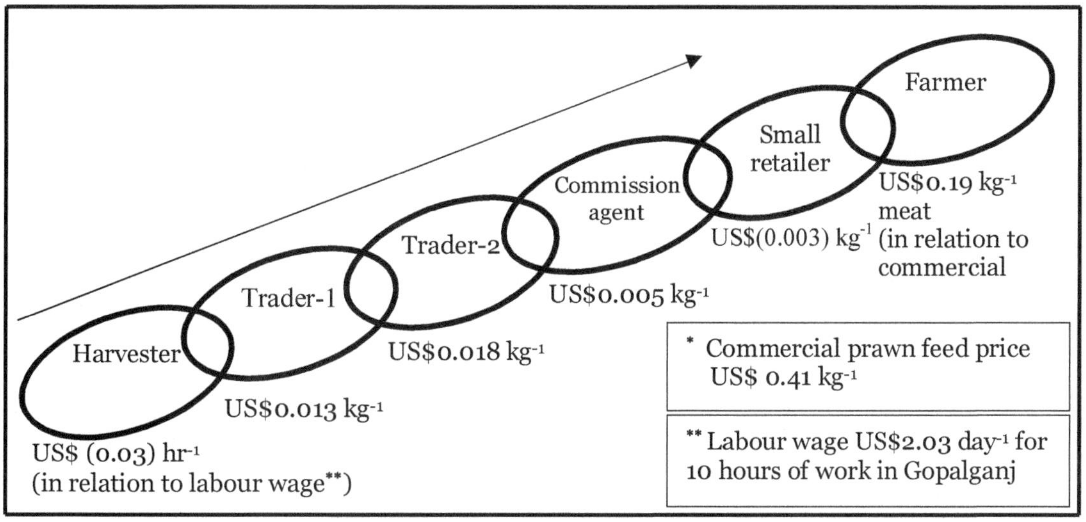

Fig. 6. Profit margins of the chain actors of snail marketing channel

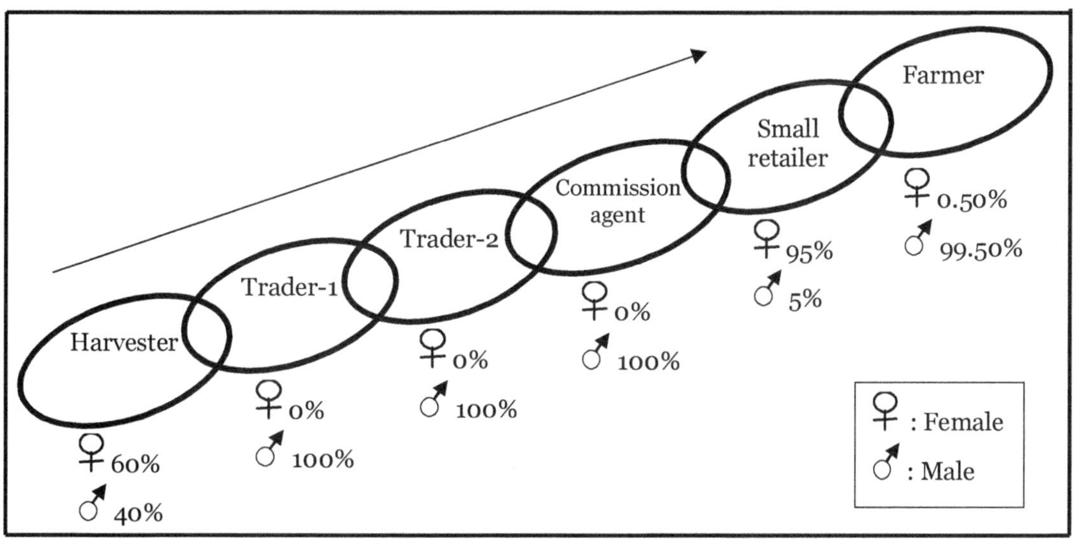

Fig. 7. Amount of *P. globosa* capture in Chanda Beel in different years

Conclusion

The snail sector has an important contribution to the national prawn farming sector. It also provides a source of income for all the men and women involved in the marketing chain, from harvesters to farmers. Concerns have, however, recently be raised about the long term sustainability of this sector at the current rate of exploitation lack of planning and improper management has led to decreases in wild stocks at an alarming rate. Due to the snail's ecological importance in limiting the overgrowth of aquatic environments, as well as its economic significance in generating livelihood options for the poor and providing low cost protein source in aquaculture, the government needs to impose policy options to protect snail habitats and populations.

Acknowledgement

The survey was done by the corresponding author Sk. Ahmad-Al-Nahid, during his Ph.D. research field work, part of the Sustaining Ethical Aquaculture Trade (SEAT) project, which is co-funded by the European Commission within the Seventh Framework Programme - Sustainable Development Global Change and Ecosystem (project no. 222889; http://www.seatglobal.eu). The authors would like to thank all respondents who spent time helping gather data for the study.

References

Ahmed, N. 2001. Socio-economic aspects of freshwater prawn culture development in Bangladesh. PhD dissertaion. Institute of Aquaculture, University of Stirling, Scotland, UK.

Ahmed, N., Demaine, H. and Muir, J.F. 2008. Freshwater prawn farming in Bangladesh: history, present status and future prospects. *Aqua. Res.* 39: 806-819.

Andrews, E.B. 1964. The functional anatomy and histology of the reproductive system of some pilid gastropod mollusks. pp. 121-139. *In*: Proc. The Malacological Society of London.

Baby, R.L., Hasan, I., Kabir, K.A. and Naser, M.N. 2010. Nutrient analysis of some commercially important molluscs of Bangladesh. *J. Sci. Res.* 2 (2): 390-396.

Banglapedia, 2006. National encyclopedia of Bangladesh. Retrieved September 23, 2011 from http://www.banglapedia.org/httpdocs/HT/S_0437.HTM.

Bombeo-Tuburan, I., Fukumoto, S. and Rodriguez, K.M. 1995. Use of the golden apple snail, cassava, and maize as feeds for the tiger shrimp, *Penaues monodon* in ponds. *Aquaculture*. 131 (1-2): 91-100.

Chowdhury, M.H. 1999. Study on thana level snail trading system. A study report of GOLDA project of CARE Bangladesh. Funded by the Department for International Development (DFID), UK.

DoF. 2010. National Fish Week Compilation. Department of Fisheries (DoF), Ministry of Fisheries & Livestock, Dhaka, Bangladesh. 120 p.

DoF. 2011. National Fish Week Compilation. Department of Fisheries (DoF), Ministry of Fisheries & Livestock, Dhaka, Bangladesh. 136 p.

Gain, P. 1998. Chanda beel: shrimps attacks snails and environment. pp. 17-20. *In:* P. Gain (ed.), Earth Touch. The Society for Environment and Human Development (SEHD), Dhaka, Bangladesh.

Gocurrency.com. 2011. Currency Converter And Exchange Rates. Retrieved September 24, 2011 from http://www.gocurrency.com/v2/historic-exchange-rates.php.

Holtzman, J.S. 2002. Rapid appraisals of commodity subsectors. *In:* Giovannucci D. (ed.), A guide to developing agricultural markets and agro enterprises. World Bank: Washington DC. USA.

Jahan, M.S., Akter, M.S., Sarker, M.M., Rahman, M.R. and Pramanik, M.N. 2001. Growth ecology of *Pila globosa* (Swainson) (Gastropoda: Pilidae) in simulated habitat. *Pakistan J. Biol. Sci.* 4 (5): 581-584.

Khan, M.A. 1998. Chanda beel: wildlife and biodiversity in peril. pp. 17-20. *In:* Gain P. (ed.), Earth Touch. The Society for Environment and Human Development (SEHD), Dhaka, Bangladesh.

Khan, M.S.I., Hussain, B.B. and Chakraborty, T.R. 2005. Snails in the Chanda beel area. IUCN Bangladesh Country Office, Dhaka, Bangladesh. 38 p.

Muir, J.F. 2003. The future for fisheries: economic performance. Fisheries Sector Review and Future Development Study. Commissioned with the association of the World Bank, DANIDA, USAID, FAO, DFID with the cooperation of the Bangladesh Ministry of Fisheries and Livestock and the Department of Fisheries, Dhaka. 172 p.

Nath, R.D., Rahi, M.L., Hossain, G.S. and Huq, K.A. 2008. Bangladesh status of fresh water snail in Khulna district. *Bangladesh Res. Pub. J.* 1 (4): 337-347.

Saha, B.K. 1998. Ecology and bio-eonomics of the freshwater edible snails of Bangladesh. PhD dissertation. Rajshahi University, Bangladesh. 162 p.

Sing, R. 1991. Seasonal variation in calorific values of *Pila globosa* and *Bellaniya bengalensis*. *J. Ecobiol.* 3 (2): 98- 101.

Sultana, P., Thompson, P.M. and Ahmed, M. 2002. Women-led fisheries management - a case study from Bangladesh. ICLARM-The World Fish Center, Malaysia and Bangladesh. pp. 89-96.

Xoing, W.W., Tian, Y.G., Zhang, Z.G., Vue, C.J., Ying, J.G., Ji, L.P. and Wen, Z.W. 1987. Tolerance of *Azolla caroliniana* and its application. Fisheries Research, International Rice Research Institute, Philippines.

SCREENING OF DIFFERENT TOMATO VARIETIES IN SALINE AREAS OF BANGLADESH

M.A. Siddiky[1,2], P.K. Sardar[3], M.M. Hossain[3], M.S. Khan[4] and M. Khabir Uddin[1*]

Abstract

A field study was conducted to screen out a number of Bangladeshi Tomato (*Lycopersicon esculentum* L.) varieties for salinity tolerance. Three levels of salinity were 2.0-4.0 dS m^{-1}, 4.1-8.0 dS m^{-1} and 8.1-12.0 dS m^{-1}. Significant varietal and or salinity treatment effects were registered on plant height, leaf area, plant growth, yield, dry matter $plant^{-1}$, Na^+ and Cl^- accumulation in tomato tissues. Variety BARI Tomato 14, BARI Hybrid Tomato 5 and BARI Tomato 2 consistently showed superior biological activity at moderate salinity (4.1-8.0 dS m^{-1}), based on dry matter biomass production thus displaying relatively greater adaptation to salinity. Under saline condition, all plant parameters of tomato varieties were reduced compared to the control except number of fruits of BARI Tomato 14, BARI Hybrid Tomato 5 and BARI Tomato 2. Thus, BARI Tomato 14, BARI Hybrid Tomato 5 and BARI Tomato 2 can be regarded as a breeding material for development of new tomato varieties for tolerance to salinity in saline areas of Bangladesh.

Keywords: Tomato Varieties, Salinity and Screening

[1]Department of Environmental Sciences, Jahangirnagar University, Dhaka, Bangladesh. [2]Horticulture Research Centre, BARI, Gazipur-1701, Bangladesh, [3]On Farm Research Division, BARI, Khulna, Bangladesh and [4]CSO (Ret'd), Soil Science Division, BARI, Gazipur-1701, Bangladesh.

*Corresponding author's email: khabir88@yahoo.com (M. Khabir Uddin)

Reviewed by Dr. Md. Mokter Hossain, Bangladesh Agricultural University, Mymensingh, Bangladesh.

Introduction

Salinity is a significant problem affecting agriculture worldwide, including Bangladesh, resulting in substantial losses in crop yield. In Bangladesh, coastal areas about 2.86 million ha covered by 30% of the total crop land of the country. Of this, nearly 1.056 million ha are affected by varying degrees of salinity (Karim *et al.*, 1990). The severity of salinity of this area increases with the desiccation of the soil. It affects crops depending on degree of salinity at the critical stages of growth and reduces yield and in severe cases, total yield is lost. It has become imperative to explore the possibilities of increasing potential of these (saline) lands for increasing production of crops. Out of coastal cultivable saline area, about 328 (31%), 274 (26%) and 190 (18%) thousand hectares of land are affected by very slight (2.0-4.0 dS m^{-1}), slight (4.1-8.0 dS m^{-1}) and moderate salinity (8.1-12.0 dS m^{-1}), respectively are scope to successfully crop production (SRDI, 2010). Suitable cultivars are required to overcome the serious limitation posed by salt affected coastal areas. Tomato (*Lycopersicon esculentum* L.) is an important vegetable in Bangladesh. Extensive research is necessary to develop growing conditions in moderate salinity to produce good vegetative growth. The tomato plant is moderately sensitive to salinity (Peralta *et al.,* 2005), although considerable differences between cultivars may be observed (Allen *et al.,* 1998). Tomato can tolerate salinity up to 2.5-2.9 dS m^{-1} in the root zone without yield losses (Sonneveld and Van der Burg, 1991). The exact salinity level may vary depending on cultivar sensitivity (Caro *et al.*, 1991) and environmental conditions (Karlberg *et al.,* 2006). Therefore, the experiment was undertaken to evaluate the ability of ten varieties of tomato for growing under different saline conditions and screen them for salinity tolerance.

Materials and Methods

The field experiment was conducted in the farmer's field at three dispersed locations, Agricultural Research Station, BARI, Khulna, in a naturally salt affected soil to screen out the performance of salinity tolerant varieties of tomato. The properties of soil of the site are presented in Table 1a. The treatment consisted of three salinity levels 2.0-4.1, 4.1-8.0 and 8.1-12.0

dS m^{-1} as shown in Table 1a, Table 1b and tomato varieties viz. BARI Tomato 2, BARI Tomato 3, BARI Tomato 4, BARI Tomato 8, BARI Tomato 9, BARI Tomato 11, BARI Tomato 14, BARI Hybrid Tomato 3, BARI Hybrid Tomato 4, and BARI Hybrid Tomato 5. The experiment was laid out in randomized complete block design with three replications. The unit plot size was 2m x 3m. Thirty day old seedlings were transplanted in 20-30 November, 2010. Tomato production technology such as fertilizer and pest management, intercultural operations were done properly. Data were taken on different yield parameters like plant height, number and weight of fruits $plant^{-1}$, root fresh weight, relative green mater value in youngest fully expanded leaves, leaf area $plant^{-1}$, and biomass of root and shoot. Na^+ and Cl^- concentration were also measured from dry mater of plant. Relative green mater value was measured with Minolta SPAD-502 chlorophyll meter. Statistical analysis of the data was performed and the means were compared at 5% probability level.

Table 1a. Properties of the naturally salt affected soil in the field experiment

Location	Soil depth (0-15 cm)			
	EC dS m^{-1}	pH	SAR (mmol L^{-1})	Texture
ARS farm, Khulna	2.0-4.0	7.3	2.1	Clay loam
Khornia, Dumuria	4.1-8.0	7.79	11.64	Clay loam
Panchpota, Dumuria	8.1-12.0	8.11	18.76	Clay loam

ARS= Agricultural Research Station, EC=Electrical Conductivity, SAR=Sodium Absorption Ratio

Table 1b. Soil salinity class and area in coastal saline belt of Bangladesh

Land classification	Salinity (dS m^{-1})	Saline area (10^3 ha)
Non saline with some very slightly saline	S_1 (2.0-4.0)	328
Very slightly saline with some slightly saline	S_2 (4.1-8.0)	274
Slightly saline with some moderately saline	S_3 (8.1-12)	189
Strongly saline with some moderately saline	S_4 (12.1-16.0)	162
Very strongly saline with some strongly saline	S_5 (>16.0)	102

Source: SRDI, 2010

Results and Discussion

Plant growth was significantly affected by different varieties as well as salinity (Tables 2a & 2b). Treatment with lower salinity gave the higher values of most plant parameters as compared to the normal salinity. Plant height, leaf area, number of fruits, fruit weight, shoot and root dry matter weight were significantly different among the salinity levels across the tomato varieties (Tables 2a & 2b). Na^+ and Cl^- accumulation in the plant tissues was significantly ($P<0.05$) influenced by the tomato varieties and salinity levels (Table 3).

The severe reduction in plant height demonstrated a consistent effect of high salinity in plant growth rate. BARI Tomato 14 was superior with respect to plant height and it was visually relatively stable and healthier even under the moderate salinity level (8.1-12.0 dS m^{-1}). Similarly, BARI Hybrid Tomato 5 and BARI Hybrid Tomato 3 and BARI Tomato 11 exhibited better performances as well as plant height among the ten tomato varieties (Table 2a). Adler and Wilcor (1987) found that salinity adversely affected the vegetative growth of tomato in respect of plant height and dry weight.

Table 2a. Response to salinity on the different yield parameter of tomato varieties

Variety	Plant height (cm)			Leaf area $plant^{-1}$ (cm^2)			Relative green mater value (%)		
	Salinity EC (dS m^{-1})								
	2.0-4.0	4.1-8.0	8.1-12.0	2.0-4.0	4.1-8.0	8.1-12.0	2.0-4.0	4.1-8.0	8.1-12.0
BT2	82.5	78.8	69.5	445.1	488.6	385.5	36.2	38.0	40.2
BT3	91.4	85.6	70.6	472.5	445.3	371.6	35.1	36.6	36.4
BT4	92.3	87.4	76.6	488.4	419.6	382.7	34.8	35.9	36.2
BT8	87.6	78.5	66.8	445.6	410.3	316.1	35.4	36.8	39.1
BT9	88.7	79.6	59.6	450.8	417.2	321.5	35.6	37.3	39.2
BT11	115.3	106.4	90.2	455.1	421.3	313.6	35.2	37.5	38.9
BT14	119.7	115.3	111.6	510.2	695.5	450.4	36.6	38.8	40.6
BHT3	114.6	104.4	100.5	445.5	413.8	361.3	35.8	37.4	39.4
BHT4	102.7	98.8	71.2	469.3	452.4	372.7	36.1	37.8	39.7
BHT5	105.6	98.2	92.4	714.6	816.3	479.2	36.4	38.2	40.4
LSD (0.05)	17.21	11.53	9.75	125.85	112.67	98.48	2.59	2.75	5.05

Note: BT2 = BARI Tomato 2, BT3 = BARI Tomato 3, BT4 = BARI Tomato 4. BT8 = BARI Tomato 8, BT9 = BARI Tomato 9, BT11 = BARI Tomato 11, BT14 = BARI Tomato 14, BHT3 = BARI Hybrid Tomato 3, BHT4 = BARI Hybrid Tomato 4 and BHT5 = BARI Hybrid Tomato 5.

Slight salinity (4.1-8.0 dS m^{-1}) stimulated expansion of leaf surface in BARI Tomato 14, BARI Hybrid Tomato 5 and BARI Tomato 2 compared to the control while leaf area was in other tomato varieties posed in same salinity level. Salinity level 8.1-12.0 dS m^{-1} caused serious leaf surface reduction in all the varieties. Adams (1991) reported that comparisons of tomato in different salinity sources had led to the conclusion that the exposure of tomato to low and moderate salinity affected fruit growth mainly through osmotic effects rather than ion specific toxicity, provided that the basic nutrient supply is balanced and adequate.

On the other hand, green mater value did not differ among the varieties, but clearly demonstrated a steady adjustment in response to moderate salinity *via* elevated values with the increasing salinity across the ten varieties (Table 2a). Shani and Dudley (2001) reported that the yield loss was occurred due to reduced photosynthesis, high energy and carbohydrate expenses in osmoregulation and interference with cell functions under saline conditions.

Table 2b. Response to salinity on number of fruit, fruit weight and total dry matter plant^{-1} of ten tomato varieties

Variety	Number of fruit plant^{-1}			Fruit weight (kg plant^{-1})			Total dry matter (g plant^{-1})		
	Salinity (dS m^{-1})								
	2.0-4.0	4.1-8.0	8.1-12.0	2.0-4.0	4.1-8.0	8.1-12.0	2.0-4.0	4.1-8.0	8.1-12.0
BT2	21	24	20	1.70	1.71	1.60	16.45	12.83	7.80
BT3	20	18	16	1.45	1.22	1.16	11.94	9.97	6.00
BT4	19	16	14	1.40	1.10	1.02	11.84	9.76	5.94
BT8	25	19	18	1.58	1.34	1.22	13.37	11.11	6.47
BT9	26	22	19	1.61	1.38	1.27	14.48	11.49	6.85
BT11	34	29	23	1.50	1.26	1.11	12.70	10.52	6.23
BT14	27	29	26	1.93	2.01	1.80	19.64	15.79	9.91
BHT3	23	20	17	1.55	1.17	1.09	13.03	10.91	6.32
BHT4	25	21	20	1.62	1.43	1.25	15.85	11.86	7.15
BHT5	26	27	25	1.85	1.90	1.75	17.93	12.73	8.59
LSD (0.05)	1.87	2.57	3.25	0.29	0.43	0.37	2.73	1.84	1.21

In salinity level 4.1-8.0 dS m^{-1} BARI Tomato 14, BARI Hybrid Tomato 5 and BARI Tomato 2 gave the highest values of fruit number and weight plant^{-1} (29, 27, 24, 2.01, 1.90 and 1.71, respectively). Whereas, variety BARI Tomato 4 gave the lowest values. There are inconsistencies in the literature regarding the contribution of fruit number to EC-induced reductions in tomato fruit yield. Eltez *et al.* (2002) reported that the number of fruits was unaffected by moderate salinity and that reduced yield was entirely due to smaller fruit. According to Olympios *et al.* (2003), the number of fruits plant^{-1} was restricted when the level of salinity in the root zone was 8 dS m^{-1} or higher. Comparing the response of different varieties to salinity, it could be clear that reduced in 8.1-12.0 dS m^{-1} salinity. Moreover, the number of fruit and weight of variety BARI Tomato 14, BARI Hybrid Tomato 5 and BARI Tomato 2 was enhanced by salinity (4.1-8.0 dS m^{-1}). This evidence could be a good sign for positive response of plants to salinity.

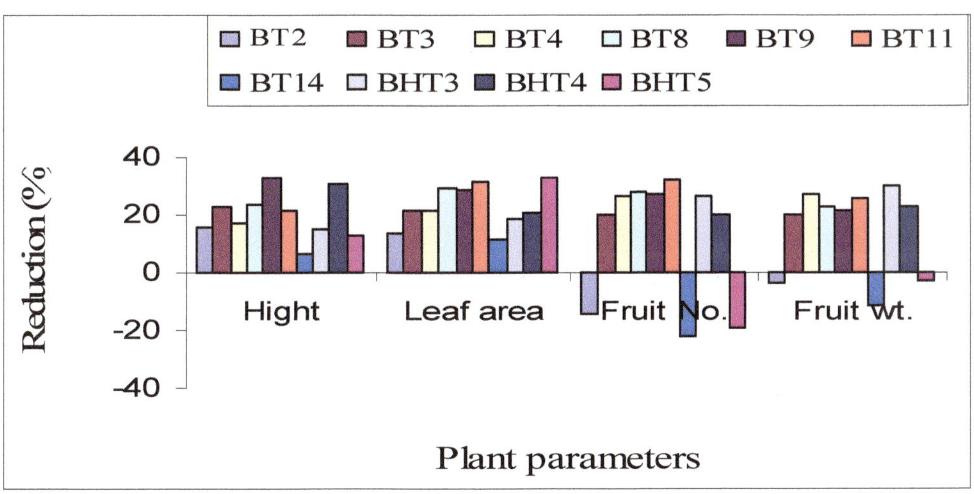

Fig. 1 Reduction in plant growth parameters due to salinity treatments

Comparing the response of different varieties to salinity, it could be seen that all plant parameters of different varieties were reduced compared to control except number of fruits of some varieties such as BARI Tomato 14, BARI Hybrid Tomato 5 and BARI Tomato 2 (Fig. 1). Moreover, the fruit fresh weight for variety BARI Tomato 14, BARI Hybrid Tomato 5 and BARI Tomato 2 was enhanced by salinity. This evidence could be a good sign for positive response of plants to salinity.

Analysis of root and shoot dry matter in the experiment suggested a moderate susceptibility of the ten tomato varieties salinity under the field condition (Fig. 2). Salinity treatment stimulated the production (or maintenance) of more roots at a slight salinity level (4.1-8.0 dS m^{-1}), with BARI Tomato 14, BARI Hybrid Tomato 5 and BARI Tomato 2 showing the greatest response. Despite their moderate susceptibility to salinity, these varieties apparently exhibited very high genetic potential across the salinity level compared to the other seven varieties. Plaut *et al.* (2004) reported that the suppressive effect of moderate salinity on tomato fruit size seems to originate from a restriction of water transport into the fruit, which results in enhanced rates of dry matter accumulation.

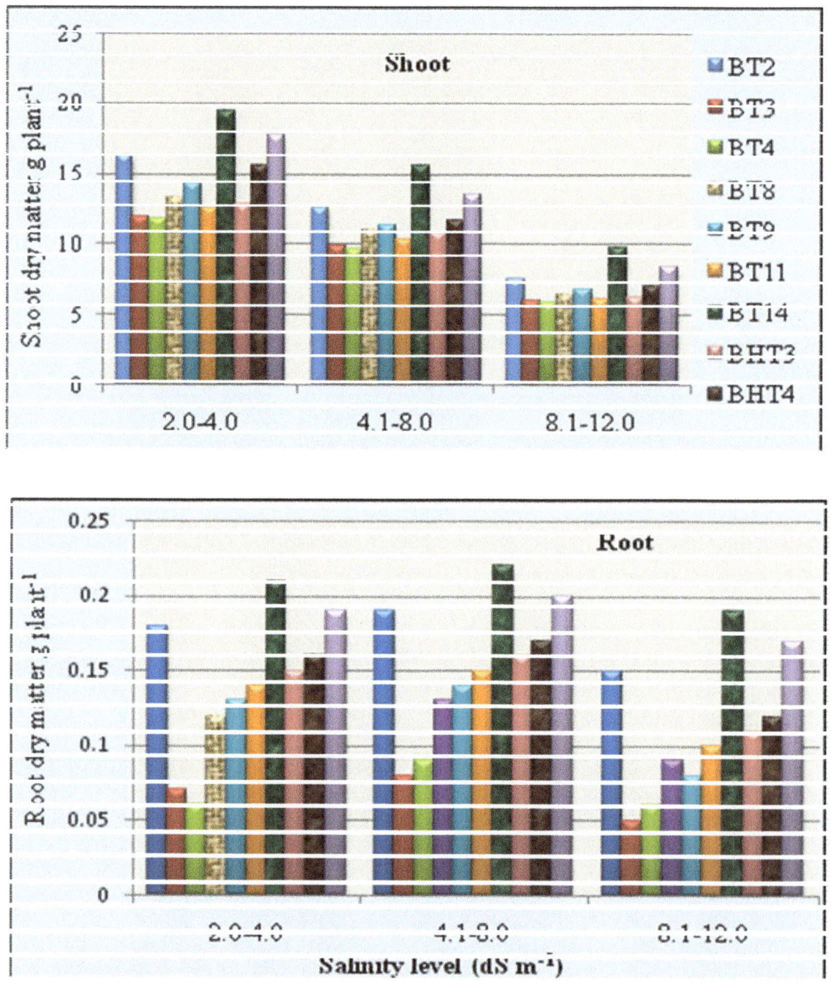

Fig. 2. Response of ten tomato varieties to different salinity based on shoot and root dry matter

Similarly, there was a considerable decrease in the root and shoot dry matter across the varieties at 8.1-12.0 ds m^{-1} (Fig. 1). BARI Tomato 14, BARI Hybrid Tomato 5 and BARI Tomato 2 exhibited superior performance in respect of both root and shoot dry matter $plant^{-1}$. On salinity treatment, dry matter production was adversely reduced in all the ten varieties, except for BARI Tomato 14, BARI Hybrid Tomato 5 and BARI Tomato 2. For instance, BARI Tomato 14, BARI Hybrid Tomato 5 and BARI Tomato 2 significantly produced more root dry matter at higher salinity treatment. The results indicated that the shoot and root dry weights decreased in saline condition, due to the moderate salinity (8.1-12.0 dS m^{-1}). Similar outcome were obtained earlier *by* Mohammad *et al.* (1998) in other tomato cultivars.

Total dry matter yields, derived from the experiments, revealed that BARI Tomato 14, BARI Hybrid Tomato 5 and BARI Tomato 2 were produced more dry matter than the other varieties in different salinity level (Table 2b). The highest plant dry matter was recorded in BARI Tomato 14. BARI Tomato 14, BARI Hybrid Tomato 5 and BARI Tomato 2 were considered relatively tolerant to the moderate salinity (8.1-12.0 dS m^{-1}), but BARI Tomato 4, BARI Tomato 3, BARI Tomato 11 and BARI Hybrid Tomato 3 were not significantly different on the basis of relative dry matter biomass. Based on this study, the four varieties are regarded as relatively weak to the moderate salinity (8.1-12.0 dS m^{-1}). Heakal *et al.* (1990) reported that dry matter yield of plant shoots decreased with increasing salinity.

Table 3. Response to salinity on Na^+ and Cl^- concentration in plant dry matter of tomato varieties

Variety	Na^+ concentration $plant^{-1}$			Cl^- concentration $plant^{-1}$		
	Salinity dS m^{-1}					
	2.0-4.0	4.1-8.0	8.1-12.0	2.0-4.0	4.1-8.0	8.1-12.0
BT2	1.0	32.5	41.9	14.5	65.8	85.8
BT3	3.3	61.3	75.9	19.2	88.7	125.5
BT4	3.7	63.2	72.1	19.0	90.3	121.4
BT8	1.8	41.6	50.8	11.4	80.4	100.7
BT9	1.6	40.3	48.4	18.5	75.2	98.4
BT11	2.6	54.4	63.9	25.1	85.5	110.3
BT14	0.5	21.7	30.2	14.2	50.1	70.2
BHT3	2.4	50.2	60.7	12.6	84.6	105.5
BHT4	2.2	45.8	58.6	9.2	70.4	93.6
BHT5	1.1	28.6	39.4	13.8	62.3	80.4
LSD (0.05)	1.25	15.68	17.63	6.23	9.88	12.83

Mineral content analysis suggested that Na^+ and Cl^- were accumulated in the plant for a majority of the tomato genotypes (Table 3). Both varietal and salinity effects, along with the interactions thereof, were detectable. Tomato variety BARI Tomato 14, BARI Hybrid Tomato 5 and BARI Tomato 2 demonstrated relatively low accumulation of the two ions (Na^+ and Cl^-) in a moderate salinity (8.1-12.0 dS m^{-1}) and low salinity (2.0-4.0 dS m^{-1}) treatment, Ghadiri *el al.* (2005) reported restricted water uptake by salinity due to the high osmotic potential in the soil and high concentrations of specific ions that may cause physiological disorders in the plant tissues and reduce yields. Salinity increases the accumulation of toxic ions such as Na^+ and Cl^- in different plant parts, tissues, cells and cell organelles. Accumulation of Na^+ and/or Cl^- takes place in the chloroplasts of higher plants which affects growth rate, and is often associated with a decrease in photosynthetic electron transport activities in photosynthesis (Kirst, 1989).

Conclusion

Soil salinity is a major constraint to economic use of land for agriculture especially in the coastal regions of Bangladesh. The results of the present study showed that fruit yield of tomato were reduced by increasing salinity. Low salinity gave better results for most plant parameters and as salinity increased there was a reduction in plant growth and yield. Variety BARI Tomato 14, BARI Hybrid Tomato 5 and BARI Tomato 2 showed an optimistic response to saline agriculture by producing more yield under slight and moderate saline conditions. Hence, variety BARI Tomato 14, BARI Hybrid Tomato 5 and BARI Tomato 2 can survive and produce good yield in coastal fields of Bangladesh.

References

Adams, P. 1991. Effects of increasing the salinity of the nutrient solution with major nutrients or sodium chloride on the yield, quality and composition of tomatoes grown in rockwool. *J. Hort. Sci.*, 66: 201-207.

Adler, P.R. and Wilcor, G.E. 1987. Salt stress, mechanical stress, or Chloromequat chloride effects on morphology and growth recovery of hydroponic tomato transplants. *J. Am. Hort. Sci.*, 112: 22-25.

Allen, R.G., Pereira, L.S. Raes, D. and Smith. M. 1998. Crop evapotranspiration: Guidelines for computing crop water requirements. Irrigation & Drainage. Paper 56. UN-FAO, Rome, Italy.

Caro, M., Cruz, V., Cuartero, J., Estañ, M.T. and Bolarin, M.C. 1991. Salinity tolerance of normal-fruited and cherry tomato cultivars. *Plant and Soil.*, 136: 249- 255.

Eltez, R.Z., Tii-zel, Y., Gi.i, L.A., Tii-zel, J.H. and Duyar, H. 2002. Effects of different EC levels of nutrient solution on greenhouse tomato growing, *Acta Hortic.*, 573: 443-448.

Ghadiri, H., Dordipour, I., Bybordi, M. and Malakourti. M.J. 2005. Potential use of Caspian Sea for supplementary irrigation in North Iran. *Agric. Water Manage.*, 79: 209-224.

Heakal, M.S., Modaihsh, A.S., Mashhady, A.S. and Metwally. A.I. 1990. Combined effects of leaching salinity and potassium content of water on growth and water use efficiency of wheat and barley. *Plant and Soil*, 125: 177-184.

Karim, Z., Ibrahim, A.M., Iqbal, A. and Ahmed, M. 1990. Drought in Bangladesh Agriculture and Irrigation Schedules for Major Crops, Bangladesh Agricultural Research Council (BARC), *Soils Public*, No. 34. 11p.

Karlberg, L., Ben-Gal, A., Jansson, P-E. and Shani, U. 2006. Modeling transpiration and growth in salinity-stressed tomato under different climatic conditions. *Ecological Modeling*, 190: 15-40.

Kirst, G.O. 1989. Salinity tolerance of eu-karyotic marine algae: Annual Review Plant Physiology. *Plant Molecular Biol.*, 40: 21-53.

Mohammad, M., Shibli, R., Ajouni, M. and Nimri, L. 1998. Tomato root and shoot responses to salt stress under different levels of phosphorus nutrition. *J. Plant Nutr.*, 21: 1667-1680.

Olympios, C.M., Karapanos, I. C., Lionoudakis, K. and Apidianakis, I. 2003. The growth, yield and quality of greenhouse tomatoes in relation to salinity applied at different stages of plant growth. *Acta Hortic.*, 609: 313-320.

Peralta, E., Knapp, S. and Spooner, O.M. 2005. New species of wild tomato *(Solanum* section *Lycoperisicon:* Solanaceae) flom; Iorthem Pem. *Systematic Bot.*, 30: 424-434.

Plaut, Z., Grava, A., Yehezke, I.C. and Matan, E. 2004. How do salinity and water stress affect transport of water, assimilates and ions to tomato fruits? *Physiologia Plantarum*, 122: 429-442.

Shani, U. and Dudley, L.M. 2001. Field studies of crop response to water and salt stress. *Soil Sci. Soc. Am. J.*, 65:1522–1528.

Sonneveld, C. and Van der Burg, A.M.M. 1991. Sodium chloride salinity in fruit vegetable crops in soilless culture. *Netherlands J. Agri. Sci.*, 39: 115-122.

SRDI. 2010. Saline soils of Bangladesh. SRMAF project, Ministry of Agriculture, Govt. Peoples Repiublic of Bangladesh. 35p.

17

REPRODUCTIVE BIOLOGY OF FEATHER BACK, CHITAL (*Notopterus chitala*, Ham.) CULTURED IN A POND OF BANGLADESH

A.H.M. Kohinoor*, D.A. Jahan, M.M. Khan, M.S. Islam and M.G. Hussain

Abstract

Studies on Gonadosomatic index (GSI), ova diameter and histology of the gonad were performed to understand reproductive biology of Feather back, Chital (*Notopterus chitala*) for a period of 6 months from January to June 2010. Thirty live chital fish were used in this study. The mean GSI values for female chital were found to range between 0.20±0.013 and 4.63±0.50. The highest GSI value was found in June. The smallest diameter of ovum was recorded 0.04 mm (January) and the largest was 4.00 mm (June). During the experimental period, the fecundity was ranged from 8,238 to 18,569 (mean 13,052±4607) in fish samples with body weight range from 1,296 to 2,360 (mean 1,742.50±474.44 g) while the relative fecundity was 5.65 to 14.33. Histological study revealed that the percentage of late perinucleolus (LPN) stage was highest in April and Cortical alveoli (CA) stage appeared from April and reached to maximum in May. Vitellogenic stage (VG) appeared in the month of May and chronologically increased through June. Vitellogenic stage (VG) of oocyte as well as highest ova diameter reached to the peak in June. The variations in the gonad weight and GSI of the female fish reached to the peak during June indicating maturity of ovary and definite spawning season.

Keywords: Reproductive biology, feather back, chital

Freshwater Station, Bangladesh Fisheries Research Institute, Mymensingh, Bangladesh

*Corresponding author's email: kohinoor41@gmail.com (A.H.M. Kohinoor)

Reviewed by Dr. Saleh M.A. Mobin, NMIT, Australia.

Introduction

Feather back, Chital (*Notopterus chitala*) is a rheophilic important freshwater fish found widely in lenthic waters. This species is widely distributed in deep and clear waters in the rivers, beels, reservoirs, haors, baors and ponds in Bangladesh (Azadi *et al.*, 1994). In recent years, the catch of this species has been declining fast due to environmental degradation (Hossain *et al.*, 2006). This fish is rich in nutritive value and commands high market price despite of the presence of a large number of intramuscular bones. In addition to the above qualities, Chital plays a significant role in regulating the population imbalance that may be caused by wild breeding of common carp, abundance of other minnows and insects in ponds under composite carp culture where strict control on the population size of the stocked fish is essential to obtain optimum production (Chaudhuri *et al.*, 1975). Some progressive farmers are thinking to culture this species in pond with carps using polyculture technique. The fry of chital is not available throughout the year for growing out in the farms for producing marketable size fish, which indicates significance for developing seed production technology of this species. Although literature is available, on the morphometry, food and feeding habits, natural breeding (Singh *et al.*, 1980; Hossain, 1999) of this fish, very scanty systematic attempts have been made to study the reproductive biology and breeding technique in captivity of this fish. This paper reports a detailed account of the reproductive biology of chital.

Materials and Methods

The study was carried out for the period of six months from January to June 2010 in the Freshwater Station, Bangladesh Fisheries Research Institute, Mymensingh, Bangladesh. Fifty brood fish of chital were stocked in a rearing pond with an area of 400 m^2 between August and September. After stocking, the brood fishes were fed with live food such as small fish, tilapia fry, small prawn etc. @ 2% of estimated body weight.

Fish Sample collection

Five female live fish samples were collected by seine net once in a month from January to June 2010 from the experimental pond. After collection, live fish were transported to the laboratory and the gonads were removed by opening the peritoneal cavity and were used for the study.

Determination of the reproductive periodicity

The methods were used to determine the reproductive periodicities of the fish are as follows:

Gonadosomatic Index (GSI)

Total body weight of selected fishes in each month was considered to calculate the mean Gonadosomatic Index (GSI). Gonadosomatic Index (GSI) was calculated according to the formula:

$$GSI = \frac{\text{Weight of gonad}}{\text{Weight of fish}} \times 100$$

Diameter of ova

After dissecting the overy, small representative part from the anterior, posterior and middle portion of the ovary was removed separately. The ova of the samples were separated using physiological saline solution (0.65% NaCl) with a petridish and spread on a glass slide to measure the diameter under a microscope with an ocular micrometer. The units of the ocular micrometer that visible in the microscope were standardized with a stage micrometer for measurement of ova diameter in micrometer (μm). Approximately 300 ova from each sample were measured and ova diameters were determined and were expressed in unit value of the ocular micrometer.

Histology of Gonad

To study the sequential histological changes, the ovaries tissues were fixed in aqueous Bouin's fluid and then routine histological process was followed for slide preparation. The gonadal tissues were serially sectioned at a thickness of 4μ and stained with haematoxylin-eosin. From the histological preparation of the gonadal tissues, the cells and nuclear diameters of cells and various oocytes were measured by ocular micrometer.

Results and Discussion

The different development stages of gonad are presented below:

Gonad observation

The oval shaped chital ovary was single lobed lying in the body cavity. The size and extent of occupancy of the body cavity were found to vary with size and maturity of the females. Ovary showed the highest weight in June indicating the mature stage. During this time, the colour of the ovary was yellowish. The immature ovary was compact but the matured one was filament like structure (Plate 1).

Plate 1. Physical appearance of Chital ovary

Gonadosomatic index (GSI)

Monthly variations in GSI values of chital are presented in Table 1 and Fig. 1. The mean GSI values of female chital were found to range between 0.20±0.013 to 4.63±0.50. The greatest GSI value was found in June, which indicates peak period of ovarian growth. It is evident from the results that the GSI values increased slowly from January to March and then sharply increased in June. Thomas *et al.* (2003) described that gonadosomatic index (GSI) found lowest in early stage and highest in later stage in fin fish and shell fish.

Table 1. Mean and standard deviation (±SD) of total body weight, total length, gonad weight and GSI (%) at successive months of female *N. Chitala*

Month	No. of fish examined	Total length (cm)	Body weight (g)	Gonad weight (g)	GSI (%)
January	4	53.90±5.90	1178±131.45	2.34±0.36	0.20±0.01
February	4	60.88±1.40	1605±114.65	3.78±0.40	0.24±0.02
March	4	65.85±0.95	2101±357.84	11.2±2.73	0.53±0.04
April	4	65.58±4.98	1855±463.41	34.52±14.28	1.80±0.45
May	4	66.7±2.66	2019.75±206.0	165.55±23.79	3.21±0.98
June	4	67.65±4.64	2041±518.19	95.50±29.86	4.63±0.50

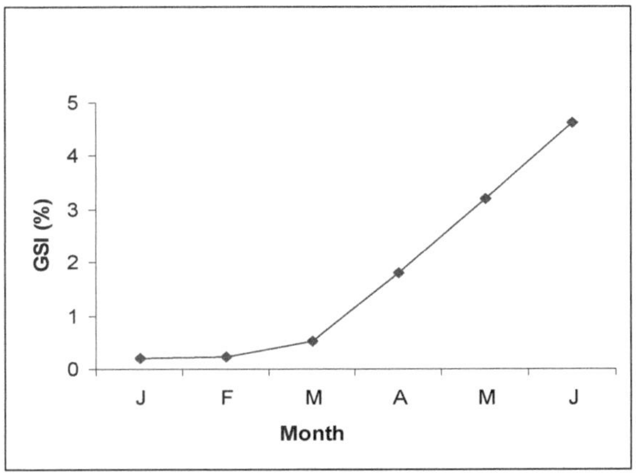

Fig. 1. Monthly trend of changes in Gonadosomatic Index (GSI) of female *N. chitala*

Diameter of ova

Increasing diameter of ova indicates the development of eggs which is related to approaching the breeding season of fish. The observed ova size at different months indicated that the chital fish become sexually matured between May and June.

For easy interpretation, the ova were grouped into two categories on the basis of size of diameter i.e. category I (0.04-0.168 mm) and category-II (1.00-4.00 mm). The smallest diameter of ovum was recorded 0.04 mm and the largest was 4.00 mm during the study period. During January to March, only category-I oocytes were found and this category was also widely distributed up to June. But in June it was comparatively lower percentage (66.04±4.73) than other months. During the observation period, mean ova diameter was 0.117±0.02 mm and 0.120± 0.03 mm in February and March, respectively (Table 2).

Table-2 also showed that two categories of ova were observed from April. The ova belonging to category II was found in April with a rapid increase in diameter in May. It was observed that the abundance of category II oocytes were highest in June (34.21±6.54%) and lowest in April (8.86±3.38%). The mean ova diameter of Category I and II, were 0.122±0.02 mm and 2.51±0.54; 0.124±0.021 and 3.16±0.89 mm; 0.129±0.02 mm and 3.21±0.47 mm in the month of April, May and June, respectively (Table 2).

It is mentioned by Rahman (1989) that chital breeds in June and having an egg size of 3.0 to 4.5 mm in diameter. While, Radheyshyam and Sarangi (2004) studied the breeding activity and egg incubation of the same species in captivity and observed the diameter of fertilized eggs ranges were 4.8-8.2 mm. Hossain *et al.*, (2006) cultured chital in ponds and found the fertilized eggs having average diameter of 4.5 mm. The results obtained in the present study are consistent with the study mentioned above.

Table 2. Category wise distribution of oocytes during observation period

Groups	Months				
	February	March	April	May	June
Category-I (mm)	0.117±0.02	0.120± 0.03	0.122±0.02	0.124±0.021	0.129±0.02
Category-II (mm)	-	-	2.51±0.54	3.16±0.89	3.21±0.47

Fecundity

Breeding performance of a fish species depends on different parameters like fecundity, spawning response, fertilization rate, hatching rate, hatching time, size of spawn, yolk sac absorption time period, larval survival rate etc. Therefore, to understand fecundity of this species a study was carried out. In this study fecundity was found to ranged from 8,238 to 18,569 (mean 13,052±4607), in fish samples with body weight from 1296 to 2360 g (mean 1742.50±474.44 g) while the relative fecundity were 5.65 to 14.33 (mean 7.99±4.24 g). Details of results are given in Table 3.

Table 3. Mean gonad weight and fecundity of *N. chitala*

Mean. body wt. (g)	Mean gonad wt. (g)	Mean relative fecundity (No. of eggs/g B. Wt.)	Mean fecundity
1742.50 ± 474.44	66.25±6.80 g	7.99±4.24 g	13052±4607
(1296 to 2360)	(59.00 to 73.00)	(5.65 to 14.33)	8238 to 18,569

The results analysis shows a nearly perfect positive linear relationship between fecundity and body weight, where the calculated values of regression coefficient, intercept and coefficient of correlation were 0.1021, + 409.69 and 0.4917, respectively (Fig. 2). Shing *et al.* (1980) observed natural spawning of chital in pond ecology and estimated the fecundity of the fish to be around 400 eggs kg^{-1} body weight. In another study, Hossain *et al.* (2006) reported that fecundity was 5761 eggs of a 4200 g female weight.

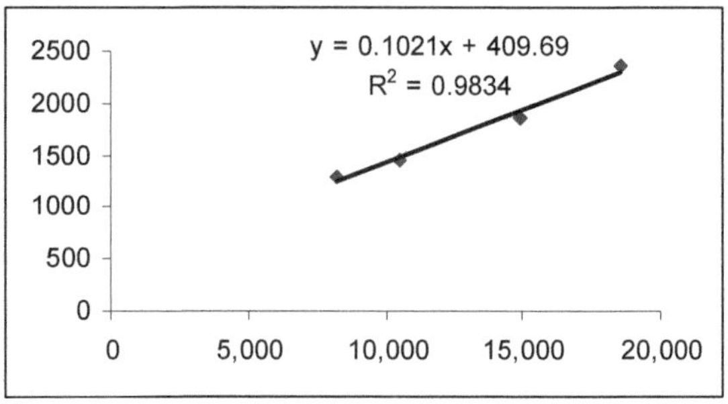

Fig. 2. Relationship between body weight and fecundity of *N. chitala*

Histology of gonad

Histological structure of gonad was examined and different stages of oocytes indicated that the oocytes did not develop synchronously. In gonad, different developmental stages of oocyte viz. oogonia (OG), early perinucleolus stage (EPN), late perinucleolus stage (LPN), cortical alveoli stage (CA), vitellogenic stage (VG) were distinguished on the basis of size appearance of nucleolus and cytoplasm. It was found that at the earlier stages of the study period (January/February), all oocytes were found in oogonial stage. Later on, the percentage of oogonia showed a decreasing trend in the successive months.

The highest percentage early perinucleolus stage (EPN) was found in May (31.00±2.65 %) with a decreasing trend in the following months. As a result, the percentage of late perinucleolus (LPN) stage was highest in April (7.67±1.53%) and then their appearance was found to decrease gradually. But the oogonia stage, early and late perinucleolus stage showed a decreasing trend from February to June. Cortical alveoli stage (CA) appeared from April and turned greatest in May (10.67±2.08%). Vitellogenic stage (VG) appeared in the month of May and chronologically increased up to June (33.67±3.51%. Atretic oocytes (AO) were not found during the study period. Month wise per cent distribution of different development stage of oocytes is shown in Table 4. No literature is available on histology of gonad of this species for comparison.

Table 4. Month wise per cent distribution of different development stage of oocytes in sample of ovarian tissue of Chital

Months	Oogonia	EPN	LPN	CA	VG
January	100	-	-	-	-
February	100	-	-	-	-
March	92.67±1.53	5.33±1.53	2.00±0.98	-	-
April	76.33±4.51	14.67±2.52	7.67±1.53	1.33±0.58	-
May	25.33±2.52	31.00±2.65	6.67±2.08	10.67±2.08	26.33±1.15
June	23.67±3.51	30.33±2.52	5.00±1.13	7.33±0.58	33.67±3.51

In the present study, five stages of oocyte were identified that were: (a) oogonia, (b) early perinucleolus stage, (c) pre-vitellogenesis, (d) vitellogenesis and (e) maturation stage. These stages of oocyte development in Chital are similar to that of *Ompok pabda, Amblypharyngodon mola, Puntius sophor, Puntius gonionotus and Chela cachius* (Zakia, 1996; Begum, 1997 and Kohinoor 2000). These developmental stages are seemed to be fairly similar to that of *Pleuronectes flesus* described by Janseen *et al.* (1995) who described eight stages of oocyte development in their study. On the other hand, Mollah (1986) observed the oocyte maturation in catfish, *Clarias macrocephalus,* and identified seven stages: oogonia, chromatin nucleus, early perinucleolus, late perinucleolus, yolk vesicle, early yolk- granule and late yolk-granule stage. In another histological study, the stage of oogonia of *L. rohita* and *C. cirrhossus* were found from the month of February while, it was fully absent in June. Jahan, (2008) observed that the EPN and LPN stages appeared from April and the vitellogenic stage (VG) in May.

Conclusion

The variations in the gonad weight and gonado somatic index (GSI) of the female chital fish reached to the peak during June indicating maturity of ovary and definite spawning season. While Vitellogenic stage (VG) of oocyte as well as highest ova diameter was observed in June also. The present study on reproductive biology of chital may contribute to the seed production technology of this fish species in captivity, which may prelude to their introduction as a potential candidate for aquaculture and paves the way to conserve this endangered native fish species.

References

Azadi, M.A., Mahamood, N. and Shafi, M. 1994. Studies on the age and growth of Chital, *Notopterus chitala* (Ham.) from the Kaptai Reservoir, Bangladesh. *Chittagong Univ. Studies.*, 18 (2): 197-205.

Begum, K. 1997. A study on the histological analysis of gonadal development, fecundity, embryonic and larval development of *Ompok pabda* (Hamilton-Buchanan 1822). M. Sc. Dissertation, Department of Zoology, Dhaka University. 108 p.

Chaudhuri, H., Chakrabarty, R.D., Sen, P.R., Rao, N.G.S. and Jena, S. 1975. A new high in fish production in India with record yield by composite fish culture in freshwater ponds. *Aquaculture*, 6: 343-355.

Hossain, Q.Z. 1999. Some observations on breeding and fry rearing of chital (*Notopterus chitala*, Hamilton) in Bangladesh. *Fishing Chimes.*, 19 (7): 13-16.

Hossain, Q.Z., Hossain, M.A. and Parween, S. 2006. Breeding biology, captive breeding and fry nursing of humped featherback (*Notopterus chitala*, Hamilton-Buchanan, 1822). *Ecoprint*, 13: 41-47

Jahan. D.A. 2008. Substitution of fishmeal protein by soybean meal protein in the diet of two Indian major carps. Ph.D. Dissertation, Department of Fisheries Biology & Genetics, Bangladesh Agricultural University, Mymensingh. 268p.

Janseen, P.A.H., Lambert, J.G.D. and Goos, H.J.T. 1995. The annual ovarian cycle and the influence of pollution on vitellogenesis in the flounder, *Pleuronectes flesus*. *J. Fish. Biol.*, 47: 509-523.

Kohinoor, A.H.M. 2000. Development of culture technologies of three small indigenous fish mola (*Amblypharyngodon mola*), punti (*Puntius sophore*) and chela (*Chela cachius*) with notes on some aspects of their biology. Ph.D. Dissertation, Department of Fisheries Management, Bangladesh Agricultural University, Mymensingh. 366p.

Mollah, M.F.A. 1986. Cycle changes in the ovary of freshwater catfish, *Clarias macrocephalus* (Gunther). *Indian J. Fish.*, 33 (1): 54-65.

Radheyshyam and Sarngi, N. 2004. Breeding for sac-fry production of feather back fish *Notopterus chitala* Hamilton) in control condition. Abstract, Souvenir Natl. Semi. Zoology and Human Welfare, 22-24 Nov. 2004 at Dr. S. P. Mukherjee Govt. Deg. College of Allahabad University, Phapharman, Allahabad, 16p.

Rahman, A.K.A. 1989. Freshwater Fishes of Bangladesh. Zoological Society of Bangladesh. 364 p.

Shing, S.B., Dey, R.K., Reddy, P.V.G.K. and Mishra, B.K. 1980. Some observation on breeding, growth, and fecundity of *Notopterus chitala*. *J. Inland Fish. Soc.* India, 12 (2): 13-17.

Thomas-Jones, E., Thorpe, K., Harrison, N., Thomas, G., Morris, C. and Hutchinson, T.H. 2003. Dynamics of estrogen biomarker responses in rainbow trout exposed to 17β-estradiol and 17α-ethinylestradiol. *Environ. Toxicol. Chem.*, 22: 3001–3008.

Zakia, A. 1996. Comparative studies on the gonadal histology and morphology of different interaspecific hybrids of *Puntius gonionotus* (Bleeker, 1849). M. Sc. Dissertation, Department of Zoology, Dhaka University, Dhaka. 138p.

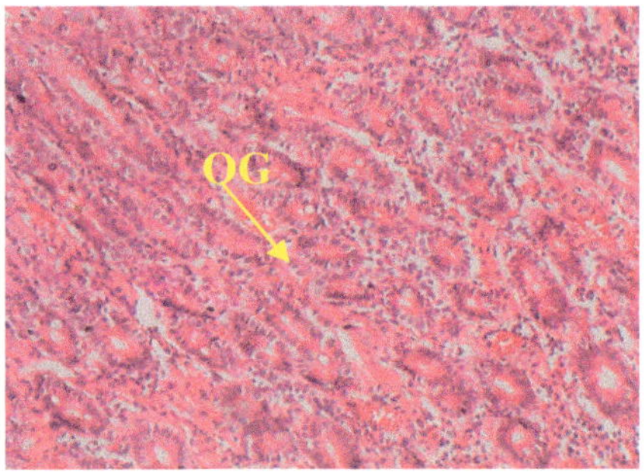

January

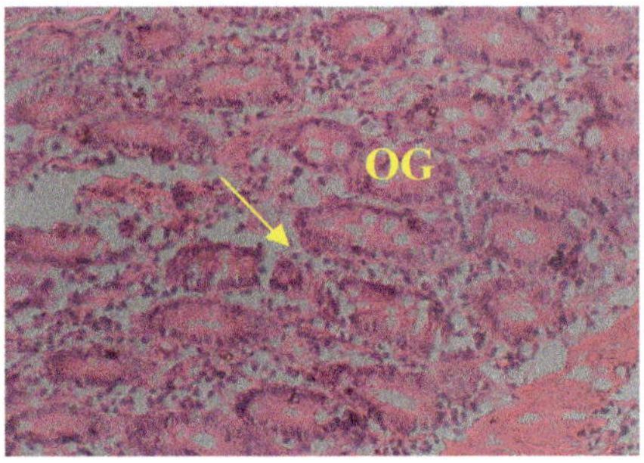

February

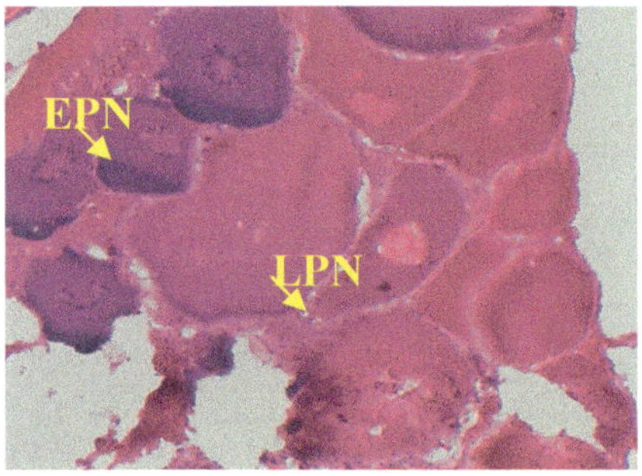

March

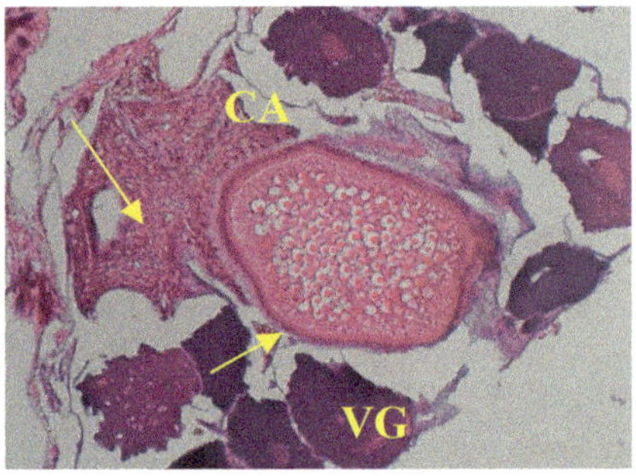

April

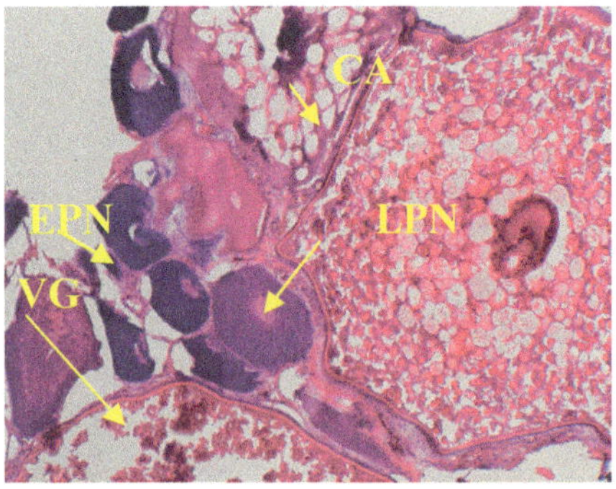

May

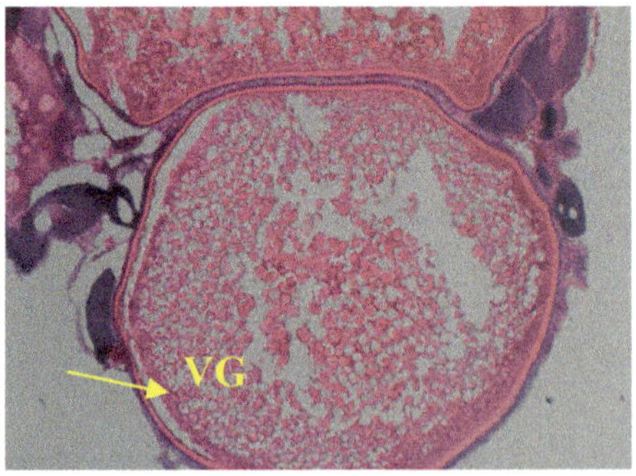

June

Plate 2. Developmental stage of Oocytes in different months

18

SUITABILITY OF DUCKWEED (*Lemna minor*) AS FEED FOR FISH IN POLYCULTURE SYSTEM

M.Z.H. Talukdar, M. Shahjahan* and M.S. Rahman

Abstract

In the present study, we conducted an experiment to evaluate the effects of duckweed (*Lemna minor*) as feed on fish production in polyculture. The experiment had 2 treatments where in treatment 1 (T1) ponds were supplied with duckweed as feed and in treatment 2 (T2) ponds were kept as control (without supply of duckweed). Average survival rates in T1 and T2 were 90 and 89%, respectively. The specific growth rates (SGR) were higher in T1. Calculated net production in T1 was 6.25 tons $ha.^{-1}$ $yr.^{-1}$ and in T2 was 2.84 tons $ha.^{-1}$ $yr.^{-1}$. The ranges of physico-chemical parameters analyzed were within the productive limit and more or less similar in all the ponds under both treatments during the experimental period. There were 24 genera of phytoplankton under 5 major groups and 10 genera of zooplankton under 3 major groups found in the experimental ponds. The net production in T1 was significantly higher than that of T2 indicated the use of duckweed as feed for fishes is economically sustainable in polyculture.

Keywords: Duckweed, polyculture, fish, feed, production, pond, water quality

Department of Fisheries Management, Bangladesh Agricultural University, Mymensingh-2202, Bangladesh.

*Corresponding author's email: shahjahanm75@yahoo.co.uk (M. Shahjahan)

Reviewed by Dr. Binay Kumar Chakraborty, Matshya Bhabon, Romna, Dhaka, Bangladesh.

Introduction

Aquaculture in Bangladesh has rapidly progressed in recent years with a contribution of 44% to the annual fish production. Among different techniques of aquaculture, polyculture is one of the most important techniques. The basic principle of fish polyculture systems rests on the idea that when compatible species of different feeding habits are cultured together in the same pond, the maximum utilization of all natural food sources takes place without harmful effects. Polyculture or mixed culture of carps has been found as an economically viable and technically sustainable in perennial water bodies (Alikhuni, 1957; Chen, 1976). The selection of fish species is very important for polyculture systems. In the present study, tilapia (*Oreochromi mossambicus*), sharpunti (*Barbades gonionotus*), grass carp (*Ctenopharyngodon idella*), catla (*Catla catla*) and mrigal (*Cirrhinus mrigala*) were selected for polyculture. These species are suitable for low inputs culture system in small ponds and ditches for their quick growth and for maximum production within short period. Bangladesh has numerous seasonal water bodies in the form of shallow ponds, ditches, roadside canals, pits in rice fields, which retain water for 4-6 months. The natural environment of Bangladesh is also suitable for growing these fish species, which can be cultured in both shallow seasonal ponds and deeper perennial ponds.

Duckweed are small floating aquatic plants which are widely available in Bangladesh and consist of four genera viz., *Lemna*, *Spirodela*, *Wolfia* and *Wolfiella* among which about 40 species have been identified (Journey *et al.*, 1991; Skillikorn *et al.*, 1993). Duckweed can easily grow abundantly with minimum cost and can be made available as much cheaper feed than other alternative plant protein sources. Recently duckweed has been accepted as protein rich (40-45% of the dry weight) feed for fish (Landolt and Kandeler, 1987; Leng *et al.*, 1995; Saha *et al.*, 1999). According to Porath and Agami (1977), the weight of grass carp could be tripled (from 100 to 300 g) within 50 days when feeding a mixture of *Lemna gibba* and *Lemna minor*. Duckweed protein has higher concentration of essential amino acids, lysine and methionine than most plant proteins and more closely resembles animal protein in that respect (Journey *et al.*, 1991). In Bangladesh, many studies have been carried out on the use of duckweed as feed for fishes in monoculture (Kohinoor *et al.*, 1993; Bornali, 2004; Haque, 2005; Uddin *et al.*, 2007; Chowdhury *et al.*, 2008), but so far, few works have been carried out in polyculture system. In the present study, we conducted an experiment to evaluate the effect of duckweed (*Lemna minor*) as feed on the fish production in polyculture system.

Materials and Methods

Duration of study

The experiment was conducted for a period of 90 days in the earthen ponds each measuring 1 decimal (40-m^2) area at Bangladesh Agricultural University, Mymensingh, Bangladesh.

Pond preparation

Before fish stocking water of the experimental ponds were drained out to eradicate all the undesirable fishes, renovated and liming was done in all the ponds at the rate of 1 kg 40 m^{-2}. Ponds were filled up with deep tube well water and fertilized with poultry dropping 10 kg, urea 100 g and TSP 100 g 40 m^{-2} as initial doses.

Stocking of fish

The experiment had two treatments each with three replications. In T_1, ponds were supplied with duckweed as supplementary fish feed and in T_2, ponds were kept as control (without supply of duckweed). Ponds were stocked at a stocking density of 151 fingerlings per decimal (40-m^2) at the ratio of tilapia: sharpunti: grass carp: catla: mrigal = 45: 38: 15: 38: 15.

Fertilization and supply of duckweed

The ponds were fertilized fortnightly with poultry dropping 10 kg, urea 60 g and TSP 90 g per 40 m^2 to grow natural food phytoplankton and zooplankton. Fresh duckweeds were supplied everyday to the ponds (T_1) at the rate of 50% of the total body weight (wet weight basis) of the fish.

Physico-chemical parameters

Various physical and chemical water quality parameters of the ponds such as water temperature (°C), transparency (cm), dissolved oxygen (mg L^{-1}), pH, free CO_2 (mg L^{-1}), total alkalinity (mg L^{-1}), PO_4-P (mg L^{-1}) and NO_3-N (mg L^{-1}) were estimated fortnightly. Water temperature was recorded with a Celsius thermometer and transparency was measured with a Secchi disc of 30 cm diameter. Dissolved oxygen was measured directly with a DO meter (Lutron, DO-5509) and a portable digital pH meter was used to measure pH. Free CO_2 and total alkalinity were determined by titrimetric method (APHA, 1992). PO_4-P (mg L^{-1}) and NO_3-N (mgL^{-1}) were determined by a Hach Kit (DR/2010, a direct reading Spectrophotometer).

Biological parameters

Biological parameters of ponds water such as phytoplankton density (cells L^{-1}) and zooplankton density (cells L^{-1}) were estimated fortnightly. The counting of plankton (both phytoplankton and zooplankton) was done with the help of Sedgwick-Rafter Counting Cell (S-R cell) under a compound binocular microscope. The plankton population was determined by using the formula of Rahman (1992). Identification of plankton (phytoplankton and zooplankton) up to generic level were made according to Prescott (1964), Needham and Needham (1963) and Belcher and Swale (1978).

Survival, growth and production of fish

Fish samples were collected with a cast net monthly to estimate the growth in length (cm) and in weight (g) and to check up the health condition of fish. At the end of the experiment, all fish were harvested through repeated netting by seine net to calculate gross and net production of fish.

The survival rate was estimated by the following formula:

$$\text{Survival rate (\%)} = \frac{\text{No. of harvested fishes}}{\text{Initial No. of fishes}} \times 100$$

Specific growth rate (SGR, percent per day) was estimated by the following formula:

$$\text{SGR (\% per day)} = \frac{\text{Loge } W_2 - \text{Loge } W_1}{T_2 - T_1} \times 100$$

Where, W_1= Initial live body weight (g) at time T_1 (day)
W_2= Final live body weight (g) at time T_2 (day)

Statistical analysis

Values are expressed as means ± standard deviation (SD). Data were analyzed by one-way analysis of variance (ANOVA) followed by Tukey's post hoc test to test for statistically significant differences between treatments. Statistical significance was set at $p < 0.05$. Statistical analyses were performed using SPSS Version 14.0 for Windows (SPSS Inc., Chicago, IL).

Results and Discussion

The present study was conducted to evaluate the suitability of duckweed as feed for fishes in polyculture of tilapia, sharpunti, grass carp, catla and mrigal. The supply of duckweed at the rate of 50% of body weight of fishes showed better production performance.

Physico-chemical parameters

The physico-chemical parameters such as water temperature (°C), transparency (cm), dissolved oxygen (mg L^{-1}), pH, free CO_2 (mg L^{-1}), total alkalinity (mg L^{-1}), PO_4-P (mg L^{-1}) and NO_3-N (mg L^{-1}) of the ponds were found to be within the acceptable ranges for fish culture (Table 1). There was no abrupt change in any parameter of the pond water during the tenure of experiment. The results were more or less similar to the findings of Wahab *et al.*, (1995), Kohinoor *et al.*, (1998), Uddin *et al.*, (2007) and Chowdhury *et al.*, (2008).

Table 1. Physico-chemical parameters (Means ± SD) of the experimental ponds during the experimental period

Parameters	Treatments			
	Treatment 1		Treatment2	
	Means	SD	Means	SD
Water temperature (°C)	28.02	1.57	27.92	1.49
Transparency (cm)	30.50	2.97	32.08	4.21
Dissolved oxygen (mgL^{-1})	6.63	0.50	6.23	0.79
pH	7.45	0.21	7.43	0.19
Free CO_2 (mgL^{-1})	2.85	0.30	2.90	0.28
Alkalinity (mgL^{-1})	75.33	7.40	74.00	8.30
Phosphate-phosphorous(mgL^{-1})	0.59	0.11	0.64	0.07
Nitrate-nitrogen(mgL^{-1})	1.80	0.08	1.54	0.18

Biological parameters

Fortnightly fluctuation of phytoplankton density (cells L^{-1}) and zooplankton density (cells L^{-1}) are shown in Table 2. The average density of phytoplankton of the ponds under T1 was 53.98 ± 4.74 (x10^3) cells L^{-1} and that of the ponds under T2 was 47.60 ± 4.07 (x10^3) cells L^{-1}. The average density of zooplankton of the ponds under T1 was 10.02 ± 0.94 (x10^3) cells L^{-1} and that of the ponds under T2 was 8.27 ± 0.76 (x10^3) cells L^{-1}. The generic status of phytoplankton and zooplankton found during the tenure of experiment are shown in Table 3. Phytoplankton and zooplankton population in number and genera were more or less similar to the findings of Uddin *et al.*, (2007) and Chowdhury *et al.*, (2008).

Table 2. Fortnightly fluctuation of phytoplankton and zooplankton densities in the ponds during the experimental period.

Parameters	Treatments	Sampling days						Means±SD
		1	2	3	4	5	6	
Phytoplankton (x10^3 cells L^{-1})	T1	47.30	49.70	54.30	60.00	55.20	57.40	53.98±4.74
	T2	47.60	51.70	40.90	45.10	51.10	49.20	47.60±4.07
Zooplankton (x10^3 cells L^{-1})	T1	9.90	10.60	8.40	10.90	10.70	9.60	10.02±0.94
	T2	7.90	7.50	9.40	8.10	9.00	7.70	8.27±0.76

Table 3. Generic status of phytoplankton and zooplankton found in the experimental ponds

Phytoplankton		Zooplankton
Bacillariophyceae	**Cyanophyceae**	**Cladocera**
Asterionella	*Anabaena*	*Daphnia*
Cyclotella	*Aphanocapsa*	*Diaphanosoma*
Diatoma	*Chroococcus*	
Fragilaria	*Gomphospaeria*	**Copepoda**
Navicula	*Microcystis*	*Cyclops*
Tabellaria	*Oscillatoria*	*Diaptomus*
Chlorophyceae	**Dinophyceae**	**Rotifera**
Actinastrum	Ceratium	*Asplanchna*
Chlorella		*Brachionus*
Closterium	**Euglenophyceae**	*Filinia*
Gloeocystis	*Euglena*	*Keratella*
Oocystis	*Phacus*	*Polyarthra*
Pediastrum		*Trichocerca*
Scenedesmus		
Ulothrix		
Volvox		

Survival, growth and production of fish

The survival rate of fishes in T_1 and T_2 were 90 and 89%, respectively (Table 3). More or less similar survival rates were observed in polyculture system (Mostaque, 1995) and monoculture of Thai sharpunti (Kohinoor *et al.*, 1993). The specific growth rate (SGR) of tilapia, sharpunti, grass carp, catla and mrigal were 0.99, 1.03, 1.51, -0.08 and 2.00 in T_1, and 0.65, 0.83, 0.58, -0.003 and 1.86 in T_2, respectively

(Table 4). SGR values in T_1 were higher than those in T_2 except catla. Catla showed negative growth rate in both treatments, most probably it could not compete successfully with other fishes. SGR values obtained in the present study are similar to those obtained by Hossain *et al.*, (1997). It can be said that the higher specific growth rate in T_1 was due to use of duckweed as feed for fishes.

Table 4. Survival, growth rate of individual fish species during the experimental period

Treatments	Species stocked	Stocking density (No. per 40 m²)	Survival rate (%)	Average initial weight (g)	Average final weight (g)	Specific growth rate
T1	Tilapia	45	82	32.67	68.92	0.99
	Sharputi	38	88	30.33	65.67	1.03
	Grass carp	15	96	74.33	231.03	1.51
	Catla	38	82	28.47	26.82	-0.08
	Mrigal	15	100	18.83	84.69	2.00
T2	Tilapia	45	82	32.67	53.11	0.65
	Sharputi	38	88	30.33	56.72	0.83
	Grass carp	15	86	74.33	115.38	0.58
	Catla	38	90	28.47	28.40	-0.03
	Mrigal	15	100	18.83	74.00	1.86

The calculated net production of fish of the ponds under T_1 was 6.25 ± 1.07 tons ha.$^{-1}$ yr.$^{-1}$ and those of the ponds under T_2 were 2.84 ± 0.85 tons ha.$^{-1}$ yr.$^{-1}$, respectively (Fig. 1). The net productions of fish were significantly higher in T_1 might be due to supply of duckweed as feed. More or less similar productions were recorded in duck weed-based polyculture system (Mazumder *et al.*, 1999; Kabir 2003). The effectiveness of duckweed as low cost supplementary feed was observed through 6 months production trial of Thai sharpunti (Kohinoor *et al.*, 1999). The production was significantly higher in ponds with supply of duckweed than that of the ponds without supply of duckweed in monoculture of tilapia (Bornali, 2004; Uddin *et al.*, 2007; Chowdhury *et al.*, 2008) and sharpunti (Haque, 2005). Considering the present and previous study, it is clear that duckweed is a suitable feed items for fishes in both monoculture and polyculture system.

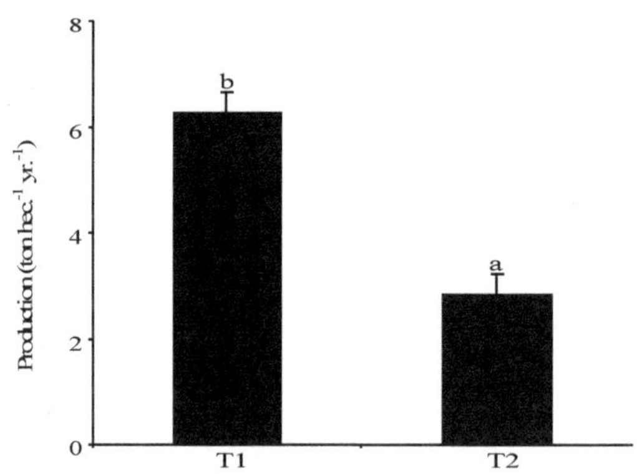

Fig. 1. Net production (means ± SD; n = 3) of fish in two treatments. Values accompanied by different letters are statistically significantly different ($p < 0.05$).

In conclusion, suitability of duckweed as feed for fishes was analyzed in polyculture system. Most of the water quality parameters of the ponds under T_1 and T_2 were more or less similar but the higher production of fish was recorded in T_1. The reason behind the higher production in T_1 was due to supply of duckweed as feed. Influence of duckweed on production of fish is positively significant indicated that duckweed might be used as preferable feed items for fishes in polyculture.

References

Alikhuni, K.H. 1957. Fish culture in India. *F.M. Bull. Indian Coun. Agric. Res.*, 20: 144.

APHA (American Public Health Association), 1992. Standard Methods for the Examination of Water and Wastewater. American Public Health Association, Washington DC. p. 874.

Belcher, H. and Swale E. 1978. A Beginner's Guide to Freshwater Algae. HMSO, London. p. 47.

Bornali, K. 2004. Effect of duckweed (*Lemna minor*) as supplementary feed on monoculture of tilapia (*Oreochromis niloticus*). MS thesis. Mymensingh: Bangladesh Agricultural University. 104p.

Chen, T.P. 1976. Aquaculture practices in Taiwan. Fishing News (Books) Ltd., West Byfleet, Surrey, England. 124p.

Chowdhury, M.M.R., Shahjahan M., Rahman, M.S. and Sadiqul Islam M. 2008. Duckweed (*Lemna minor*) as supplementary feed in monoculture of nile tilapia, *Oreochromis niloticus*. *J. Fisheries & Aquatic Sci.,* 3: 54-59.

Haque, M.S. 2005. Use of duckweed (*Lemna minor*) as supplementary feed in monoculture of sharpunti (*Puntius gonionotus*). MS thesis. Mymensingh: Bangladesh Agricultural University. 94 p.

Hossain, M.A., Ahmed, M. and Islam, M.N. 1997. Mixed culture of fishes in seasonal ponds through fertilization and feeding. *Bangladesh J. Fisheries Res.,* 1: 9-18.

Journey, T., Skillicorn, P. and Spira, B. 1991. Duckweed aquaculture: A new aquatic farming system for developing countries. Emena Technical Department, World Bank.

Kabir, A.N.M.A. 2003. Use of duckweed (*Lemna minor*) as feed for fishes in polyculture. MS thesis. Mymensingh: Bangladesh Agricultural University. 97p.

Kohinoor, A.H.M., Akteruzzaman, M. and Shah, M.S. 1993. Red tilapia production potential under low input management in Bangladesh. 3rd Asian Fisheries Forum, Octobor 26-30, World Trade Center, Singapore.

Kohinoor, A.H.M., Islam, M.L., Wahab, M.A. and Thilsted, S.H. 1998. Effect of mola (*Amblypharyngodon mola* Ham.) on the growth and production of carps in polyculture. *Bangldesh J. Fisheries Res.,* 2: 119-126.

Kohinoor, A.H.M., Islam, M.S., Begum, N. and Hussain, M.G. 1999. Production of thai sharpunti (*Barbodes gonionotus* Bleeker) in polyculture with carps using low-cost feed. *Bangldesh J. Fisheries Res.,* 3: 157-164.

Landolt, E. and Kandeler, R. 1987. The family of Lemnaceae a monographic study. Veroffentlichungen Des Geobotanischen Institute der Edg. Tech. Hochschule, Stiftung Ruebel, Zuerich. 638p.

Leng, R.A., Stambolie, J.H. and Bell, R. 1995. Duckweed: A potential high-protein feed resource for domestic animal and fish. AAAP Conf. Proc. Bali. 1: 103-114.

Mazumder, D., Kohinoor, A.H.M., Islam, T. and Tripathi, S.D. 1999. Low-cost, weed-based sustainable aquaculture system for seasonal water-logged rice-fields. *Bangldesh J. Fisheries Res.,* 22: 79-83.

Mostaque, A. 1995. Studies on the mixed culture of fishes in seasonal ponds through fertilization and feeding. MS thesis. Mymensingh: Bangladesh Agricultural University. 107p.

Needham, J.G. and Needham, P.R. 1963. A Guide to Study of Freshwater Biology. 5th Edn., Holden-Day, Inc., San Francisco, 106 p.

Porath, D. and Agami, M. 1977. Enhancement of protein production in fish ponds with duckweed (*Lemnaceae*). *Israel J. Bot.,* 26: 51-51.

Prescott, G.W. 1964. Algae of Western Great Lakes area. Wm. C. Brown Co. Dubuque, IOWA, 946 p.

Rahman, M.S. 1992. Water Quality Management: Aquaculture. BRAC Prokashana, Mohakhali, Dhaka, Bangladesh. 84 p.

Saha, J.K., Rahmatullah, S.M. and Mazid, M.A. 1999. Optimization of stocking density of duckweed, *Wolffia arhiza* (Linn.) and *Lemna* sp. *Bangldesh J. Fisheries Res.,* 7: 161-168.

Skillikorn, P., Spira, W. and Journey, T. 1993. Duckweed aquaculture. A New Aquatic Farming System for Developing Countries. The World Bank, Washington DC, 68 p.

Uddin, M.N., Rahman, M.S. and Shahjahan M. 2007. Effects of duckweed (*Lemna minor*) as supplementary feed on monoculture of GIFT strain of tilapia (*Oreochromis niloticus*). *Progressive Agriculture,* 18: 183-188.

Wahab, M.A., Ahmed, Z.F., Islam, M.A. and Rahmatullah, S.M. 1995. Effect of introduction of common carp, *Cyprinus carpio* (L.) on the pond ecology and growth of fish in polyculture. *Aquaculture Research,* 26: 619-628.

19

RESPONSE OF DIFFERENT COMBINATIONS OF MANURE AND FERTILIZERS TO THE YIELD OF RAPE CROP

S.P. Baliyan[1]*, P.S. Baliyan[2] and K.S. Madhava Rao[3]

Abstract

This study was an attempt to determine the effect of different combinations of manure and fertilizers on the yield of rape crop so as to improve yield by recommending the best combination of manure and fertilizers to the rape growers in Botswana. Rape crop was planted on farmers field in a complete randomized block design where nine different combinations of manure and fertilizers, namely, chicken manure (Ch); Ch and NPK (2:3:2); Ch and Fe (Iron); Ch, NPK and Fe; Ch and Urea; Ch, NPK and Urea; Ch, Urea and Fe; Ch, NPK, Urea and Fe; and control (No application of manure and fertilizer) were considered. Each fertilizer combination was replicated four times in a total of thirty six plots in the layout of the experiment. The data on rape yield was collected. The leaves of rape were harvested three times a month and the rape yield (in kg) was recorded in a harvest sheet. A two way analysis of variance (ANOVA) was performed using the SPSS software. The results indicated that fertilizer combination of chicken manure, NPK and Urea produced the highest yield of rape which was 2.61 times higher than the yield produced by the control plots (where no manure and fertilizer was applied) and therefore, a fertilizer combination of chicken manure, NPK and Urea was recommended to the rape growers. Suggested future studies included the comparative profitability of different fertilizer combinations as well as the effects of different doses of the organic and inorganic fertilizers on rape production.

Keywords: Fertilizers Response, Fertilizer Combinations, Organic and Inorganic Fertilizers, Rape Production Improvement, Soil Fertility

[1]Department of Agriculture, Livingstone Kolobeng College, Gaborone, Botswana
[2]Department of Accounting and Finance, University of Botswana, Gaborone, Botswana
[3]Department of Statistics, University of Botswana, Gaborone, Botswana

*Corresponding author's email: spbaliyan@yahoo.com (S.P. Baliyan)

Introduction

Vegetables are important protective food and highly beneficial for the maintenance of health and prevention of diseases including malnutrition deficiencies as they contain high quality of minerals, vitamins and fiber. Rape (*Brassica napus)* is a major leafy vegetable crop grown during the times of the year when the temperature is low (Acquaah, 2005; Decoteau, 2000). Rape production among small holder farmers is important for its nutritional and income generation (FAO, 1997; Acquaah, 2005). The rape is grown by most of the small scale farmers in Botswana but they are not harvesting enough and good quality yield and therefore, the farmers have not been able to to meet out the local demand which force to import vegetables from South Africa. There are various constraints resulting in low production of vegetables including rape in Botswana which include poor soil fertility, water scarcity, poor cultivation skills, pest and disease attack, poor availability of inputs and harsh climate (Baliyan and Kgahti, 2009). The soils in Botswana have been broadly characterized into poor sandy soils which are lacking in nutrients and require improvement by the addition of organic matter and chemical fertilizers for successful production of vegetable crops like rape (Belien, 1997 and Braun, 1975). Growing of vegetables requires knowledge and skills of both crop production and business practices (Williams, 1984). In order to harvest high yields of quality produce, there is a need to plan each production step and follow all cultural practice including application of proper fertilizers.

The small scale farmers in Botswana are not aware of the use and the effect of different manure and fertilizers on the rape production and therefore are not practicing any manure and fertilizer application strategies (Staring, 1979; Staring, 1979). Rape requires thirteen essential nutrients elements for its proper growth as it is the case for most other green leafy vegetables. All these elements are equally essential and cannot be replaced by other nutrients. With a complete lack or insufficient supply of nutrients, normal rape growth cannot take place (Mengel and Kirkby, 2001). Soils in Botswana are poor which leads to poor growth and yield of the rape. Different types of fertilizers (organic and inorganic) are available to improve the soil fertility. Organic fertilizers such as compost, chicken manure and kraal manure are commonly used in growing of vegetables including rape. Since rape crop has a high nutritional requirement, organic fertilizers cannot constantly supply all the nutrients to its requirement and therefore, inorganic fertilizers have to be used to ensure proper growth and yield of rape crop. Organic and inorganic fertilizers can be combined and applied to boost the constant supply of the essential nutrients but their response on the yield of rape is not well known. Therefore, this experiment was conducted to find out the most effective combination of manure and fertilizer for rape crop so as to convenience rape growers on the effect of application of different fertilizers.

The objective of this study was to determine the effect of different combinations of manure and fertilizers on the yield of rape crop and therefore to recommend the best combination of manure and fertilizers to the rape growers in Botswana. In order to fulfill the objectives of the study, the null and the alternative research hypotheses of interest were stated as follows:

H_0 :The nine combinations of fertilizers have equal effects on the total yield of rape

H_1 :The nine combinations of fertilizers do differ in their effects on the total yield of rape

H_0^* :The four blocks have equal effect on the total yield of rape

H_1^* :The four blocks do differ in their effects on the total yield of rape

Materials and Methods

The seeds of a recommended variety of rape for Botswana (English Giant) were planted in a seed bed during the winter season and then four weeks old seedlings were transplanted in the prepared thirty six experimental plots. The plot size was 6 × 2.4 meter with a spacing of 60 × 40 cm (Row to row × Plant to plant) and thus 4 rows of 6 meter length were planted in each plot. A Complete Randomized Block Design was adopted for the experiment and the treatments were allocated through randomization using a random sampling method in each of the four blocks. Nine different combinations of manure and fertilizers namely chicken manure (Ch), Ch and NPK (2:3:2); Ch and Fe (Iron); Ch, NPK and Fe; Ch and Urea; Ch, NPK and Urea; Ch, Urea and Fe; Ch, NPK, Urea and Fe; and control (No application of manure and fertilizer) were considered as treatments for the experiment (Table 1). Thus each treatment was replicated four times among the four blocks giving a total of thirty six plots in the layout of the experiment.

Table 1. Different fertilizer combinations (treatments) used in the experiment

Treatments	Combination of fertilizers
Treatment 1	Chicken manure (Ch)
Treatment 2	Chicken manure, NPK (2:3:2)
Treatment 3	Chicken manure, Fe (Iron)
Treatment 4	Chicken manure, NPK, Fe
Treatment 5	Chicken manure, Urea
Treatment 6	Chicken manure, NPK, Urea
Treatment 7	Chicken manure, Urea, Fe
Treatment 8	Chicken manure, NPK, Urea, Fe
Treatment 9	Control (No application of any manure and fertilizer)

All the manure and fertilizers were applied at the suggested rates using recommended methods of their application (Table 2). All the recommended management and cultural practices such as watering, weeding and pest and disease control were strictly followed throughout the experimentation period.

Table 2. Rates and application methods of manure and fertilizers used in the experiment

Manure/Fertilizer	Quantity applied (Kg ha^{-1})	Application method	No. of Application
Chicken manure	20000	Basal	One
NPK	1000	Basal	One
Urea	100	Top dressing	every 3 weeks
Iron (iron chelates)	500 mg	Foliar spray	4 sprays

The data on rape yield was collected. The rape crop was harvested three times a month giving a total of nine harvests. Only the two central rows of rape in each plot were harvested as a precautionary measure to avoid the leaching effect of the nutrients from the neighboring plots which may have affected the yield of rape and therefore the results of the study. The yield obtained from each plot for each harvest was weighed and recorded in a pre-structured harvest sheet. The total yield of rape for different treatments was calculated by adding the yield obtained from all the harvests of rape crop from all the respective treatments plots. Data were entered in SPSS software and a Two Way Analysis of Variance (ANOVA) was performed to determine the significance of difference between the means of rape yield from different combinations of fertilizers (treatments) and effect of blocking the experimental field. The treatment means were separated by the Least Significant Difference (LSD Turkey test) at alpha ≤ 0.05.

Results and Discussion

The output of Two Way Analysis of Variance (ANOVA) is presented in Table 1. The two way analysis of variance reveals that the treatments differ significantly $(p-value < 0.0001)$, while there are no significant block effects $(p-value = 0.126)$. In other words, the nine fertilizer combinations (treatments) namely, Ch; Ch NPK; Ch Fe, Ch NPK Fe; Ch Urea; Ch, NPK Urea; Ch, Urea, Fe; Ch, NPK, Urea, Fe; and Control contribute differently to the total yield of rape. However, the total yield does not seem to be impacted by how the experimental fields were laid out in terms of blocks in the study design which indicates that the differences in the rape yield was only because of the effects of the different combinations of fertilizers considered in the experiment.

Table 3. Influence of different combinations of fertilizers on the yield of rape

Source	SS	df	MS	F	Sig
Blocks	89.450	3	29.817	2.102	0.126
Treatments	1036.292	8	129.537	9.133	0.0001
Error	340.388	24	14.183		
Total	1466.130	35			

The multiple comparisons based on LSD principle reveal that the following seventeen combinations of treatments out of the $\binom{9}{2} = 36$ combinations are significant at 5% level.

(Ch, Ch NPK), (Ch, Ch Urea), (Ch, Ch NPK Urea), (Ch, Ch Urea Fe), (Ch, Ch NPK Urea Fe), (Ch NPK, Ch NPK Urea), (Ch NPK, Control), (Ch Fe, Ch NPK Urea), (Ch Fe, Control), (Ch NPK Fe, Ch NPK Urea), (Ch NPK Ch, Control), (Ch Urea, Ch NPK Urea), (Ch Urea, Control), (Ch NPK Urea, Ch Urea Fe), (Ch NPK Urea, Ch NPK Urea), (Ch NPK Urea, Control) and (Ch NPK Urea Fe, Control).

The Figure 1 shows the plot of estimated marginal means of total yield of rape over the four replications (blocks).

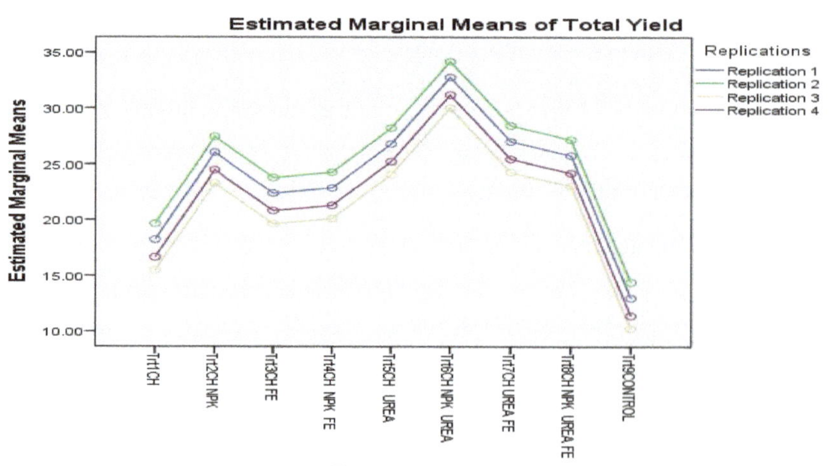

Fig. 1. Estimated marginal means of total yield of rape over the four replications (blocks)

It is observed from Figure 1 that the fertilizer combination (Ch NPK Urea) emerges as the best fertilizer combination to be replicated in the blocks to maximize the rape yield as this fertilizer combination has produced the highest yield of rape (128.1kg) followed by Ch Urea Fe and Ch Urea which produced total rape of 105.2 kg and 104.3 kg, respectively. The treatment Ch produced the lowest total yield (70 kg) among the eight combinations of fertilizer treatments. As expected the Control (no application of fertilizers) resulted in the lowest total yield of 49.1 kg of rape because no nutrients were supplied to the rape crop. These results reflected that NPK is an important source of nutrients for improving the yield of rape because Ch Urea Fe and Ch Urea have produced almost the same yield of rape where as the Ch NPK Urea fertilizer combination has not only produced the highest yield (128.1 kg) but also the yield difference between Ch NPK Urea and the two fertilizer combinations of (Ch Urea Fe and Ch Urea) varied from 22.9 kg to 23.8 kg. Therefore, it was concluded the application of NPK fertilizer is responsible for this difference in rape yield. However, it should also be noted that NPK is one of the expensive fertilizers for farmers in Botswana and therefore profitability analysis of rape production with and without NPK fertilizer should be conducted so as to maximize the income from rape production.

Conclusion and Recommendations

Results of this study indicated that the application of the manure and fertilizers can not only increase the yield of rape but excess application of nutrients (especially nitrogen) also reduces the yield of rape. The results have indicated that chicken manure, NPK and urea emerged the best fertilizer combination because it produced the highest yield of rape. Therefore, the rape growers were recommended to use the best combination of manure and fertilizers to obtain the higher yield of rape crop. This study provides directions for conducting experiments in future in the areas of:

- Response of different combinations of manure and fertilizers to the yield of rape crop on different types of soils. *To find out the best treatment for getting the highest yield of rape in different types of soils.*
- Response of different doses of the treatments used in this experiment to the yield of rape crop: *To find out the best dose of the treatments for getting the highest yield of rape.*
- Response of different doses of the treatments used in this experiment to the quality of rape crop: *To find out the best dose of the treatments for getting the best quality yield of rape.*
- Comparative economic profitability of fertilizer combinations of Ch NPK Urea, in rape production: *To find out the best profitable combination of the fertilizers for obtaining the highest income and profit in rape production.*

Acknowledgement

The corresponding author would like to acknowledge the support from the Department of Agricultural Research, Ministry of Agriculture, and Republic of Botswana for providing the opportunity to conduct this experiment during his tenure as an Agricultural Research Officer. He would also like to thank Dr. G.C. Wiles for the immense technical support during the conduction of this research.

References

Acquaah. 2005. Horticulture: Principles and Practices (3rd Ed.) Pearson Education, Inc: New Jersey. 817 p.

Baliyan, S.P. and Kgathi, D.L. 2009. Production and marketing problems in small scale horticultural farming in Botswana. Proceeding of the XVI International Symposium on Horticultural Economics and Management. *Acta Hort.* (ISHS). 831: 31-40.

Belien, J.S. 1977. Research Agronomist's Interim Report, Technical Document No. 7, UNDP FAO Project BOT/72/019: Swamp and Dryland Soils of the Okavango.155 p. Braun, B.H. 1975. Investigations on the soils at Motopi and Moshu, Technical Note No. 3, UNDP FAO Project BOT/72/019. 144 p.

Decoteau, D.R. 2000. Vegetable crops. Prentice-Hall, Inc, New Jersey. 464 p.

FAO. 1997. Agriculture, food and nutrition for Africa: A resource book for teachers of agriculture. FAO Publishing managerial group, Rome, Italy. 308 p.

Mengel, K. and Kirkby, E.A. 2001. Principles of plant nutrition (5th Ed.) Kluwer Academic Publishers, Dordrecht. 849 p.

Staring, G. 1979. Results of agronomic trials during 1978/79, Field document no. 16, UNDP FAO Project BOT/72/019: Swamp and Dryland Soils of the Okavango. 16 p.

Staring, G. 1980. Soil, Water and Crop Production in Ngamiland, Field Document No. 15, UNDP FAO Project BOT/72/019: Swamp and Dry land Soils of the Okavango. 33 p.

Williams, R.D. 1984. Institute of Food and Agricultural Sciences, University of Florida. Gainesville, 32611. 105 p.

EFFECT OF TRAINING AND VISIT SYSTEM ON PROFESSIONALIZATON OF EXTENSION AGENTS IN OSUN STATE AGRICULTURAL DEVELOPMENT PROGRAMME OF NIGERIA

J.O. Akintonde[1*], O.A. Akinboye[1], C.O. Farayola[2] and O. S. Akintola[1]

Abstract

The study was carried out to examine the effect of training and visit system on professionalization of extension agents in Osun State Agricultural Development Programme of Nigeria. Seventy-five data were collected by intervieweing extension agents constituted the sample size and were interviewed through the aid of using a well-structured questionnaire and information obtained were analyzed using frequency counts, percentages and person correlation coefficient was used for testing the research hypothesis. The findings revealed that 80% of the respondents were between the age group of 30 -50 years, spent between 10 – 30 years in service (84%) and are married (98.7%). A large proportion (56%) of the respondents holds higher national diploma in general agriculture, exhibited favorable attitude towards the farmers (93.3%), specials in general agriculture (34.7%) and required training in one or more areas of agriculture (69.3%). Analysis performed shows that all the characteristics had positive relationship with effect of training and visit on professionalization of extension agents except years of service, which had a negative relationship though all characters were statistically insignificant. Hence, the more the numbers of years spent on the job the more experience you would be. Which indicated that personal characteristics of the respondent influenced the professionalization of the extension agents through training and visit system?

Keywords: Training, Visit, Professionalization, Extension, Nigeria

[1]Department of Agricultural Extension and Rural Development, Ladoke Akintola University of Technology, P.M.B.4000, Ogbomoso, Oyo State, Nigeria.
[2]Agricultural Development Management, Department of Agricultural and Rural Management Training Institute (ARMTI), P.M.B. 1343, Ilorin Kwara State, Nigeria,

*Corresponding author's email: joakintonde@yahoo.com (J.O. Akintonde)

Reviewed by Dr. Monzur Morshed, South China Agricultural University, China.
Dr. Md. Abdul Jalil Mridha, Bangladesh Rice Research Institute, Gazipur, Bangladesh.

Introduction

Agricultural extension refers to the sets of activities of communication, information, demonstration and technical training geared towards transforming and disseminating to farmers new or improved technologies (IFAD, 2001). In addition, it is a means of increasing agricultural productivity. It deals directly with improvement of agriculture. Its prime objective is to help farmers developm and increase production by persuading them to adopt improved technical practices, develop skills, knowledge and attitude favorable to change in the farmers and their families. Ti enables them to benefit from research and technology and its ultimate aim is to raise their existence through improved living standard (Ogunfiditimi and Ewuola, 1995). Extension agents are the people charged with the responsibilities of bringing about changes within a target system. They assist farmers by increasing their awareness of improved agricultural technology as well as improving their decision-making capabilities.

The training and visit (T&V) system is an extension management system that as developed for the World Bank and it was aimed at upgrading the technical content of field extension activities while making agents activities more predictable and thus more accessible to farmers. The idea was to increase the effectiveness of agricultural extension service through comprehensively structured training, delivery and administrative system. During the post independence era, agricultural extension agents were saddled with both educational and non-educational duties such as supply of input.

Many of these systems were built with insufficient attention to skill development of field agents. Most of the extension agents were inadequately trained and had low morale because of poor career prospect, poor mobility, delay in claims and settlement and low level of education (Birkhaeuser *et al.*, 1991).

Training and visit system provides an organization structure and detailed mode of operation that is well informed. Extension agents visit farmers regularly and transmit message relevant to production needs and problems faced by the farmers are quickly, feedback to research scientist for relevant and immediate solution (Ogunfiditimi and Ewuola, 1995). The success of Training and Visit System is premised on the following underlying principles;

- Fully trained professional staff
- A single line of command
- Regular farm visit, time bound work and training programme
- Orientation towards working with farmers on their fields
- Regular and continuous training of all staff members to continuously upgrading their skills and
- Effective flow of information among research, extension and farmers (Gapson, 1990).

Moreover, Training and Visit System provide for relevant extension staff, regular training to upgrade their professional competence and serve the technological needs of the farmers, which involve the systematic application of well-known management principles with a view to professionalizing the extension service (Ogunfitidimi and Ewuola, 1995). Professionalization is the transformation of many occupations into profession. It is the capacity of extension staff to identity production problems in the field, recommend appropriate measures to solve them and train farmers on how to use the measures on their farms. Similarly, professionalization in agricultural extension can be defined as throwing of ideas from journalism, audio-visual, communication and advertising as well as from political and religious propaganda (Sappo, 1993). The idea behind professionalization is to increase the status of the extension agents' employee to extension professionals. In many agricultural sectors where professionalization is, implemented extension agents are also given opportunities to be collaborates with institutions outside the farms especially universities. This partnerships help extension agents foster a sense of contribution to dissemination of information and educational efforts beyond the local areas. Professionalization adds greater integrity, flexibility and authority to extension agents (IFAD, 2001).

This study was conducted to ascertain the effect of training and visit extension system on the professionalization of extension agents in Osun State Agricultural Development Programmes with the aim of identifying the personal characteristics of the extension agents as well as ascertaining the training needs of extension agents under the T & V System. It was also hypothesized that no significant relationship exists between the respondents' personal characteristics and the effect of T & V System on the professionalization of extension agents.

Methodology

The study was conducted in Osun State Agricultural Development Programme (OSSADEP). Osun State occupied a land mass of approximately 8,02 km^2 and it is bounded in the west by Oyo State, Ondo and Ekiti State in the east, Kwara State in the north and Ogun State in the south. There are 30 local government areas in the study area all of which falls under 3 zones of ADP for administrative purpose and the zones are Iwo, Ife-Ijesa and Osogbo.

The state observed 2 seasons namely dry and rainy seasons and it is covered by secondary forest savanna hence farming is the traditional occupation of the people in the study area through other income generating activities can be found on the area. The main cash crops in the area are cocoa and palm produce while food crops such as yam, rice, cassava, millet, plantain etc are grown in the area. Apart from being rich in agricultural produce, it is also endowed with a number of mineral resources such as gold, clay limestone and granite. The state also has great potentials for growth in economic activities.

The population of the study consist of all extension agents currently in the services of Osun State Agricultural Development Programme i.e. OSSADEP. The study area consist of three ADP zones namely Iwo, Ife-Ijesa and Osogbo. A multistage sampling technique was employed in the course of the study. Iwo agricultural zone consist of two extension blocks while both Ife-Ijesa and Osogbo zone respectively consist of six extension blocks. Firstly the 2 blocks in Iwo ADP was purposively selected which consist of 16 cells while 3 blocks each from both Osogbo and Ife-Ijesa ADP zone was selected through simple random sampling hence giving a total of 8 extension blocks i.e. 32 cells x 2=64 cell. Each blocks consist of 8 extension cells thus given a total of 80 extension cells and all the extension agents in the 80 cells constitute the sample size but in the course of the administration of the instruments, only 75 questionnaires was found properly filled by the respondents.

Well-structured questionnaire was administered on the respondents and both descriptive and inferential statistics were used in analyzing the data. Descriptive statistics used include frequency counts, percentages and cumulative percentage, while inferential statistic used to test for relationship was Pearson product moment correlation (PPMC).

Results and Discussion

It could be observed on table 1 that majority of the respondents (80%) were below the ages of 50 years while 18.7% were the ages of 50 years and above and only 1.3% of the sampled respondents were less than 30 years of age. This implies that majority of the respondents are young and active thus were able to face and cope with the rigours of extension works. The table further shows that majority of the respondents (84%) have spent between 10-30 years in the services of Osun State Agricultural Development Programme while 14.7% accounted for those respondents that have spent between 31-40 years and the remaining 1.3% represent the respondent that have spent less than 10 years as an extension agent.

Table 1. Frequency and percentage distribution of respondents according to their socioeconomic characteristics

Socioeconomic characteristic	Frequency	Percentage
Age (years)		
Below 30	1	1.3
30 – 39	25	33.3
40 – 49	35	46.7
50 and above	14	18.7
Years of service		
Less than 10	1	1.3
10 – 20	46	61.3
21 – 30	17	22.7
31 – 40	11	14.7
Marital status		
Single	1	1.3
Married	74	98.7
Educational attainment		
N.C.E	1	1.3
N.D (or OND)	21	28.0
H.N.D	42	56.0
B.Sc.	11	14.7
Areas of specialization		
General agriculture	26	34.7
Crop production	22	29.3
Fishery	4	5.3
Livestock production	5	6.7
Agricultural mechanism	5	6.7
Horticulture	6	8.0
Soil science	4	5.3
Agricultural extension	3	4.0
Attitude toward farmers		
Unfavourable	5	6.7
Favourable	70	93.3
Training needs		
Drug administration	6	8.0
Livestock production	15	20.0
Agroforestry	4	5.4
Crop production	2	2.7
Agricultural business/marketing	4	5.4
Agricultural mechanism	11	14.6
Fishery	7	9.3
Plant protection	26	34.6

Source: Field survey, 2009

This implies that the extension agents becomes more experienced and skillful based on the distribution of the numbers of years they have spent in service (i.e. the more the number of years in service, the more experienced the respondent) hence would be more familiar with the technicalities of T & V system of extension service.

Similarly, it was observed that majority (98.7%) of the respondents were married while the remaining 1.3% of the respondents were single. This implies that majority of the respondents are responsible, have some dependants and capable of concentrating on their jobs hence increasing their professional skills leading to effective discharge of their duties and responsibilities. In addition, it was observed that 56% of the sampled respondents are Higher National Diploma (HND) holders, 28% of them are Ordinary National Diploma (OND) holders while 14.7% represent those respondents who are Bachelor of Science (B.Sc.) degree holders and only 1.3% accounted for the respondent who have National Certificate of Education (NCE). This implies that all the respondents in the service of OSSADEP are educated hence may have been exposed to different kinds of training in their various training institution that may enhance skill development.

The table further revealed that 34.7% of the respondents specializes in general agriculture while 29.3% are specialist in crop production and only 2.7% of the respondents have more than one area of specialization. In addition, 10.6% specializes in fish production and soil science respectively while livestock production and agriculture mechanization accounted for 13.4% of the total respondents. Only 4% specializes in agricultural extension services. This implies that all the respondents have different area of specialization though may have knowledge in areas outside their discipline. Similarly, it was observed that majority of the respondents (93.3%) have positive attitude towards the farmers while only 6.7% of the respondents accounted for those with negative attitude towards the farmers. This implies that there is mutual understanding and cooperation between the extension agents and the farmers, which enable them to discharge their duties efficiently and effectively.

Further analysis shows that majority of the respondents (69.3%) required training in different areas of agriculture like drug administration, livestock production, Agro-forestry, crop production, fishery, agricultural mechanization etc while 6.7% of them required training in all the areas of agriculture and 9.3% required training in two or more areas. The remaining 14.7% of the respondents accounted for those who gave no responses to the question asked. This implies that majority (85.3%) of the sampled respondents still require improved training in one or more areas of agriculture for adequate professionalization in their field of discipline for effective performance in carrying out their duties. This finding in corroborated by Olatidoye (1999), when he opined that appropriate training enables an agent to develop into a competent professional and to live up to this role, he is expected to be in close touch with relevant scientific developments and latest research information on all the sub-sector of agriculture is his immediate environment hence he must possess the ability to identify production constraints and suggest appropriate remedial measures which creates credibility not only for himself but also for the institution he represent.

Hypothesis testing

Data presented on table 2 shows that there was no significant relationship between respondents' personal characteristics and the effect of T & V on the professionalization of the extension agent. Form the correlation test with the following correlation values: age (r = 0.08), educational level (r = 0.019), marital status (r = 0.0059), religion (r = 0.013), years in service (r = 0.020) and area of specialization (r = 0.155); all the values are insignificant. This implies that the personal characteristics of the respondents does not in any way influence the professionalization of the extension agents through training and visit hence the initial hypothesis is hereby accepted.

Table 2. Relationship between respondents' personal characteristics and effect of training of visit system on professionalization of extension agents

Personal characteristics	Correlation value	Remark
Age	0.058	NS
Years in service	-0.020	NS
Marital status	0.059	NS
Educational attainment	0.019	NS
Area of specialization	0.155	NS
Religion	0.013	NS

At 5% significant level

Recommendations

Based on the findings of the study, the follow recommendations were made;

1. There should be regular training of extension agents through fortnightly training by efficient subject matter specialists.
2. All the essential infrastructural facilities that will aid in proper training of the extension agents should be put in place at all the three tiers of government.
3. The constraint associated with training, visit system of extension should be adequately addressed, and the authorities concerned should proffer appropriate solution.

Conclusions

This study examines the effect of training and visit system of extension on the professionalization of extension agents in the services of Osun State Agricultural Development Programme. The following conclusions were made based on the findings of the study.

1. Majority of the respondents were adults between the ages of 30-50 years, have spent between 10-30 years in service, married and were all educated.
2. A large proportion of the respondents specialized in general agriculture, exhibit positive and favorable attitude towards the farmers and need training in different areas of agriculture.
3. There was no significant relationship between the personal characteristics of the respondents' effect of training and visit system on professionalization of extension agents because all the correlation values obtained were insignificant.

References

Birkhaeuser, D., Evenson, R.G and Feder, G. 1991. The Economic Impact of Agricultural Extension. University of Chicago. 609 p.

Gapson, D. 1990. What is Research in the Context of National Agricultural Research System. KSA lecture ISNAA, Wagening Holland, p. 11.

International Fund for Agricultural Development (IFAD). 2001. Agricultural Extension and Support for farmers' innovation in Western and Central Africa. Edweb, p. 2.

Ogunfitidimi, T.G and Ewuola, S.O. 1995. The Synthesis of Comparative Agricultural Extension System. EMM Press ltd, Ibadan, p. 42.

Olatidoye, K. 1999. Extension Training, Features, Objectives and Principles of Unified Agricultural Extension System. Unpublished Seminar Material in Osun State Agricultural Development Programme. pp. 2-4.

Sappo, H. 1993. "Profile of Rural Poverty on Africa". The State of World Rural Poverty. Published by New York University Press and Intermediate Technology Publication.

21

BIOEFFICACY OF PLANT EXTRACTS TO CONTROL CERCOSPORA LEAF SPOT OF MUNGBEAN (*Vigna radiata*)

M.N. Uddin[1], M.A. Bakr[2], M.R. Islam[3], M.I. Hossain[4] and A. Hossain[5]

Abstract

The experiment was conducted at Bangladesh Agricultural Research Institute farm, Joydebpur, Gazipur during March to July 2007 to evaluate the bioefficacy of some plant extracts in controlling *Cercospora* leaf spot of mungbean. Six indigenous plant species i.e. Neem leaves extract (1:4 w/v), Garlic cloves extract (1:5 w/v), Biskatali leaves extract (1:4 w/v), Alamanda leaves extract (1:6 w/v), Arjun leaves extract (1:4 w/v) and Debdaru leaves extract (1:5 w/v) were used in this experiment. The experiment was laid out in RCBD with seven treatments and four replications. Data on disease incidence, severity, yield contributing characters and yield of mungbean were recorded. Naturally, infection of the disease was considered in this experiment. The lowest disease incidence (7.33%) at 60 DAS was found in T1. Lowest and similar disease severity (PDI= 4.55) was found in T2 and T3 at the same DAS. Neem extract treated plots gave better response in yield (1.26 t ha^{-1}) and all the yield contributing parameters like inflorescences plant^{-1} (13.45), tallest plant (51.44 cm), the maximum number of pods plant^{-1} (26.81), length of pod (8.56 cm), number of seeds pod^{-1} (12.64) and 1000 seeds weight (27.33 g) followed by T2 and T3. The highest disease incidence (26.50%) and disease index (13.65%) were recorded in treatment T7 at 60 DAS. Yield and all yield contributing factors were lowest in same treatment. The results of the experiment suggested that the use of neem leaves extracts are effective for minimizing *Cercospora* leaf spot incidence, severity and increasing yield of mungbean.

Keywords: Mungbean, *Vigna radiate,* Cercospora Leaf Spot, Control, Plant extracts

[1]Agriculture Extension Officer, Department of Agricultural Extension, MOA, Bangladesh
[2]Ex-CSO, Plant Pathology Division, Bangladesh Agricultural Research Institute, Gazipur, Bangladesh
[3]Professor, Department of Plant Pathology, Sher-e-Bangla Agricultural University, Dhaka, Bangladesh
[4]Training, Implementation and Monitoring Officer (TIMO), IFDC, Bangladesh
[5]Assistant Professor, Department of Plant Pathology, Sythet Agricultural University, Bangladesh
*Corresponding author's email: mnumahmud@gmail.com (M.N. Uddin)

Introduction

Mungbean (*Vigna radiata*), belongs to the family leguminosae and sub family papilionaceae, is one of the most important pulse crops having global economic importance as a dietary ingredient of the staple food in tropical and subtropical region. It is grown three times in a year covering 21862 ha with an average yield of 0.82 t ha^{-1} (BBS, 2009). It can fix atmosphere nitrogen through symbiotic relationship with soil bacteria and improve the soil fertility (Yadav *et al.*, 1994). Mungbean has been considered as a "poor men's protein" (Mian, 1976). Apart from 26% protein, it also contains 51% carbohydrate, 10% moisture, 4% minerals and 3% vitamins (Khan, 1981). In Bangladesh, total sixteen diseases of mungbean have been recorded (Fakir, 1983 and Bakr, 1993) which are responsible for lower yield of mungbean. Among the diseases, *Cercospora* leaf spot is a serious disease of mungbean (Verma and Sandhu, 1992) and causes yield losses of up to 58% (Lal *et al.*, 2001). It is widely distributed all over the country where mungbean is cultivated. The causal organisms of this disease are *Cercospora cruenta*, *C. conescens*, *C. kikuchii* and *C. caracallae*. Among these, *Cercospora cruenta* is the most prevalent species (Talukder, 1974). The initial symptoms of the disease appear as water soaked spot on leaves. As spots become older may coalesce together, causing enlarged dead area on the infected leaves. Heavy infections of *Cercospora* can cause premature defoliation of the mungbean plant. Sometimes the leaves may become malformed and wrinkled. Maturity is delayed in the diseased plants resulting poor pod formation. Seeds that developed on severely infected plants are small and immature (Poehlman, 1991). Different approaches such as spray of chemical fungicides (Singh and Singh, 1978), spray of different plant extracts and use of resistant variety are tried to control *Cercospora* leaf spot of mungbean. Though the released varieties of mungbean in Bangladesh are described as resistant variety but most of them show susceptible reaction towards this disease and the susceptibility varied due to variation of weather situation during trail before release and commercial cultivation. Indiscriminate and frequently use of chemicals affects the soil health. Harmful chemical substances enter into the food chain that ultimately causes serious health hazards. Eco-friendly management like use of plant extract against plant diseases control is however, a recent approach to plant diseases management and it has drawn the special attention of the plant pathologist all over the world. Most of the plant extracts are cost effective and do not have harmful effects on beneficial soil microorganisms. In Bangladesh, only a few attempts have been made to evaluate plant extracts against plant diseases (Ashrafuzzaman and Hossain, 1992; Hossain *et al.*, 1993; Ashrafuzzaman and Khan, 1992; Suratuzzaman *et al.*, 1994). Many researchers reported that plant extracts have antifungal properties and potentially used against many plant diseases (Singh and Dwivedi, 1987; Tariq and Magee, 1990; Lakshmanan *et al.*, 1990). It can help to avoid environmental pollution caused by chemicals and thus become most rewarding one to our existing socio-economic conditions and environmental threats. In the light of above background, the present piece of research work has been undertaken to estimate the effects of different botanicals on the incidence and severity of the *Cercospora* leaf spot of mungbean, yield and yield contributing characters of *Cercospora* leaf spot infected mungbean.

Materials and Methods

The experiment was conducted in the farm of Bangladesh Agricultural Research Institute (BARI), Joydebpur, Gazipur, Bangladesh during the period from March to July 2007. In this research work, BARI moog-3 was used as test crop where as BARI moog-3 is resistant to Cercospora leaf spot and Yellow mosaic virus of mungbean. Seeds were collected from pulse crop research division of BARI, Joydebpur, Gazipur. Seven different treatments including untreated control were explored in the experiment. Treatments were T_1 = Neem (*Azadirachta indica*) leaves extract (1:4 w/v), T_2 = Garlic (*Allium sativum*) cloves extract (1:5 w/v), T_3 = Biskatali (*Polygonum hydropiper*) leaves extract (1:4 w/v), T_4 = Alamanda (*Allamanda cathertica*) leaves extract (1:6 w/v), T_5 = Arjun (*Terminalia aurjuna*) leaves extract (1:4 w/v), T_6 = Debdaru (*Polyalthia longifolia*) leaves extract (1:5 w/v), T_7 = Untreated (control). The experiment was laid out in Randomized Complete Block Design (RCBD) with four replications. The unit plot size was 4.0 m × 3.0 m. The distance between two blocks and two plots were 1.0 m and 0.5 m, respectively. The beds were prepared manually after ploughing and laddering. Then the seeds were sown in the bed in line sowing with 30 cm line-to-line distance. Fertilizers and manures were applied as per recommendation of BARI. Watering and intercultural operations were done when necessary. Plant extracts were collected from Bangladesh Agricultural Research Institute, Joydebpur, Gazipur. The extracts were prepared by using the method of Ashrafuzzaman and Hossain (1992). After washing weighted leaf, the big leaves were cut into small pieces. For getting extract, weighted plant parts were blended and then distilled water was added into the jug of the blender. The pulverized mass was squeezed through 3 folds of fine cotton cloth. For getting 1:4 (w/v) ratio 400 ml of distilled water was added with 100 g of plant parts, 1:5 (w/v) ratio 500 ml of distilled water was added with 100 g of plant parts, 1:6 (w/v) ratio 600 ml of distilled water was added with 100 g of plant parts. Plant extracts were applied in the field as foliar spray. Spraying was done 3 times at 7 days interval starting from vegetative growth. The plants were naturally infected. The disease incidence and severity was recorded at 30, 40, 50 and 60 DAS (Days after Sowing). For disease incidence, the infected plants were identified by comparing it symptoms critically with those already published by Ahmed (1985). The incidence of *Cercospora* leaf spot was calculated as follows:

Percentage Infected plants

$$= \frac{\text{Number of infected plant in each plot}}{\text{Total number of plants in each plot}} \times 100$$

Disease severity were calculate by using (0-9) disease severity score of Metha and Mondal (1978) and determined as PDI (Percent Disease Incidence) by using following formula (Krisna Prasad *et al.*, 1979).

Percent Disease Index (PDI)

$$= \frac{\text{Sum of total ratting}}{\text{Total number of observation} \times \text{Highest grade in the scale}} \times 100$$

The severity of *Cercospora* leaf spot disease was recorded according to the grade of Metha and Mondal (1978) describe in the following table:

Numeric score	Symptom severely on plants at maximum flowering and pod formation stage
0	No visible symptoms on leaves, plant growth, flowering and pod formations are normal
1	10% leaf area infection
3	11-30% leaf area infection
5	31-50% leaf area infection
7	51-70% leaf area infection
9	71% and above leaf area infection

Data on plant height (cm), number of primary branches plant^{-1}, number of inflorescences plant^{-1}, number of pods plant^{-1} were collected from standing crops and data on pod length (cm), number of seeds pod^{-1}, 1000 seed weight (g), yield plot^{-1} (kg), yield (ton) hectare^{-1} were recorded after harvesting. The data obtained for different characters were analyzed to find out the significant differences among the treatments. The analysis of variance was performed by using MSTAT program. The significance of the difference among the treatment means was estimated by DMRT (Duncan's Multiple Range Test) at 5% level of probability (Gomez and Gomez, 1984).

Results and Discussion

A gradual increasing trend for both disease incidence and severity was found with the increasing of DAS. There were no significant variations in *Cercospora* disease incidence and disease severity (0-9 scale) for different treatments at 30 DAS (Table 1). Significant differences were recorded in disease incidence and disease severity (0-9 scale) for different treatments at 40 DAS. The highest disease incidence (13.42% infected plant) and disease severity (4.85) were recorded from treatment T_7 which was close to T_6 with disease incidence (5.08% infected plant) and disease severity (2.45).The diseases incidence and diseases severity were increased gradually up to 39.72% and 116.50% respectively compare with 50 DAS. The lowest disease incidence (3.67% infected plant) and disease severity (1.95) were recorded from treatment T_1 which was statistically similar with treatment T_3 (3.92% infected plant) and (2.10). The diseases incidence and diseases severity were increased gradually up to 58.86% and 56.41% respectively compare with 50 DAS. Statistically significant differences were recorded in disease incidence and disease severity for different treatments at 50 DAS. The highest *Cercospora* disease incidence (18.75% infected plant) and disease severity (10.50) were recorded from treatment T_7 which was nearly to T_6 (8.33% infected plant) and diseases severity (3.75). The diseases incidence and diseases severity were increased gradually up to 41.33% and 30.00% respectively compare with 60 DAS.At 50 DAS the lowest disease incidence (5.83% infected plant) and disease severity (3.05) were recorded from treatment T_1 where as the diseases incidence and diseases severity were increased gradually up to 25.72% and 58.36% respectively compare with 60 DAS. At 60 DAS different treatments showed a remarkable variation in *Cercospora* disease incidence and disease severity. The highest disease incidence (26.50% infected plant) and disease severity (13.65) were recorded from treatment T_7 followed by T_6 (10.33% infected plant) and disease severity (5.00). On the other hand the lowest disease incidence (7.33% infected plant) was recorded from treatment T_1 which was statistically similar with treatment T_3 (8.08% infected plant) and the lowest disease severity (4.55) was recorded from treatment T_2 and T_3. Among the different plant extracts neem extracts was most effective. Bishkathali and garlic were also effective than alamonda, arjun and debdaru leaves extracts. Miah *et al.* (1990); Ahmed (1985) reported that neem extract had potential for controlling *Cercospora* leaf spot in mungbean. It has been observed that neem leaves extract resulted significant reduction of *Cercospora* leaf spot of mungbean over untreated (control) (BARI, 2007; Khan, 1981; Mehta and Mondal, 1978).

Table 1. Disease incidence and severity influenced by different plant extracts used for the control of *Cercospora* leaf spot of mungbean (cv. BARI moog 3) recorded at different date

Treatments	Incidence (% Plant infected)				Severity (0-9 Scale)			
	30 DAS	40 DAS	50 DAS	60 DAS	30 DAS	40 DAS	50 DAS	60 DAS
T_1	1.75	3.67 c	5.83 d	7.33 d	1.33	1.95 d	3.05 c	4.83 b
T_2	1.58	4.25 bc	6.83 cd	8.42 cd	1.30	2.15 bcd	3.25 bc	4.55 b
T_3	1.58	3.92 bc	6.25 d	8.08 d	1.18	2.10 cd	3.10 c	4.55 b
T_4	1.75	5.00 b	7.83 bc	9.67 bc	1.40	2.40 bc	3.60 bc	4.95 b
T_5	1.83	4.50 bc	7.67 bc	8.83 bcd	1.25	2.35 bc	3.25 bc	4.70 b
T_6	1.83	5.08 b	8.33 b	10.33 b	1.28	2.45 b	3.75 b	5.00 b
T_7	1.67	13.42 a	18.75 a	26.50 a	1.30	4.85 a	10.50 a	13.65 a
CV (%)	12.14	12.89	7.75	8.55	18.66	7.96	8.48	7.10

Different letter (s) in the same column showed the significant different at 0.05 level of probability

T_1 = Neem (*Azadirachta indica*) leaves extract (1:4 w/v), T_2 = Garlic (*Allium sativum*) cloves extract (1:5 w/v), T_3 = Biskatali (*Polygonum hydropiper*) leaves extract (1:4 w/v), T_4 = Alamanda (*Allamanda cathertica*) leaves extract (1:6 w/v), T_5 = Arjun (*Terminalia aurjuna*) leaves extract (1:4 w/v), T_6 = Debdaru (*Polyalthia longifolia*) leaves extract (1:5 w/v), T_7 = Untreated (control).

A little variation were recorded for plant height due to application of different plant extracts used in this experiment for controlling *Cercospora* leaf spot of mungbean (Table 2). The tallest plant (51.44 cm) was recorded from treatment T_1 which was statistically similar with T_2, T_3, T_4, T_5, T_6 (50.05 cm, 50.82 cm, 49.87 cm, 49.86 cm, 49.19 cm, respectively). On the other hand, the shortest plant (44.75 cm) was recorded from treatment T_7. The highest number of inflorescences plant^{-1} (13.45) was recorded from treatment T_1 which was identical with T_3 (13.05) (Table 2). Alternatively, the lowest number of inflorescences plant^{-1} (10.02) was recorded from the treatment T_7. Williams *et al.* (1968); Poehlman (1991); Roy and Malathi (2001) reported the similar effect on number of inflorescences per plant and plant height of mungbean. The highest number of primary branches per plant (5.38) was recorded from treatment T_1 which was more or less statistically similar with treatment T_2- T_6 (Table 2). Conversely, the lowest number of primary branches per plant (3.62) was recorded from the treatment T_7. The maximum number of pods per plant (26.81) was recorded from the treatment T_1 (Table 2). On the other hand, the minimum number of pods per plant (21.07) was recorded in treatment T_7. The maximum pod length (8.56 cm) was recorded from the treatment T_1 which was statistically similar with treatment T_3 (8.51 cm) (Table 2). On the contrary, the minimum pod length (6.08 cm) was recorded from the treatment T_7 followed by T_6 treatment (6.85 cm) which was identical T_4. The maximum number of seeds pod^{-1} (12.64) was recorded for treatment T_1, which was statistically similar with treatment T2, and T_3 (12.04 and 12.00 respectively) (Table 3). On the other, hand the minimum number of seeds per pod (10.26) was recorded from the treatment T_7 which was statistically similar with treatments T_6, T_5 and T_4 (10.82, 10.96 and 11.00 respectively). Increase of number of seeds per pod^{-1} over control showed variation for different treatments in the present experiment (Table 3). The maximum increased number of seeds pod^{-1} (23.20%) was recorded from the treatment T_1 and the minimum was recorded in treatment T_6 (5.46%). The maximum 1000 seeds weight (27.33 g) was recorded in treatment T_1 which was statistically similar with treatments T_2, T_3, T_4, T_5, and T_6 (26.15 g, 27.26 g, 26.02 g, 26 g and 26.46 g, respectively) (Table 3). Oppositely, the minimum 1000 seeds weight (23.73 g) was recorded from the treatment T_7. Increases of 1000 seeds weight over control in percentages showed variation for different treatments under the present experiment. The maximum increase 1000 seeds weight (15.17%) was recorded from the treatment T_1 and the minimum was recorded from treatment T_6 (9.57%) (Table 3).

Table 2. Effects of different treatments on yield contributing characters of *Cercospora* leaf spot infected mungbean (cv. BARI moog-3)

Treatments	Plant height (cm)	Number of inflorescence s plant^{-1}	Number of primary branches plant^{-1}	Number of pods plant^{-1}	Pod length (cm)
T_1 = Neem leaves extract (1:4 w/v)	51.44 a	13.45 a	5.38 a	26.81 a	8.56 a
T_2 = Garlic cloves extract (1:5 w/v)	50.05 a	12.51 bc	5.00 ab	25.42 ab	8.04 b
T_3 = Biskatali leaves extract (1:4 w/v)	50.82 a	13.05 ab	5.34 a	26.00 ab	8.51 a
T_4 = Alamanda leaves extract (1:6 w/v)	49.87 a	12.00 cd	4.74 b	25.03 ab	7.01 d
T_5 = Arjun leaves extract (1:4 w/v)	49.86 a	12.35 cd	4.94 b	25.00 ab	7.63 c
T_6 = Debdaru leaves extract (1:5 w/v)	49.19 a	11.85 d	4.71 b	24.26 b	6.85 d
T_7 = Untreated (control)	44.75 b	10.02 e	3.62 c	21.07 c	6.08 e
CV (%)	2.95	3.36	5.18	5.45	2.91

Different letter (s) in the same column showed the significant different at 0.05 level of probability

The maximum yield plot^{-1} (1.51 kg) was recorded from the treatment T_1 i.e. by applying neem extract which was statistically similar with treatment T_3 (1.45 kg) i. e. biskatali leaf extracts (Table 3). In contrast, the minimum yield plot^{-1} (1.21 kg) was recorded from the treatment T_7 where no extract was applied. Increase of yield per plot over control showed differences for different treatments in this experiment. The data of yield plot^{-1} was computed to yield hectare^{-1}. The maximum yield (1.26 ton) was recorded for treatment T_1 which was statistically similar with treatments T_3 and T_2 (1.21 and 1.20 ton) (Table 3). On the other hand, the minimum yield per hectare (1.01 ton) was recorded from the treatment T_7. The maximum increased yield hectare^{-1} (24.75%) over control was recorded from treatment T_1 and the minimum was recorded from treatment T_6 (12.87%). Among the different plant extracts, neem leaves extracts was more effective and bishkathali also had similar effects (Table 3). Islam *et al.* (2006) also informed that seed treatment with bishkathali extract increased 29.74% yield of wheat over control.

Table 3. Number of pods plant^{-1}, pod length, seed yield and increase of seed yield as influenced by different treatments used for the management of leaf spot of mungbean (cv. BARI moog-3)

Treat ments	Number of seeds pod^{-1}	Increase of seeds pod^{-1} over control (%)	1000 seeds weight (g)	Increase of 1000 seeds weight over control (%)	Yield (kg plot^{-1})	Yield (ton ha^{-1})	Increase of yield hectare^{-1} over control (%)
T_1	12.64 a	23.20	27.33 a	15.17	1.51 a	1.26 a	24.75
T_2	12.04 a	17.35	26.15 a	10.20	1.44 b	1.20 ab	18.81
T_3	12.00 a	16.96	27.26 a	14.88	1.45 ab	1.21 ab	19.80
T_4	11.00 b	7.21	26.02 a	9.65	1.42 bc	1.18 b	16.83
T_5	10.96 b	6.82	26.00 a	9.57	1.40 bc	1.17 b	15.84
T_6	10.82 b	5.46	26.46 a	11.50	1.37 c	1.14 b	12.87
T_7	10.26 b	--	23.73 b	--	1.21 d	1.01 c	--
CV(%)	5.51	--	3.25	--	3.41	3.41	--

In a column figures having dissimilar letter(s) differ significantly at 0.05 level of probability.

T_1 = Neem (*Azadirachta indica*) leaves extract (1:4 w/v), T_2 = Garlic (*Allium sativum*) cloves extract (1:5 w/v), T_3 = Biskatali (*Polygonum hydropiper*) leaves extract (1:4 w/v), T_4 = Alamanda (*Allamanda cathertica*) leaves extract (1:6 w/v), T_5 = Arjun (*Terminalia aurjuna*) leaves extract (1:4 w/v), T_6 = Debdaru (*Polyalthia longifolia*) leaves extract (1:5 w/v), T_7 = Untreated(control).

Conclusion

The overall results suggested that the use of neem leaves extracts are effective for minimizing *Cercospora* leaf spot incidence, severity and increasing yield of mungbean than other treatments. The enthusiastic researchers can be motivated to explore other plant extracts and spell out the extract concentration for control *Cercospora* leaf spot of mungbean.

References

Ahmed, H.U. 1985. Disease problem of pulse and oilseed crops in Bangladesh. Presented at the First Biennial Conference of the Bangladesh Phytopathological Society, 13-14. April, 1985, BARI, Joydebpur, Gazipur, Bangladesh. pp. 13-14.

Ashrafuzzaman, H. and Hossain, I. 1992. Antifungal activity of crude extracts of plants against *Rhyzoctonia solani* and *Bipolaris sorokiniana*. *Proc. BAU. Res. Prog.* 6: 188-192.

Ashrafuzzaman, M.H. and Khan, A.R. 1992. Antifungal activity in vitro of some plant extracts on *Rhizoctonia solani*. *Bangladesh J. Sci. Res.* 10 (2): 243-244.

Bakr, M.A. 1993. Plant protection of mungbean in Bangldesh. pp. 177-184. *In:* Proceedings of the seminar on mungbean in South Asia, 11-15 March, 1991, New Delhi, India (Earskine, W. and Saxena, M.C.Eds), ICARDA, Alleppo, Syria Kit.

BARI. 2007. Research Report for 2006-2007. Bangladesh Agricultural Research Institute, Joydebpur, Gazipur.

BBS. 2009. Bangladesh Bureau of Statistics, Statistical Yearbook of Bangladesh, Statistics Division, Ministry of Planning, Government of the People's Republic of Bangladesh, Dhaka.

Fakir, G.A. 1983. Pulses disease and their control (in Bangla), Mymensingh, Bangladesh Agricultural University. p. 14.

Gomez, K.A. and Gomez, A.A. 1984. Statistical Procedure for Agricultural Research (2nd ed.). Int. Rice Res. Inst., A Willey Int. Sci., Pub. pp. 28-192.

Hossain, I., Ashrafuzzaman, H. and Khan, M.H.H. 1993. Biocontrol of *Rhizoctonia solani*. *BAU Res. Prog.* 7: 264-269.

Islam, M.A., Aminuzzaman, F.M., Islam, M.R. and Zamal, M.S. 2006. Seed treatment with plant extract and vitavax-200 in controlling leaf spot with increasing grain yield of wheat. *Int. J. Sustain. Agril. Tech.* 2 (8): 15-20.

Khan M.R.I. 1981. Nutritional quality characters in pulses. pp. 199-206. *In:* Proceedings of National workshop on pulses, BARI, Gazipur.

Krisna Prasad, K.S., Siddanamatch, A.S. and Hedge, R.K. 1979. Development of peanut rust disease in Karnatak state. *India Plant Div. Report.* 63 (8): 692-695.

Lakshmanan, P., Mohan, S. and Jeyarajan, R. 1990. Antifungal properties of some plant extracts against *Thanatephorus cucumeris*, the causal agent of color rot disease of *Phaseolus aureum*. *Madras Agric. J.* 77 (1): 1-4.

Lal, G., Kim, D., Shanmugasundaram, S. and Kalb, T. 2001. Mungbean production. AVRDC. World Vegetable Center. Tainan, Shanhua: AVRDC-The World Vegetable Center. p. 6.

Mehta, P.P. and Mondal, K.K. 1978. Field screening of groundnut cultivars against rust of tikka. *Indian Phytopath.* 31: 259-260.

Miah, A., Ahmed, M.U., Sharma, N.R., Ali, A. and Miah, S.A. 1990. Antifungal activity of some plant extracts. *Bangladesh J. Botany.* 19 (1): 5-20.

Mian, A.L. 1976. Grow more pulse to keep your pulse well: An essay of Bangladesh Pulse, Department of Agronomy, Bangladesh Agricultural University, Mymensingh. pp. 11-15.

Poehlman, J.M. 1991. The mungbean. Westview Press. BOULDER. pp. 169-274.

Roy, A. and Malathi, V.G. 2001. Molecular cloning of cowpea golden mosaic geminivirus and its relationship with s yellow mosaic geminivirus. *Tropical Agril. Res.* 13: 341-352.

Singh, D.V. and R.R Singh, 1978. Field evaluation of fungicides for the control of Cercospora leaf spot of green gram. *Pesticides.* 12: 28–9.

Singh, R.K. and Dwivedi, R.S. 1987. Fungitoxicity of different plants. *National Academy Sci.* 10 (3): 89-91.

Suratuzzaman, M., Hossain, I. and Fakir, G.A. 1994. Control of seed borne fungi of two rice varieties with some plant extracts. *Prog. Agric.* 5 (1): 11-15.

Talukder, M.J. 1974. Plant diseases of Bangladesh. *J. Agric. Res.* 1: 61-86.

Tariq, V.N. and Magee, A.C. 1990. Effect of volatiles from garlic bulb extracts on *Fusarium oxysporum* fsp. *lycopersici. Mycological Res.* 94 (5): 617-620.

Verma, M.M. and Sandhu, S.S. 1992. Mungbean yellow mosaic disease. pp. 28-37. *In:* Proceeding of an International Workshop, Bankok, Thailand, 2-3 July 1991.

Williams, F.J., Grewal, J.S. and Amin, K.S. 1968. Serious and new diseases of pulse crops in India in 1966. *Pulse Diseases Report.* 52: 302-304.

Yadav, D.S., Panwar, K.S. and Singh, V.K. 1994. Management of pulse csrops in sequential cropping. p.27. *In:* Recent advances in pulses research (Asthana, A.N. and Ali, M., eds.). Kanpur, India: Indian Society of Pulses Research and Development, Indian Pulses Research Institute.

COMBINED EFFECT OF ORGANIC MANURE AND POTASSIUM ON GROWTH AND YIELD OF ONION CV. BARI PIAZ-I

H.K. Barman[1], M.N. Siddiqui[2], M.A.Siddique[1], M.S. Roni[3] and M. Nuruzzaman[1]

Abstract

An experiment was conducted to find out the combined effect of cowdung and potassium on the growth and yield of onion cv. BARI piaz-I at Horticulture Farm, Bangladesh Agricultural University, Mymensingh during the period from December 2010 to March 2011. The experiment was laid out in a randomized complete block design with three replications. The two factors experiment had four levels of cowdung, viz., 0, 5, 10 and 20 tons ha^{-1} and four levels of potassium, viz. 0, 50, 150 and 250 kg K ha^{-1}. Doses of cowdung and potassium showed significant variation in respect of all the parameters studied. The combination of 10 tons cowdung and 250 kg K ha^{-1} gave the tallest plant (46.60 cm), the highest number of leaves $plant^{-1}$ (6.40), the highest length of bulb (3.27 cm), the highest diameter of bulb (4.83 cm), individual weight of bulb (51.23 g), dry matter content (12.66%) and yield of bulb ha^{-1} (12.83 tons); whereas the control treatment gave the shortest plant (38.15 cm), lowest number of leaves $plant^{-1}$ (5.68), diameter of bulb (3.41), individual weight of bulb (35.65g) and gave lowest bulb yield ha^{-1} (9.16 tons).

Keywords: Cowdung, Potassium, Onion, Bulb Yield

[1]Department of Horticulture, Bangladesh Agricultural University, Mymensingh, Bangladesh
[2]Department of Biochemistry and Molecular Biology, Bangabandhu Sheikh Mujibur Rahman Agricultural University, Gazipur-1706, Bangladesh
[3]Department of Horticulture, Bangabandhu Sheikh Mujibur Rahman Agricultural University, Gazipur-1706, Bangladesh
*Corresponding author's email: nuralam_hstu@yahoo.com (M.N. Siddiqui)

Introduction

Onion (*Allium cepa* L.) belonging to the family Alliaceae is one of the most important spice as well as vegetables crops in the world including Bangladesh. Among the spices crops grown in Bangladesh onion ranks top in respect of production and second in respect of area. It is a short duration and quick growing having various uses such as vegetables, spices and medicinal. The main edible portion is the bulb, which is a modified organ consisting of thickened fleshy scale leaves and stem plate (Jones and Mann, 1963). In Bangladesh onion has been an integral part of the people's daily diet and its use is very common in almost all food preparations (Hossain and Islam, 1994). Medicinal value of this spice crop is helpful for human being and it has preservative.

The demand of onion bulbs and its seeds are increasing with the ever increasing population. To meet the national requirement, Bangladesh officially imported 47 thousand tones of onion in 1999 with a cost of about US$ 10.3 million. The total production of onion in Bangladesh in far less than the quantity required. The total requirement is 1030 thousand metric tons and 330 thousand metric tons is imported annually (Anonymous, 2005). There are many causes for such limited production, which include production of onion only in the winter season and use of sub-optimal production inputs at the farmers' level. As a result, huge amount of onion bulb is imported and smuggled in from neighboring countries like India, Myanmar and Pakistan.

Onion production is greatly influenced by agronomic practices (Mondal *et al.*, 1986). The production of onion in Bangladesh can be increased either by extending the land area under cultivation or by increasing the yield of the crop. The $hectare^{-1}$ yield of onion can be increased by efficient use of manures and fertilizers (Katwale and Saraf, 1994). Organic manure improves soil physical chemical properties that are important for plant growth (Synman *et al.*, 1998). Organic fertilizers has positive effect on root growth by improving the root rizosfer conditions (structure, humidity, etc.) and also plant growth is encouraged by increasing the population of microorganisms (Shaheen *et al.*, 2007). Organic fertilizers contain plant nutrients. Organic acids which occur in decomposition increases the benefits of nutrients (Anonymous, 2010). Cowdung is important organic manure in Bangladesh. The farmers of our country today is burned as fuel it and is this lost to agriculture. Cowdung manure may be bulky (nutrient contents are very low per unit volume) and it improve physical conditions of the soil and supplies N-P-K and other nutrients to plants. Now a day's gradual deficiency in soil organic matter and reduced yield of crop are an alarming factor and burning issue for the farmers and agriculturists.

Judicial application of organic and inorganic fertilizers is important for the production of onion. Among the nutrients, potassium plays the most important role for photosynthetic activity of leaf, as it helps in translocation of food (Rai, 1981). Potassium helps in root development, and exerts a balancing role in multi nutrient fertilizer application (Brady, 1995). Among the yield contributing factors, application of proper doses of potassium is of great importance (Magruder and Allard, 1937; Sotomayer, 1975). The higher doses of potassium showed positive effect to attain lower percentage of rotten bulb and weight loss (Dilruba, 2001). Adequate potassium levels are especially important in improving bulb quality and storage life (Ali *et al.,* 1994).

Considering the above points, the main objective of this present experiment was to study the effect of different levels of organic manures and potassium on the growth and bulb yield of onion cv BARI-I plants.

Materials and Methods

The experiment work was conducted at the Horticulture Farm of Bangladesh Agricultural University, Mymensingh, during the period from December, 2010 to March, 2011 in an attempt to evaluate the effect of different levels of cowdung and potassium on growth and yield of onion cv. BARI piaz-I. The soil of experiment area was medium high in topography and sandy loam in texture belonging to the Old Brahmaputra Flood Plain Alluvial Tract under the AEZ number 9 (UNDP, 1988). The research work was carried out with onion variety BARI piaz-I. The seeds were collected from the Spices Research Centre of Bangladesh Agricultural Research Institute (BARI), Gazipur. The experiment was consisted of 16 treatment combinations and was laid out in the Randomized Complete Block Design with three replications. An area of 153 square meter of uniform fertility with good irrigation and drainage channels was divided into three equal blocks, representing the replications. A 75 cm space between the blocks and 50 cm between the plots was provided. The size of a unit plot was 1.2 m X 1.0 m. The experiment had two factors and was designed to study the effect of different cowdung and potassium levels on growth and yield of onion cv. BARI piaz-I). The experiment consisted of the following treatments:

Factor A: four levels of cowdung

C_0 : 0 ton cowdung ha^{-1} (control)
C_1 : 5 ton cowdung ha^{-1}
C_2 : 10 ton cowdung ha^{-1}
C_3 : 20 ton cowdung ha^{-1}

Factor B: four levels of potassium (K)

K_0 : 0 kg K ha^{-1} (control)
K_1 : 50 kg K ha^{-1}
K_2 : 150 kg K ha^{-1}
K_3 : 250 kg K ha^{-1}

In this study, manures and fertilizers were applied according to the experimental treatments. The seedlings were always kept under careful observation. Necessary intercultural operations were done throughout the cropping season for better growth and development of the plants. The crop was harvested when 75% of the tops had fallen over (Shalalby *et al.,* 1991). The data on plant height, number of leaves plant^{-1}, length of bulb, diameter of bulb, individual weight of bulb, and dry matter percentage of bulb and yield of bulb hectare^{-1} were recorded. The percentage of bulb dry matter was calculated by the following formula:

$$\% \text{ Dry matter of bulb} = \frac{\text{Constant weight of bulb}}{\text{Fresh weight of bulb}} \times 100$$

The collected data were statistically analyzed using MSTAT software. Difference (LSD) taking the probability level 5% and 1% as the minimum unit of significance and where higher levels of significance were found, they were indicated in results (Gomez and Gomez, 1984).

Results and Discussion

Plant height

The combined effect of cowdung and potassium was statistically significant. The tallest plant (46.60 cm) was observed from the treatment combination of 10 tons cowdung ha^{-1} and 250 kg K ha^{-1} the shortest plant (39.77 cm) was obtained from the control dose. The height of plant increased gradually with the increased level of cowdung up to 10 tons ha^{-1} and potassium up to 250 kg K ha^{-1} plant height will be increased (Table 1). Our findings were full agreement with the result of Singh and Dhankhar (1988).

Number of leaves

The combined effect of cowdung and potassium on the number of leaves plant^{-1} was significant (Table 1). The maximum number of leaves plant^{-1} (6.40) was recorded from the treatment combination 20 tons cowdung ha^{-1} with 250 kg K ha^{-1}, while the minimum number of plant leaves $plant^{-1}$ (5.68) was obtained from the treatment combination of 0 tons cowdung ha^{-1} with 0 kg K ha^{-1} at 70 DAT (Table 1). The result points out that the number of leaves $plant^{-1}$ was highest (6.40) with the treatment containing the highest dose of cowdung 20 tons ha^{-1} and 250 kg K ha^{-1} respectively. Singh and Dhankhar (1989) also observed similar results. From this study, it was found that the number of leaves $plant^{-1}$ was directly related with the application of optimum levels of cowdung and potassium.

Table 1. Combined effect of cowdung and potassium on plant height and number of leaves $plant^{-1}$

Treatment	Plant height (cm) at different DAT						No. of leaves $plant^{1}$ at different DAT					
	20	30	40	50	60	70	20	30	40	50	60	70
C_oK_o	7.20	11.37	20.16	24.48	30.70	38.15	1.81	2.10	3.20	3.22	3.58	5.68
C_oK_1	7.62	12.23	21.56	25.48	32.06	39.77	2.00	2.20	3.10	3.36	3.76	5.90
C_oK_2	7.54	12.23	21.37	25.90	31.45	39.85	1.98	2.20	3.20	3.33	3.73	5.86
C_oK_3	7.63	12.34	21.95	25.98	32.17	40.09	2.00	2.23	3.26	3.36	3.76	5.93
C_1K_o	7.68	12.40	21.99	26.14	32.55	41.19	2.00	2.23	3.24	3.36	3.86	5.90
C_1K_1	7.76	12.46	22.64	26.37	33.06	41.72	2.03	2.26	3.28	3.36	3.93	5.96
C_1K_2	7.68	12.49	22.24	26.68	32.93	41.38	2.00	2.26	3.26	3.40	3.90	5.96
C_1K_3	7.89	12.51	22.66	26.74	33.12	41.90	2.03	2.28	3.30	3.40	3.93	6.03
C_2K_o	7.70	12.40	22.14	26.21	32.41	40.64	2.00	2.23	3.20	3.36	3.83	5.93
C_2K_1	7.79	12.54	22.63	26.51	32.60	41.77	2.00	2.26	3.28	3.40	3.83	6.00
C_2K_2	7.86	12.49	22.19	26.31	32.95	41.59	2.03	2.26	3.23	3.36	3.86	5.96
C_2K_3	7.92	12.59	22.79	26.56	33.00	46.60	2.03	2.30	3.26	3.40	3.86	6.40
C_3K_o	7.92	12.66	22.88	26.87	33.40	42.76	2.04	2.30	3.33	3.46	3.90	6.06
C_3K_1	8.14	12.69	23.00	26.94	33.57	42.89	2.06	2.33	3.33	3.43	3.96	6.06
C_3K_2	8.20	12.79	23.00	27.04	33.86	42.99	2.08	2.36	3.36	3.46	4.00	6.10
C_3K_3	8.67	13.65	25.04	29.07	35.74	41.92	2.41	2.45	3.47	3.58	4.25	6.00
LSD at 5%	0.08	0.14	0.10	0.13	0.27	0.29	0.01	0.01	0.01	0.01	0.03	0.04
LSD at 1%	0.11	0.18	0.13	0.17	0.37	0.39	0.02	0.01	0.02	0.02	0.04	0.06
Level of significance	**	**	**	**	**	**	**	**	**	**	**	*

C_o= 0 ton cowdung ha^{-1}
C_1= 5 ton cowdung ha^{-1}
C_2= 10 ton cowdung ha^{-1}
C_3= 20 ton cowdung ha^{-1}

K_o= 0 kg K ha^{-1}
K_1= 50 kg K ha^{-1}
K_2= 150 kg K ha^{-1}
K_3= 250 kg K ha^{-1}

** = Significant at 1% level of probability
* = Significant at 5% level of probability

Length of bulb

The effect of cowdung and potassium combinedly showed significant variation with respect of bulb length. The maximum bulb length (3.27 cm) was obtained from the plant that received 10 tons cowdung ha^{-1} and 250 kg K ha^{-1}. The lowest bulb length (2.80 cm) was found from the control (Table 2).

Diameter of bulb

There was non-significant variation due to the combination effect of cowdung and potassium application. The diameters of the bulbs were varied from 3.41 cm to 4.83 cm. The maximum diameter (4.83 cm) was found from the treatment combination of 10 tons cowdung ha^{-1} and 250 kg K ha^{-1} that was statistically different from the other combination. The lowest bulb length (3.41 cm) was found from the control treatment (Table 2).

Table 2. Combined effect of cowdung and potassium on length of bulb, diameter of bulb, individual weight of bulb, dry matter % of bulb, yield of bulb hectare^{-1}

Treatment	Length of bulb (cm)	Diameter of bulb (cm)	Individual weight of bulb (g)	Dry matter of bulb (%)	Yield of bulb hectare^{-1}
C_0K_0	2.80	3.41	35.65	11.26	9.16
C_0K_1	2.88	3.71	39.63	11.33	9.41
C_0K_2	2.57	3.68	38.70	11.41	11.16
C_0K_3	2.87	3.78	40.00	11.50	9.66
C_1K_0	2.97	3.80	40.23	11.55	9.75
C_1K_1	2.88	3.86	41.46	11.61	10.83
C_1K_2	2.81	3.84	41.06	11.61	10.50
C_1K_3	2.97	3.92	42.23	11.96	10.91
C_2K_0	3.01	4.30	43.10	11.83	11.08
C_2K_1	2.99	4.37	45.30	12.21	11.75
C_2K_2	3.04	4.36	43.20	12.05	11.33
C_2K_3	3.27	4.83	51.23	12.66	12.83
C_3K_0	2.92	3.88	40.50	11.66	10.35
C_3K_1	2.88	4.12	41.90	11.81	10.50
C_3K_2	2.88	3.92	41.73	11.77	10.08
C_3K_3	2.87	4.13	42.10	11.88	10.83
LSD at 5%	0.03		0.99	0.07	0.02
LSD at 1%	0.05		1.32	0.10	0.03
Level of significance	**	NS	**	**	**

NS = Not significant, ** = Significant at 1% level of probability, * = Significant at 5% level of probability

Individual weight of bulb

The combined effect of cowdung and potassium had significant effect on the individual weight of bulb as well as significant influence on their interaction effect. However, the highest bulb weight (51.23 g) was found from the treatment combination of 10 tons cowdung ha^{-1} and 250 kg K ha^{-1} and the lowest (35.65 g) was observed from control (Table 2). This result is an agreement with that of Singh and Dhankhar (1988) who found similar result on effect of cowdung and potassium on fresh weight of onion bulb.

Dry matter percentage of bulb

There was significant variation of dry matter content of bulb on the effect of different levels of cowdung and potassium. However, it was found that 10 tons cowdung ha^{-1} and 250 kg K ha^{-1} produced the maximum dry matter (12.66%) and minimum dry matter (11.26%) was recorded under control treatment (Table 2). It was observed that the highest doses of cowdung and potassium as combined cause barrier effect to uptake water by plant.

Yield of bulb per hectare

The variation in yield at harvest due to the combined effect among different doses of cowdung and potassium was observed to be significant. The maximum yield of bulb (12.83 tons ha^{-1}) was observed with 10 tons cowdung ha^{-1} and 250 kg K ha^{-1} and the minimum yield of bulb (9.16 tons ha^{-1}) was obtained from control treatment (Table 2). Mandaria and Khan (2003) observed similar results and they reported that application of cowdung and potassium showed the best performances in terms of growth and yield.

The present findings revealed that application of 10 ton cowdung ha^{-1} and 250 kg K ha^{-1} might be recommended for proper growth and yield of onion cv. BARI piaz-I to get more yield under the Bangladesh Agricultural University, Horticulture Farm conditions.

References

Ali, M.Y., Haq, M.F. and Rahman, M.S. 1994. Effect of fertilizer doses on the yield and storability of onion. *Bangladesh J. Agril. Sci.* 23 (2): 7-12.

Anonymous. 2005. Action plan for increasing the production of spices crop to further 25%, 2005. Ministry of Agriculture, Govt. of the Peoples Republic of Bangladesh. p. 14.

Anonymous.2010.http://www.slidefinder.net/t/life_your_soil/introductionsoil_microbiology/1116901. environmental soils.cas.psu.edu.

Brady, N.C. 1995. The Nature and Properties of Soils. Hall of India Pvt. Ltd., New Delhi. p. 369.

Dilruba, S. 2001. Effect of different doses of nitrogen and potassium on growth, yield and storage of onion. MS thesis, Department of Horticulture, Bangladesh Agricultural University, Mymensingh. 58 p.

Gomez, K.A. and Gomez, A.A. 1984. Statistical Procedure for Agricultural Research. (2nd Ed.) John Willey & Sons, New York. pp. 28-192.

Hossain, A.K.M.A. and Islam, J. 1994. Status of Allium production in Bangladesh. *Acta Hort.* 358: 33-36.

Jones, H.A. and Mann, L.K.1963. Onion and Their Allies. Leonard Hill (books) Ltd., London. p. 32.

Katwale, T.R. and Saraf, R.K. 1994. Studies on response of onion to varying levels of fertilizer doses during monsoon season in Satpura plateau. *Orissa J. Hort.* 22 (1-2): 95-97.

Magruder, R. and Allard, H.K. 1937. Bulb formation in some American and European varieties of onions as affected by length of day. *J. Agric. Res.* 54: 716-752.

Mandira, C. and Khan, E.B. 2003. Effect of nitrogen and potassium on growth and yield and yield attributes of onion. ICAR-Research Conference, India, 14 (1-2): 9-11. [Cited from CAB Abst., 2202/08-2003/10].

Mondal, M.F., Brewster., J.L., Morris, F.E.L. and Butler, H.A. 1986. Bulb development in onion (*Allium cepa* L.). Effect on plant density and sowing date in field conditions. *Ann. Bot.* 58 (2): 187-195.

Rai, M.M. 1981. Principles of Soil Science. MacMillan India Limited, Calcutta. pp. 179-182.

Shaheen. A., Fatma, M., Rizk, A. and Singer, S. M. 2007. Growing onion plants without chemical fertilization. *Res. J. Agr. Biol. Sci.*, 3 (2): 95-104.

Shalalby, G.L., El-Muraba, A.L., Kandeel, N.M. and Gamie, A. M. 1991. Effect of some cultural practices on onion bulb production from sets at three planting dates, direction of ridges and cultivars. *Assiut J. Agril. Sci.* 22 (2): 103-121.

Singh, J. and Dhankhar, S.B. 1989. Effect of nitrogen, potash on growth, yield and quality of onion. *Indian Agric.* 32 (3): 163-170.

Singh, J., Dhankhar, B.S. and Ram, V. 1988. Effect of nitrogen, potash on growth, yield and quality of onion. *Veg. Sci.* 16 (2): 136-144.

Snyman, H.G., Jong, D.E. and Aveling, T.A.S. 1998. The stabilization of sewage sludge applied to agricultural land and the effects on maize seedlings. *Water Sci. Tech.* 38 (2): 87-95.

Sotomayer, R.J. 1975. Garlic fertilization. Progress Agricola, Chile. 7 (1): 34. [Cited from *Hort. Abstr.* 47 (1): 53, 1977].

UNDP. 1988. Land Resources. Appraisal of Bangladesh for Agricultural Development Report 2: Agro-ecological Region of Bangladesh. FAO, Rome, Italy. 577 p.

HUMAN FACTORS THAT DERAIL EXTENSION SERVICES DELIVERY IN DEVELOPING COUNTRIES: IMPLICATIONS FOR POLICY

E.N. Ajani* and E.A. Onwubuya

Abstract

The paper examines human factors derailing extension services delivery in developing countries. Poor management of rewards and incentives; quality of work life; poor assessment system; limited quality of human resource; inadequate extension support training materials; inconsistencies of government programmes; inadequate staff training and poor extension programme evaluation were viewed as major constraints. Considering the role of extension to agricultural development, the paper recommends that efforts should be made by the various stakeholders in agricultural extension services to improve the capability of human resources involved in extension services and highlights the need for government agricultural programmes to be consistent to enable the beneficiaries of such programmes derive the benefits that will boost productivity in agriculture.

Keywords: Human Factors, Extension Services Delivery, Developing Countries

Department of Agricultural Extension, Faculty of Agriculture, University of Nigeria, Nsukka, Nigeria

Corresponding author's email: lyneajani@gmail.com (E.N. Ajani)

Introduction

Agricultural extension services delivery play an important role in agricultural development and improves the standard of living of rural dwellers. Extension education serves the needs of educating farmers on ways of improving their farm production and life style. They are educated on how to purposefully mobilize their resources, rationally allocate them and effectively utilize them to achieve their preselected goals. Vijayaragavan and Singh (1999) observed that extension organizations in developing countries face the major problems of professional incompetence and inadequate motivation among their employees.

Rao and Pareek (2004) stressed that many of the agricultural extension departments of the developing countries do not have a well-defined system of human resource management. This affects the entire extension organization and subsequently weakens the effectiveness of extension services delivery. They further noted that, there is need for the increasing importance in the area of human resource management in extension organizations to remove the respective impediments among their employees for effective job performance.

The paper therefore reviews human factors derailing extension services delivery in developing countries. It relied heavily on literature and personal observations.

Human factors derailing extension services delivery in developing countries

There are major human factors which derail effective extension services delivery. These factors include:

Poor management of rewards and incentives

This is one of the aspects of human resource management which needs special attention in extension organizations. It entails the development of reward systems which attract, retain and motivate extension personnel as well as provide training and promotional opportunities. Extension organizations in Asian and African countries have poor reward system (Vijayaragavan, 2000; Swanson *et al.,* 1998). Similar studies conducted by Wiggins (1999) disclosed that the extension agents are not only poorly paid but are paid late after reminders of visits to the headquarters about their payment. The public extension services are run by government thus operating under rules and regulations of public administration. The rules do not have provisions for rewarding superior performance or for wage payment system based on merit.

Under such administrative set up, promotion criteria are based on seniority and length of service rather than merit. Vijayaragavan and Singh (1999) in their studies viewed that, among many of the government departments particularly, the agricultural departments (including extension service department); there are low public esteem and poor payment structure. A clear job description, performance standards, and performance appraisal will help in evaluating extension work and rewarding people for meritorious service. For example, extension workers on the basis of their performance can be sent for higher education. Non-monetary rewards such as recognizing the good ideas of field workers or awarding honorable titles will also help in improving performance. Extension personal may also be encouraged to form professional societies to develop and communicate high standards, as well as to recognize superior performance. Thus the bureaucratic structure and administration of the extension services is a basic hindrance to designing a better reward system for extension staff. This hinders effective extension delivery system.

Unimproved quality of work life

Human resource development emphasized individual development through training and proper supervision. However, with the increasing complexity of extension organizations and the society, it is realized that not only training and proper supervision play important roles in extension development, but there is also need for improving the quality of work life through making the extension job more satisfying by involving the employees in work decision making process (Pareek, 2003). Studies by Thamtani and Singh (2000; 2002) showed that the work environment of the extension organizations in developing countries is poor and needs improvement to enhance effective extension delivery and agricultural development.

Poor reward and assessment systems

The reward of extension staff in case of outstanding performance or hard work and payment of other necessary allowances is far reached in developing countries (Warren, 2003). The assessment of agricultural extension workers favors inputs given out for training farmers rather than the output of extension work by the extension agents. Qualitative assessment and effective supervision is not carried out by extension administrators. Extension workers therefore, are not stimulated to achieve quality outputs or hardly become innovative, creative and take initiative towards extension services delivery. This limits extension services delivery and agricultural development.

Poor quality of human resource

At most levels of agricultural extension, limited human resource is always a constraint to the proper functioning of extension system. Zinnah *et al.* (1999) stressed that one of the major problems impeding the effectiveness of agricultural extension services in Asia and Africa is the low level of trained extension staff. FAO (1996) stated that, shortage of trained extension staff makes it impossible for the teaming population of farmers in rural areas to be effectively reached with the expected information on improved farming methods. A ration of one extension agent to eight hundred farmers (1: 800) is practical at the moment in the developing countries.

Agricultural extensionists constitute the least trained group of staff in developing nations' agricultural organizations. It is viewed that their initial formal training is usually inadequate in most cases, where in-service training is provided, it is often ad-hoc in nature and not responsive to the changing situation of extension tasks.

Inadequate extension support training materials

Most extension organizations in developing countries suffer from relevant and practical extension training materials to support field activities of their extension workers. Many extension workers rely primarily on their inter-personal communication skills which limit their scope of coverage (Adhikaraya, 1996). Inability to provide enough, specifically designed and relevant extension training support materials like communication gadgets such as mobile phones, automatic video camera, and public address system, video machines, tape recorder etc., limit the scope of extension work.

Inadequate training materials increase the work load of extension staff and reduce the quality and area of coverage of their extension messages. Experience has shown that extension worker's motivation, enthusiasm, confidence and credibility reduce when training materials are not made available to them. Poor provision of the necessary training materials as noted above in developing countries result to low quality extension services delivery.

Inconsistency of government programmes

According to Zinnah *et al.* (1999), there is low consistency within and outside the programmes of various public extension services especially those related to agricultural and community development. Strategies are not synchronized and there is often conflict, the extension workers are forced to use top-down approach despite the diversity and complexity of conditions and problems in rural areas.

The dominance of mechanistic top-down approach which entails "wait for order from the highest echelon" as the case in agricultural extension makes the extension services delivery weak and decrease the zeal in extension workers. This results to weak extension services delivery by extension workers to farmers.

Inadequate training and development

The training of extension personnel contributes directly to the development of human resource within extension organizations. Training programmes need to be directed towards maintaining and improving job performance and develop skills further (Stoner and Freeman, 2004). Training has to start with identification of training needs through job analysis, performance appraisal and organizational analysis. Once the training needs of extension personal have been identified, the next step is to organize training programmes. Methods such as games, role playing, simulation exercises, and case can be used in extension organizations to create learning situations based on experience (Lynton and Pareek, 1990). Extension agents need training not only in the technological aspects of introduced innovations but also in human relations, problem solving and basic aspects of management to effectively interact with farmers.

In most cases, the desired training of the extension agents is weak especially the practical aspects even when made available in developing countries. This slows down the pace of extension services delivery.

Poor extension programme evaluation

Inadequate extension programme evaluation impedes extension services delivery. Stufflebeam *et al.* (1974); Warren (2003) and Ajayi (1995) in their separate studies emphasized that, very few extension agents attach importance to project evaluation. Most of the extension workers prefer administrative duties and largely forget about evaluating the extension programmes that are carried out in their areas of operation for further action. Hence evaluation is a driving force that keeps a constant touch/ check on every aspect of extension organization, to ensure maximum efficiency, its programmes and activities if not taken seriously will contribute to failure of extension services delivery.

Conclusion

Human resource planning forecasts the future personnel, their training needs, financial requirements and managerial implications in extension organizations. The rapid changes in technological needs of farmers, market situations and competitive environment has made planning for human resource a challenging task for extension organizations.

Priority actions should always be taken by the developing countries to improve their agricultural sector and economy in general. It is hoped that, the governments of developing countries should always examine properly human factors with a view to remove all the bottlenecks in their agricultural extension services to pave way for agricultural development. They should not depend entirely on expatriates and foreign donor agencies to correct weaknesses in their own domestic policies. This could be achieved through the assessment of future needs of human resource strength based on the analysis of present and future human resource requirement policies and growth trends of the developing countries.

Recommendations

1. Special attention should be given to the development of specific reward system for extension workers by both public and private extension organizations. This will attract, retain and motivate extension personnel.
2. Extension personnel should be involved at all levels of programme/ project decision of extension agencies (both public and private) to carry all staff along. This will enhance the employee's working performance.
3. Extension personnel under training should be well trained to achieve maximum job efficiency. The training of quality extension personnel based on field situations in large number will narrow the gap between extension agents-farmers contact ratio, which is currently 1: 800.
4. Government programmes related to agricultural and community development should be consistent to avoid subsequent disruption in extension delivery efforts.
5. The extension agencies (public and private) should concentrate on both input and output assessment of extension personnel rather than concentrating only on assessment of inputs invested.
6. Reward of superior performance and prompt payment of salaries and other allowances to extension workers should be encouraged.
7. Extension support and training materials should be provided at all times for use by extension agents to enhance their performance.
8. The dominance of mechanistic top-down approach of government towards extension services delivery efforts should be discouraged.
9. Constant evaluation of all extension programmes should be given priority by all extension workers to ascertain the success of programmes introduced.

References

Adhikaraya, R. 1996. Strategic extension campaign: Increase cost effectiveness and farmers participation in applying agricultural technologies. (http://www.fao.org/docrep/w5830e/w5830 eoo.htm), pp. 53-58.

Ajayi, A.R. 1995. Training needs of village extension workers in Osun State Agricultural Development Projects. *Nigerian J. Agril. & Teachers Edu.* 4 (2): 1-10.

FAO (Food and Agricultural Organization). 1996. Training for agricultural and rural development. Economic and Social Development series No. 54. Rome, Italy.

Lynton, R. and Pareek, U. 1990. Training for development. Sage Publishers, New Delhi, India. pp. 14-16.

Pareek, U. 2003. Making organizational roles effective. Tata McGraw-Hill Publishers, New Delhi, India. pp. 3-5.

Rao, T.V. and Pareek, U. 2004. Designing and managing human resource systems. Oxford and IBH Publishing Company, New Delhi, India. p. 15.

Stoner, J.A.R. and Freeman, R.E. 2004. *Management*. New Delhi. Prentice Hall.

Stufflebeam, D.L., Forey, J.F. and Williams, J.G. 1974. *Education, Evaluation and Decision Making*. Peacock Publishers, INC. Itasca, Illinois. pp. 11-17.

Swanson, B.E., Farmer, B.J. and Bahal, R. 1990. The Current status of agricultural extension World wide. *In:* B.E. Swanson (ed.) Report of the Global Consultation on Agricultural Extension. pp. 43-76.

Thamtani, A. and Singh, K.M. 2000. Perceptual variations of individuals on organizational environment. *Indian J. Ext. Edu.* 25-58.

Thamtani, A. and Singh, K.M. 2002. Inter-unit variation in organizational environment of developing countries. *Manpower J.*, 28-30.

Vijayaragavan, K. and Singh, K.M. 1999. Pay Administration in Agricultural Departments. (http://www./fao.org/docre/w5830e/5830 eoh.html), pp. 16-22.

Vijayaragavan, K. 2000. Agricultural administration in India. Concept Publishing Company, New Delhi, India. p. 40.

Warren, P. 2003. The context of extension in agriculture and rural development. (http://www./fao.org/docrep/w5830e/w58 30e04 html).

Wiggins, S. 1999. Agricultural policy and extension: The African experience. *In:* G.E Jones (ed.), *Investing in Extension Programmes*. London: Elsevier Applied Services Publishers. pp. 99-106.

Zinnah, M.M., Steele, R.E. and Mattocks, D.M. 1999. From margin to mainstream: working out strategies for agricultural extension in developing countries. (http://www./ fao.org /sd/index en.html).

ASSESSMENT OF USE OF INDIGENOUS MAIZE STORAGE PRACTICES AMONG FARMERS IN ANAMBRA STATE, NIGERIA

E.N. Ajani* and E.A. Onwubuya

Abstract

The study assessed the use of indigenous maize storage practices among farmers in Anambra State, Nigeria. Data for the study was collected by the use of interview schedule from a sample of 60 respondents. Percentage, mean score and standard deviation were used for analyzing data collected for the study. Results revealed that the farmers were using indigenous technologies such as baskets, aerial (over fire) in the kitchen, bare floors, among others in storing maize. The respondents indicated that they stored maize cobs undehusked in order to overcome wastage and obtained information about indigenous storage of maize from parents and fellow farmers. Major problems militating against effective storage of maize include: attack of pests such as rodents and weevils, diseases, termite attack and use of poor quality storage materials. The respondents indicated that use of materials free from termite, clearing of surroundings against fire disaster; use of durable materials treated with insecticides will help to solve the problems. The study recommends that provision of appropriate and affordable storage structures should be made available to the maize farmers in order to avoid wasting of the produce under storage. This will help to ensure food security among rural farm households.

Keywords: Indigenous, Knowledge, Storage, Practices, Maize, Farmers, Anambra State, Nigeria

Department of Agricultural Extension, Faculty of Agriculture, University of Nigeria, Nsukka, Nigeria

*Corresponding author's email: lyneajani@gmail.com (E.N. Ajani)

Introduction

Maize is a major food for most households in Nigeria and the main source of income and employment for the majority of rural households. Food security and welfare of the farming population are dependent on the productive capacity of maize farmers (Wekesa *et al.*, 2003). The importance of appropriate and readily available post-harvest storage practices for agricultural crops cannot be overemphasized in any development plan for increased food production and enhancement of food security. Adequate storage of farm produce remains paramount for preservation of farm produce for future consumption. Maize is the most important cereal in the world after rice and wheat (Eleweanya *et al.*, 2005). It is used as staple food in developing countries, particularly in the tropics. It also serves as raw materials for many finished products.

A lot of mature, ready-to-harvest and harvested crops, especially maize are lost to spoilage, contamination, mould and pests in the field and on storage. Severe pest's infestation may cause losses of up to 50%. The use of pesticides for control of pests is effective but not economical (Echezona and Iloba, 2005). Apart from not being economical, pesticides tend to have severe side effects on the environment (Grant and Tingle, 2002).

According to Uguru (1996), storage techniques or methods can be traditional or improved methods. Collinson (2001) stated that there is urgent need for intensified efforts to be geared towards provision of adequate and efficient storage facilities in order to avoid wastage of farm produce.

Maize production among small scale farmers in Nigeria may have declined due to agricultural reforms. The reforms which commenced in 1980s and is on-going stemmed from the structural adjustment programmes. This led to cut back in government expenditure including removal of subsidies on farm inputs. As a result, greater economic burden was placed on the small scale farmers (Nyangito and Karugia, 2006; Mochoge and Zziwa, 2004). In addition, the open maize market resulting from the agricultural reforms caused fluctuations in maize returns, exploitation by middlemen and low product prices (Kodhek, 2005). The Agricultural reforms thus pose greater risk for small scale maize farmers who are risk averse to adopt improved agricultural practices (Ogada, 2009).

Lack of suitable high yielding variety as well as poor knowledge about production practices are ascribed as main reasons for low productivity of maize in Nigeria. The productivity of maize per unit area could be increased by adopting recommended scientific and sustainable management practices using a suitable high yielding variety.

Maize farmers in Anambra state are using certain local methods and techniques such as storing over fire in the kitchen, use of earthen pots, gourds, among others in preserving maize to be used for future consumption. The study therefore answers the following questions:

- What are the indigenous maize storage practices used by maize farmers in the study area?
- What forms of storage are used for maize? What are the reasons for storing maize?
- What are the problems militating against effective storage of maize?

The specific objectives were to:

i. identify indigenous maize storage practices used by farmers;
ii. ascertain reasons for storing maize; and
iii. ascertain constraints to effective storage of maize among farmers.

Methodology

The study was carried out in Anambra State, Nigeria. There are four agricultural zones in the state, namely; Aguata, Anambra, Awka and Onitsha. Two zones namely: Aguata and Anambra were purposively used for the study. The State shares boundary with Enugu State on the north, Delta State on the south, Kogi State on the west and Imo State on the east. It has an estimated population of 4.18 million and land area of approximately 5,025 sq. km (NPC, 2006). Majority of the farmers in the state are involved in the production of arable crops such as yam, cassava, cocoyam, maize, vegetables and raising of farm animals like sheep, goat, and poultry. Major perennial crops grown in the state include: oil palm, mango, oil bean, pear, breadfruit, among others. The population of the study comprised farmers in the two agricultural zones. Anambra zone is made up of four (4) extension blocks comprising 45 circles, while Aguata zone is made up of six (6) extension blocks, comprising 45 circles. Two blocks and six circles were selected respectively from the zones and circles using a simple random sampling technique. In each of the circles selected, ten farmers were selected randomly for the study. In general, the study comprised four (4) blocks and twelve (12) circles, giving a total of one hundred and twenty (120) farmers. The interview schedule used for data collection was divided into four sections based on the objectives. The first section sought information on socio-economic characteristics of the farmers. Respondents were asked to indicate their age, marital status, level of education, size of household, head of household, sources of farm labour, farming occupational status, size of land holding (hectares) and other non-farm occupations. The second section sought information on indigenous maize storage practices which include: storing over fire in the kitchen, storing on the floor, use of cribs, gourds, and earthen pots, among others. Section three sought information on reasons for storing maize. The respondents were asked to indicate reasons for storing maize. This includes: future consumption, source of income, avoidance of wastage and generate money during off-season periods. The fourth section dealt with problems militating against effective storage of maize. The respondents were asked to indicate the problems using a Likert-type scale of to a great extent (3), to some extent (2), to a little extent (1) and to no extent (0). The mean was 1.5; this was used to determine the cut off point. Data for the study were collected using interview schedule/ questionnaire. Percentage, mean score and standard deviation were used for data analysis.

Results and Discussion

Socio-economic characteristics of respondents

Data in Table 1 show that about 14% of the respondents were within the age bracket of 20-29 years and 25.0% of them were aged 30-39 years and 40-49 years, respectively. About 21.0% of the farmers were within the age of 50-59 years while 14.2% of them were aged 60 years and above. This indicates that majority of the farmers were in most productive ages, hence greater involvement in maize production.

Entries in Table 1 show that majority (69.2%) of the farmers were married, while about 13% and 17.5% were single and widowed, respectively. This implies that most of the respondents have household's hence greater involvement in maize production as well as practicing storage practices for future consumption for household members.

About 20.0% of the respondents never had any form of formal education. Greater proportion (30.0%) had attended primary school while 27.5% and 22.5% attended secondary school and tertiary institutions, respectively. Education is very important to the farmers; this will help them to participate in the farming operations as managers by storing and marketing maize, maintaining computer records, making purchases and helping with long term planning (Taylor, 1997).

About 44.0% of the respondents had a fairly large family size of 6-10 members, 40.8% of them had between 1 and 5 members, while 10.0% and 5.0% had 11-15 and 16-20 members, respectively. Thus, the large family size of 6-10 members constitutes the family labor which most of the respondents rely upon in carrying out certain tasks in the maize farm.

Majority (64.2%) of the farmers were heads of households while 35.8% of them had their husbands as heads of household. This shows that in the absence of men as heads of households, women are becoming increasingly involved in agriculture. This helps to make their families economically stronger as well as ensuring food security. FAO (2002) points out that there is a great increase in the proportions of households headed by women this had resulted to women taking more responsibilities in agricultural production. According to UNESCO (1991), women are the heads of one third of the world's households.

Most (70.8%) of the farmers depended on their family members for provision of labour used in the farm; about 6.0% and 4.2% of them obtained assistance from their relations and friends, respectively. Hired labour was rated 51.7% while about 1.3% depend on exchange labour. This implies that most of the labour used in their farms comes from household members.

Most (55.8%) of the farmers engaged in part-time farming while 44.2% were on full-time basis. Those on part–time had other non-farm occupations. They were mostly involved in petty-trading and this helps them to sustain their families during off-season periods.

Majority (50.8%) of the farmers had access to 1-2 hectares of land while about 20.0% had 3-4 hectares of land. Also, 27.5% had less than 1 hectare of land while 1.7% had more than 5 hectares. This finding shows that greater proportion of the farmers had enough farmland for maize production; hence access to land is not seen as a major obstacle for their farm work. The findings are in agreement with Ali-Olubandwa *et al.* (2010) who observe that farmers grow maize on less than three acres of land.

A greater percentage (70.1%) of the farmers was traders, 7.5% were civil servants, 4.5% were hair dressers while 10.4% were teachers. This result shows that farmers were seriously involved in other occupations. This is to enable them meet up with their family responsibilities, since most of them are heads of households.

Table 1. Distribution of the respondents according to socio-economic characteristics (n= 120)

Characteristics	Frequency	Percentage
Age (years)		
20-29	17	14.2
30-39	30	25.0
40-49	30	25.0
50-59	26	21.7
60 and above	17	14.2
Marital status		
Single	16	13.3
Married	83	69.2
Widowed	21	17.5
Level of education (years)		
No formal education	24	20.0
Primary school	36	30.0
Secondary school	33	27.5
Tertiary institutions (NCE, Polytechnic, University	27	22.5
Size of household		
1-5	49	40.8
6-10	53	44.2
11-15	12	10.0
16-20	6	5.0
Head of household		
Husband	43	35.8
Wife	77	64.2
Sources of farm labour*		
Household members	85	70.8
Relations	8	6.7
Friends	5	4.2
Hired labour	62	51.7
Exchange labour	16	13.3
Farming occupational status		
Full-time	53	44.2
Part-time	67	55.8
Farm size (hectares)		
<1	33	27.5
1-2	61	50.8
3-4	24	20.0
5 and above	2	1.7
Other non- farm occupations		
Trading	47	70.1
Civil service	5	7.5
Hair dressing	3	4.5
Teaching	7	10.4
Catering	1	1.5
Tailoring	3	4.5
Artisan	1	1.5

*Multiple responses

Indigenous maize storage practices used by the farmers

A greater percentage (41.7%) of the respondent's stored maize in crib, 18.3% stored over fire in the kitchen, 16.7% stored on the floor, among others. About 3.0% indicated that they did not store at all. This implies that they sold maize when they were fresh and had no need for storage. It could equally be that they lacked adequate storage facility and instead of allowing the maize to spoil under storage they had to sell them immediately after harvesting. The findings also indicate that the respondents had various methods of storing maize for future consumption.

Table 2. Percentage distribution of the respondents according to indigenous methods used for storing maize (n= 120)

Storage methods	Percentage
Storing over fire (aerial storage)	18.3
Floor	16.7
Gourds	1.9
Earthen pots	6.5
Solid wall bins	11.7
Crib	41.7
Do not store at all	3.3

Forms of storing maize

Data in Table 3 indicate that 85.0% of the respondents stored undehusked maize, 75.0% stored dehusked maize cob, 46.7% stored shelled maize grains, while 25.0% process into flour before storage. This implies that the respondents did not store maize in forms which can be used for immediate consumption. This could be attributed to the fact that those stored in that form cannot last for a long period of time.

Table 3. Percentage distribution of respondents based on forms of storing maize (n= 120)

Forms of storage*	Percentage
Dehusked cob	75.0
Shelled grains	46.7
Flour	25.0
Undehusked cob	85.0

*Multiple responses

Sources of information on indigenous maize storage techniques

Table 4 indicates that fellow farmers (57.1%) were the major source of information on indigenous maize storage techniques. Other sources of information include: neighbours (33.0%), parents (29.7%), and extension agents (21.7%), among others. This implies that the respondents obtained information mostly from informal sources. The finding agrees with Anyanwu *et al.* (2002) who report that farmers receive their farm information from non-professional inter-personal sources more often than professional sources.

Table 4. Percentage distribution of respondents based on sources of information on indigenous maize storage techniques (n= 120)

Sources of information*	Percentage
Fellow farmers	57.1
Extension agents	21.7
Parents	29.7
Relations	15.0
Neighbours	33.0

*Multiple responses

Reasons for storing maize

The major reasons for storing maize among farmers include: household consumption (60.0%), generate income (40.0%), avoidance of wastage (36.7%) and acquisition of money during off season period. This implies that the respondents were storing maize in order to feed members of their household. This could be attributed to the fact that maize is a major staple food consumed by most households which can be used for preparing pap, flour and other diets.

Table 5. Percentage distribution of respondents according to reasons for storing maize (n= 120)

Reasons*	Percentage
Household consumption	60.0
Generate income	40.0
Avoid wastage	36.7
Acquire money during off season period	13.3

*Multiple responses

Constraints to effective storage of maize

The major constraints to effective storage of maize include: poor finance (M= 3.1), poor knowledge of preservation materials (M= 3.0), poor access to proper storage facilities (M= 2.8) and inadequate knowledge of better storage methods (M= 2.5). Other constraints were moldiness of stored produce (M= 2.3), lack of market for produce (M= 2.1), high costs of farm inputs (2.0), among others. This implies that the respondents were highly constrained by infrastructural problems. The findings agree with Ajani and Igbokwe (2011) who report that the major constraints to crop production among farmers were lack of farm inputs such as fertilizer, herbicides, etc. However, the findings disagree with Wekesa *et al.* (2003) who state that technological information is not a major constraining factor to maize production in Kenya.

Table 6. Distribution of respondents according to constraints to effective storage of maize

Constraints	Mean score (M)	Standard Deviation (SD)
Poor knowledge of preservation materials	3.0	0.445
Poor access to proper storage facilities	2.8	0.553
Pests and rodents attack	1.9	0.745
Inadequate knowledge of better storage methods	2.5	0.719
Poor finance	3.1	0.667
Moldiness of stored produce	2.3	0.509
High cost of farm inputs such as fertilizer and herbicides	2.0	0.520
High cost of labour	1.8	0.653
Lack of market for produce	2.1	0.523

Strategies for effective storage of maize

The respondents indicated strategies that will help to alleviate the problems they encounter in maize production. They include: availability of storage facilities (93.5%), better storage methods (87.0%), provision of funds (84.4%), and use of pesticides (61.1%), among others. This implies that the farmers need to be supplied with appropriate storage technologies in order to minimize storage losses. This will in turn enhance household food security.

Table 7. Percentage distribution of respondents according to strategies for effective storage of maize

Strategies	Percentage
Good knowledge of preservation materials	38.2
Availability of storage facilities	93.5
Better storage methods	87.0
Use of pesticides	61.1
Adequate storage structures	10.2
Provision of funds	84.4

Conclusion and Recommendations

Maize farmers were using indigenous storage practices such as baskets, aerial (over fire) in the kitchen, bare floors, among others in storing the packages disseminated. The farmers should be facilitated in order to adopt improved agricultural practices by providing them with soft loans, which they can use to buy farm inputs. maize. They indicated that they stored maize cobs undehusked in order to overcome wastage emanating from storage.

Efforts should be made by the government to encourage farmers to adopt improved maize storage practices in order to reduce losses emanating from storage. The educational level of the farmers should be considered when coming up with extension packages and methods to ensure maximum adoption of

References

Ajani, E.N. and Igbokwe, E.M. 2011. Implications of feminization of agriculture on women farmers in Anambra State, Nigeria. *J. Agril. Ext.* 15: 31-39.

Ali-Olubandwa, A.M, Odero-Wanga, D., Kathuri, N.J. and Shivoga, W.A. 2010. Adoption of improved maize production practices among small scale farmers in the agricultural reform era: The case of Western Province of Kenya. *Spring.* 17 (1): 25-27.

Anyanwu, A.C., Agwu, A.E. and Umeweni, C.A. 2002. Sources of agricultural information used by women farmers in Orumba North Local Government Area of Anambra State, Nigeria. *ASSET Series.* 2 (1): 97-104.

Collinson, M. 2001. Institutional and professional obstacles to a more effective research process for small holder agriculture. *Agricultural Systems.* 69 (12): 27-36.

Echezona, B.C. and Iloba, B.C. 2005. Maize earthworm attack as influences by varying maize spatial arrangement and population in maize soybean mixture. *Agroscience Journal.* 4 (1): 60-65.

Eleweanya, A.N., Uguru, M.I., Eneobong, E.E. and Okocha, P.I. 2005. Correlation and path coefficient analysis of grain yield related characters in maize (*Zea mays*) under umudike conditions of South Eastern Nigeria. *Agroscience Journal.* 4 (1): 24-28.

FAO (Food and Agriculture Organization). 2002. Special report: FAO/WFP crop and food supply assessment mission to Malawi. (http://www.fao.org/waicent/ fao info).

Grant, I.F. and Tingle, C.C.D. 2002. Ecological monitoring methods for the assessment of pesticides impact in the tropics, Chatham, UK, National Resource Institute. p. 266.

Kodhek, G.A. 2005. *Contemporary issues determining the future of Kenyan agriculture: An agenda for policy and research.* Retrieved from http://www.yahoo.Agenda_pol_resrch(1) html.

Mochoge, B. and Zziwa, S. 2004. Agricultural and food security challenges in the IGAD region. A paper presented at the NEPAD/ IGAD regional conference *Agricultural Successes in the Greater Horn of Africa.* Nairobi, Kenya.

NPC (National Population Commission). 2006. National Population census figure, Abuja, Nigeria.

Nyangito, H.O. and Karugia, J.T. 2006. *The impact of recent policy changes on the agricultural sector and public agricultural research in Kenya.* Retrieved from http//:www.Glob_ch05.pdf.

Ogada, M.J. 2009. The role of production risk in farm technology adoption. *Environment for Development Policy Brief.*

Taylor, M. 1997. Women's work: Modern women rewrite a farm wife's job description. *Top Producer.* 14 (4): 8–13.

Uguru, M.I. 1996. *Crop Production Tools, Techniques and Practices*, Nsukka. Fulladu publishing company, Nigeria. pp. 14-17.

UNESCO, 1991. Wekesa, E., Mwangi, W., Verkuijl, H., Danda, K. and De Groote, H. 2003. Adoption of maize production technologies in the coastal lowlands of Kenya, International Maize and Wheat Improvement Center (CIMMYT). pp. 1-34.

ABUNDANCE OF MACROZOOBENTHOS IN RELATION TO BOTTOM SOIL TEXTURAL TYPES AND WATER DEPTH IN AQUACULTURE PONDS

N. Nupur, M. Shahjahan*, M.S. Rahman and M.K. Fatema

Abstract

The present experiment was conducted to evaluate the effects of bottom soil textural classes and different water depths on abundance of macrozoobenthos in aquaculture ponds. Three treatments, i.e., ponds bottom with sandy loam (T1), with loam (T2) and with clay loam (T3) were considered in this experiment. Samples were collected from three different depths (60.96 cm, 106.68 cm and 152.40 cm) with three replications. The ranges of water quality parameters were suitable for the growth of macrozoobenthos during the experimental period. Similarly, chemical properties of soil were also within suitable ranges and every parameter showed comparatively higher ranges in T2. Eight genera were recorded belonging to major groups of Chironomidae, Oligochaeta, Mollusca and Ceratoponogonidae. The highest population densities of Oligochaeta (1200±4.25 per m^2), Chironomidae (1422±4.88 per m^2), Ceratopogonidae (399±1.56 per m^2) and Mollusca (977±2.24 per m^2) were found in T2. The population densities of macrozoobenthos showed fortnightly variations in all the treatments. Among the three depths, significantly highest densities of macrozoobenthos were recorded in 106.68 cm in every treatment. The mean abundance of macrozoobenthos was significantly highest in T2. The present study indicates that loamy soil pond bottom along with water depth 106.68 cm is suitable for the growth and production of macrozoobenthos in aquaculture ponds.

Keywords: Benthos, Aquaculture, Soil Textural Class, Pond, Water Quality

Department of Fisheries Management, Bangladesh Agricultural University, Mymensingh-2202, Bangladesh

*Corresponding author's email: mdshahjahan@bau.edu.bd (M. Shahjahan)

Introduction

The benthos is defined as that assemblage of animal living in or on the sediments and dependent upon the decomposition cycle of organic matters that are the basic food supply in any aquatic ecosystem (Brinkhurst, 1974). Different kinds of macrobenthos are very important food organisms for bottom-feeding fishes; they are important in sediment-water interactions; the presence and abundance of benthos indicates normal limnological conditions of the water bodies; and they are also occasionally used as food by fishes of upper water and mid-water and column water. The growth of fish depends largely upon the availability of food organisms namely plankton and benthos. Benthos is one of the very important food items of fishes especially the bottom feeders. About 60% of total food items of bottom-feeder fishes come from insect larvae (Habib *et al.*, 1984; Ali *et al.*, 1987). Moreover, the benthic fauna, through their burrowing and feeding activities play a significant role in the exchange of nutrients, dissolved gases and other materials between sediments and overlying water. The productivity of a water body largely depends upon the plankton and benthos (Habib *et al.*, 1984).

The nutrients status of both soil and water plays a significant role in the growth and abundance of aquatic organisms, especially plankton and benthos. The chemical properties (nutrients status) have some growth promoting effect on the various species of benthos fauna (Habib *et al.*, 1984). On the other hand, nutrient status of soil depends on the type of soil texture. In soil science, the USDA (Donahue *et al.*, 1990) defines twelve major soil textural classes. Loam soils generally contain more nutrients and humus than sandy soils. However, there is no study on the effects of bottom soil textural types on growth and abundance of benthic fauna so far. Productivity of benthos as well as productivity of water body depends on the kind of textural types of pond bottom-soil along with limnological and ecological conditions. On the other hand, water depth is another important factor that control light intensity, dissolved oxygen, benthic fauna etc. in a water body. In the present study, it was evaluated the effects of bottom soil textural types and water depth on abundance of macrozoobenthos in aquaculture pond.

Materials and Methods

Experimental design

The experiment was conducted in the ponds situated at the campus of Bangladesh Agricultural University, Mymensingh during August to November 2011. The experiment had three treatments with three replications, i.e., ponds bottom with sandy loam (T1), with loam (T2) and with clay loam (T3). Samples were collected from three different depths (D1= 60.96 cm, D2= 106.68 cm and D3= 152.40 cm) with three replications.

Water quality parameters

Various physical and chemical water quality parameters of the ponds such as water temperature (°C), transparency (cm), dissolved oxygen (mg L^{-1}), free CO_2 (mg L^{-1}), total alkalinity (mg L^{-1}), PO_4-P (mg L^{-1}) and NO_3-N (mg L^{-1}) were estimated fortnightly following standard methods.

Chemical parameters of pond bottom-soil (sediment)

Chemical properties of pond bottom-soil (sediment) such as pH, available phosphorus (ppm), total nitrogen (%), organic carbon (%) and organic matter (%) were estimated fortnightly using the standard method (Sattar and Rahman, 1987).

Study of macrozoobenthos

The samples were collected fortnightly randomly from 3 depths of water of each of the ponds by lowering an open Ekman dredge on bottom mud. Then collected benthos with mud was put on a fine mesh-screen net fixed with a small steel frame and mud was washed in the pond water and the bottom materials were passed through a 0.2 mm mesh of screen net for a preliminary separation of benthos and large particles from mud and water. The benthic organisms remained on the screen-net were collected by means of fine forceps and kept into small plastic bottles containing 5% formalin for preservation.

Macrozoobenthos population density (no. per m^2) = no. found in bottom-mud collected by Ekman dredge × 44.44 = no. per 225 cm^2 × 44.44. Where, area of open mouth of Ekman dredge = 225 cm^2

The preserved organisms were then transferred to a petri dish and washed with tap water to remove the remaining washable detritus and mud. Then the samples were cleaned by means of distilled water and the organisms were separated from each other with the help of sorting needless and fine forceps, which were again sorted into major taxonomic groups by means of hand lens (magnifying glass, 65 mm, Optical Instrument Co. China) and low power microscope (magnification:10×2.5) wherever necessary. Then the organisms were counted and identified according to different taxonomic groups according to Needham and Needham (1963), Maitland and Heedspith (1974), Wetzel and Lickens (1979).

Statistical analysis

Values are expressed as means ± standard error of the mean (SEM). Data were analyzed by one-way analysis of variance (ANOVA) followed by Tukey's post hoc test to assess statistically significant differences among the different sampling days, different depths and different treatments. Statistical significance was set at $p<0.05$. Statistical analyses were performed using SPSS Version 14.0 for Windows (SPSS Inc., Chicago, IL).

Results

Water quality parameters

The results of water quality parameters are shown in Table 1. Throughout the study period, water temperature (°C), transparency (cm), dissolved oxygen (mg L^{-1}), free CO_2 (mg L^{-1}), total alkalinity (mg L^{-1}), PO_4-P (mg L^{-1}) and NO_3-N (mg L^{-1}) were within the productive ranges and showed no abrupt changes during the experimental period in all the treatments.

Table 1. Water quality parameters (Means ± SEM; n = 3) of the ponds during the experimental periods

Parameters	Treatments		
	Treatment 1	Treatment 2	Treatment 3
Water temperature (°C)	27.32 ± 3.13	27.00 ± 3.07	27.20 ± 3.20
Air temperature (°C)	27.50 ± 2.88	27.50 ± 2.88	27.50 ± 2.88
Transparency (cm)	32.30 ± 0.91*	16.00 ± 1.15	15.30 ± 1.11
Dissolved oxygen (mgL^{-1})	7.86 ± 0.24	7.21 ± 0.70	5.86 ± 0.48
Free CO2 (mgL^{-1})	3.86 ± 0.90	3.43 ± 1.62	4.00 ± 0.82
Total alkalinity (mgL^{-1})	82.72 ± 8.28	149.14 ± 9.05*	51.00 ± 6.86
Phosphate-phosphorous (mgL^{-1})	2.20 ± 0.58	2.68 ± 0.23	1.73 ± 0.46
Nitrate-nitrogen (mgL^{-1})	3.11 ± 0.55	3.58 ± 0.19	2.78 ± 0.33

* Indicates the significant difference among the treatments

Chemical parameters of pond bottom-soil (sediment)

In the present study, the ranges of pH, organic carbon (%), organic matter (%), available phosphorus (ppm) and total nitrogen (%) of pond bottom-soil in the aquaculture ponds varied from 6.8 ± 0.26 to 7.47 ± 0.21, 0.75 ± 0.10 to 0.98 ± 0.06, 1.34 ± 0.16 to 1.66 ± 0.14, 13.36 ± 1.58 to 25.87 ± 0.88 and 0.07 ± 0.01 to 0.09 ± 0.01, respectively (Table 2). All the parameters were within the suitable ranges and showed no abrupt changes during the experimental period in all treatments.

Table 2. Chemical parameters of pond bottom-soil (means ± SEM; n = 3) during the experimental periods

Parameters	Treatments		
	Treatment 1	Treatment 2	Treatment 3
pH	7.03 ± 0.14	7.13 ± 0.16	7.32 ± 0.13
Organic carbon (%)	0.78 ± 0.07	0.94 ± 0.08	0.84 ± 0.13
Organic matter (%)	1.34 ± 0.13	1.64 ± 0.12	1.41 ± 0.20
Available phosphorus (ppm)	13.86 ± 1.25	25.76 ± 0.89	16.10 ± 1.12
Total nitrogen (%)	0.07 ± 0.01	0.07 ± 0.01	0.08 ± 0.02

Abundance of macrozoobenthic fauna

During the study period, 8 genera of macrozoobenthos were recorded belonging to the major groups of Chironomidae, Oligochaeta, Ceratoponogonidae and Mollusca (Table 3). The individual groups of macrozoobenthic fauna (means ± SEM; n = 3) in the aquaculture ponds during the experimental periods are shown in Table 4.

Table 3. Species composition of macrozoobenthos found in the aquaculture ponds during the experimental periods

Oligochaeta	Chironomidae	Ceratoponogonidae	Mollusca
Brachiura sowerbyi	*Chironomus* sp.	*Culiciodes* sp.	*Viviparous bengalensis*
Peloscolex ferox	*Pentaneura* sp.	*Amphizoa* sp.	
Aelosoma sp.			

Table 4. Individual groups of macrozoobenthos (means ± SEM; n = 3) in the aquaculture ponds during the experimental period

Fortnightly sampling day	Treatments	Macrobenthos (number per m^2)			
		Oligochaeta	Chironomidae	Ceratopogonidae	Mollusca
	T1	533 ± 2.05^b	756 ± 2.35	-	311 ± 1.25
1	T2	1111 ± 3.25^a	1022 ± 3.85	222 ± 1.31	533 ± 1.34
	T3	755 ± 2.25^{ab}	799 ± 2.33	177 ± 1.28	-
	T1	311 ± 1.25^b	1111 ± 3.11	133 ± 1.74	178 ± 1.22^b
2	T2	711 ± 2.22^a	1155 ± 4.75	355 ± 1.85	888 ± 2.21^a
	T3	844 ± 2.35^a	888 ± 2.95	177 ± 1.65	-
	T1	489 ± 1.85^b	1111 ± 3.88	-	311± 1.26^b
3	T2	799 ± 2.56^a	1288 ± 4.95	399 ± 1.56^a	888 ± 2.25^a
	T3	711 ± 2.33^a	1111 ± 4.12	44 ± 0.25^b	-
	T1	178 ± 1.44^b	578 ± 2.21	-	133 ± 1.23^b
4	T2	755 ± 2.24^a	977 ± 2.23	266 ± 1.25	755 ± 2.12^a
	T3	266 ± 1.56^b	844 ± 2.35	311 ± 1.66	-
	T1	267 ± 1.65^b	333 ± 1.66^b	133 ± 1.22	444 ± 1.27
5	T2	1200 ± 4.25^a	1199 ± 3.44^a	266 ± 1.23	622 ± 1.95
	T3	711 ± 3.32^a	933 ± 2.77^{ab}	133 ± 1.20	-
	T1	356 ± 1.52^b	667 ± 2.22^b	44 ± 0.25^b	222 ± 1.21^b
6	T2	888 ± 2.26^a	1244 ± 4.21^a	399 ± 1.28^a	888 ± 2.26^a
	T3	533 ± 1.88^b	933 ± 2.24^{ab}	311 ± 1.26^a	-
	T1	444 ± 1.55^b	1067 ± 3.28	-	267 ± 1.75^b
7	T2	1022 ± 4.22^a	1422 ± 4.88	222 ± 1.21	977 ± 2.24^a
	T3	799 ± 2.23^b	1022 ± 4.22	222 ± 1.21	-

Values with different superscripts are significantly different among treatments ($p<0.05$)

The highest number of Oligochaeta (1200 ± 4.25 per m^2) was found in T2 and the lowest number (178 ± 1.44 per m^2) in T1. The highest number of Chironomidae (1422 ± 4.88 per m^2) was found in T2 and the lowest number (333 ± 1.66 per m^2) was found in T1. The highest number of Ceratoponogonidae (399 ± 1.28 per m^2) was found in T2 and the lowest number (0 per m^2) was found in T1 and T3. The highest number of Mollusca (977 ± 2.24 per m^2) was found in T2 and the lowest number of Mollusca (0 per m^2) was found in T1 and T3. On the other hand, the total number of macrozoobenthos (means ± SEM; n = 3) ranged from 888 ± 2.24 to 1783 ± 3.54, 2655 ± 4.24 to 3644 ± 5.23 and 1322 ± 2.89 to 2044 ± 3.63 per m^2 in the ponds of T1, T2 and T3, respectively (Fig. 1).

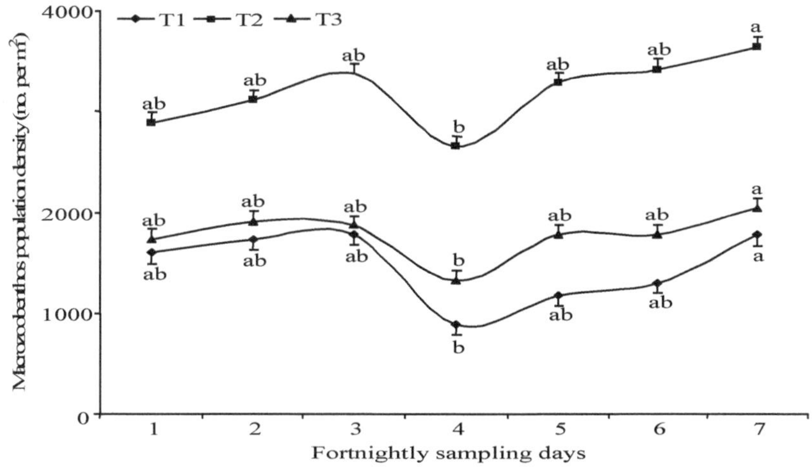

Fig. 1. Fortnightly variations in abundance (means ± SEM; n = 3) of total macrozoobenthos in the experimental ponds under three treatments during the study period. Values accompanied by different letters are statistically significantly different ($p<0.05$).

The macrozoobenthos densities showed fortnightly variations. Among different depths, significantly highest densities of macrozoobenthos were recorded in 106.68 cm in every treatment (Fig. 2). Based on mean value (Fig. 3), it was observed that macrozoobenthos showed its highest density in T2 (3211 ± 9.97 per m^2) followed by T3 (1775 ± 7.20 per m^2) and T1 (1464 ± 7.35 per m^2).

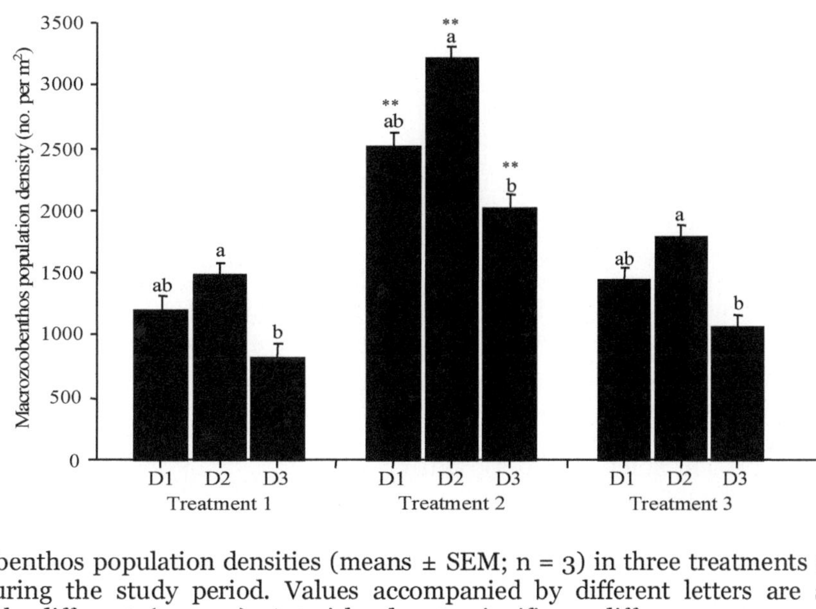

Fig. 2. Macrozoobenthos population densities (means ± SEM; n = 3) in three treatments at different depths during the study period. Values accompanied by different letters are statistically significantly different ($p<0.05$). Asterisks denote significant differences among treatments (**$p<0.05$). D1 = 60.96 cm depth, D2 = 106.68 cm depth, D3 = 152.40 cm depth.

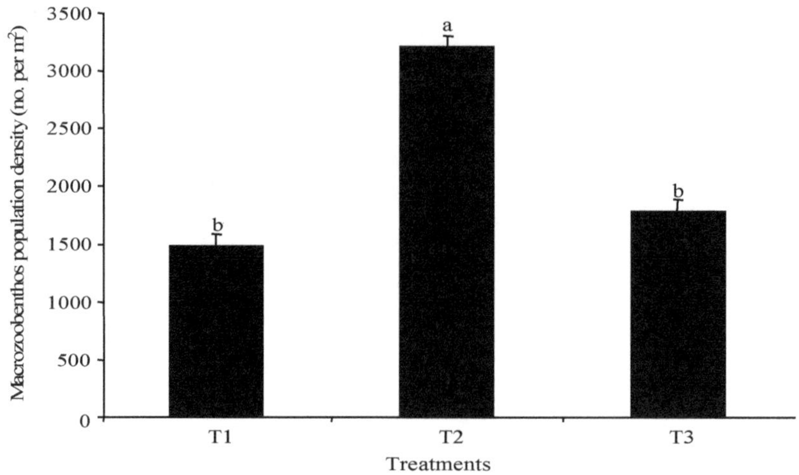

Fig. 3. Macrozoobenthos population densities (means ± SEM; n = 3) in different treatments during the study period. Values accompanied by different letters are statistically significantly different ($p<0.05$).

Discussion

The present study was conducted to evaluate the effects of bottom soil textural conditions and water depth on the abundance of macrozoobenthos in aquaculture pond. The highest abundance of macrozoobenthos found in the ponds bottom with loam and water depth 106.68 cm indicates that loamy bottom soil along with water depth 106.68 cm is suitable for the growth and production of macrozoobenthos in aquaculture pond.

The water quality parameters of the experimental ponds were within the productive ranges for the growth of macrozoobenthos during the experimental period. Within limit productive ranges of such water quality parameters have been observed by several authors (Uddin *et al.*, 2007; Chowdhury *et al.*, 2008; Rahman *et al.*, 2012; Talukdar *et al.*, 2012) in the aquaculture ponds of BAU area. Similarly, the ranges of pH, organic carbon (%), organic matter (%), available phosphorus (ppm) and total nitrogen (%) of pond bottom-soil in the experimental ponds were within the suitable ranges for the growth and production of macro-benthos (Table 2). Hossain (2000), Kohinoor (2002), Akter (2006) and Siddika *et al.* (2012) have observed more or less similar results.

A total number of 8 genera of macrozoobenthos belonging to Chironomidae (3), Oligocheta (2), Ceratopogonidae (2) and Mollusca (1) were recorded in the present study (Table 3). Other studies recorded were 23 genera belonging to 14 families in a marsh (Saha *et al.*, 2006), 29 genera under arthropods, annelids and mollusks in Vajiralongkom and Srinakarin Reservoir (Sonsupap *et al.*, 2007), 13 genera in lagoon of Nigeria (Edokpayi *et al.*, 2010), and 18 genera in two Phyla, Mollusca and Arthropods in Majidun river, Nigeria (Esenowo and Ugwumba, 2010).

In the present study, significantly highest number of Oligochaeta (1200±4.25 per m^2), Chironomidae (1422±4.88 per m^2), Ceratopogonidae (399±1.56 per m^2) and Mollusca (977±2.24 per m^2) were found in T2 (Table 4). Bais *et al.* (1992) and Hossain (2000) have observed more or less similar results. The densities of macrozoobenthos showed fortnightly variations (Fig. 1) but in every sampling day highest number were observed in T2. Interestingly, among three depths of 60.96 cm, 106.68 cm and 152.40 cm, significantly highest densities of macrozoobenthos were recorded in 106.68 cm depths in every treatment and it might be due to favourable light intensity in this depth. On the other hand, the mean abundance of macrozoobenthos was significantly highest in T2 (Fig. 3). The occurrence of these dominant groups of macrozoobenthos in T2 might be due to the suitable ecological conditions of the ponds that favoured the growth of these groups. Chemical properties of soil were comparatively higher in T2 may be an important cause to the higher abundance of macrozoobenthos population in the present study. This argument also supported by several authors (Ali *et al.*, 1987; Verneaux *et al.*, 2004; Kailasam and Sivakami, 2004) who found the significant effect of chemical properties on the macrozoobenthos growth and production. Moreover, loamy soils generally contain more nutrients and humus. Humus is a temporary intermediate product left after considerable decomposition of dead plants and animals, which might be support food for macrozoobenthos. This result indicated that pond bottom with loamy soil is more suitable for growth and production of macrozoobenthos in aquaculture ponds.

In conclusion, suitability of bottom soil textural types and water depth on macrozoobenthos population density were analyzed in aquaculture pond. Most of the water quality parameters of the ponds were more or less similar and within productive limit and chemical properties of soil were within suitable ranges. The mean abundance of macrozoobenthos density was significantly highest in T2. Significantly, highest densities of macrozoobenthos were recorded in 106.68 cm depths in every treatment. This findings of the study indicates that loamy soil bottom along with water depth 106.68 cm is suitable for the growth and production of macrozoobenthos in aquaculture ponds.

References

Akter, N. 2006. Effect of bottom soil properties on the abundance of benthic fauna in nursery ponds. MS Thesis. Department of Aquaculture, Bangladesh Agricultural University, Mymensingh, Bangladesh. 74 p.

Ali, M.M., Rahmatullah, S.M. and Habib, M.A.B. 1987. Abundance of benthic fauna in relation to metrological and physical-chemical factors of water. *Bangladesh J. Agril. Sci.* 12: 239-247.

Bais, V.S., Yatheesh, S. and Agrawal, N.C. 1992. Benthic macro-invertebrates in relation to water and sediment chemistry. *J. Freshwater Biol.* 4: 183-191.

Brinkhurst, R.O. 1974. A guide for the Identification of British Aquatic Oligochaeta, Freshwater Biological Association, Scientific Publication no. 22 & 55. 65 p.

Chowdhury, M.M.R., Shahjahan, M., Rahman, M.S. and Sadiqul Islam, M. 2008. Duckweed (*Lemna minor*) as supplementary feed in monoculture of Nile tilapia, *Oreochromis niloticus*. *J. Fish. & Aqua. Sci.* 3: 54-59.

Donahue, R.L., Miller, R.W. and Shickluna, J.C. 1990. Soils: An Introduction to Soils and Plant Growth. Prentice-Hall of India Private Ltd, New Delhi. 110001. 667 p.

Edokpayi, C.A., Olowoporoku, A.O. and Uwadiae, R.E. 2010. The hydrochemistry and macrobenthic fauna characteristics of an urban draining creek. *Int. J. Biodiver. & Conser.* 2: 196-203.

Esenowo, I.K. and Ugwumba, A.A. 2010. Composition and abundance of macrobenthos in Majidun River, Ikorodu Lagos state, Nigeria. *Res. J. Biol. Sci.* 5: 556-560.

Habib, M.A.B., Haque, A.K.M. and Islam, M.A. 1984. Status of macro-benthos in two ponds in Bangladesh Agricultural University Campus, Mymensingh. *Bangladesh J. Agril. Sci.* 11: 39-46.

Hossain, M.Y. 2000. Effects of biophosphorus in organic and inorganic fertilizers on water quality parameters and biological production. MS Thesis. Department of Fisheries Management, BAU, Mymensingh. 24 p.

Kailasam, M. and Sivakami, S. 2004. Effect of thermal effluent discharge on benthic fauna of Tuticorin Bay, South East coast of India. *Indian J. Marine Sci.* 33: 194-201.

Kohinoor, A.H.M. 2002. Development of culture technology of three small indigenous fishes mola (*Amblypharyngodon mola*), punti (*Puntius gonionotus*) and chela (*Chela cachius*) with notes on some aspects of their biology. PhD Thesis. Department of Fisheries Management, Bangladesh Agricultural University, Mymensingh. 363 p.

Maitland, P.S. and Heedspith, P.M.G. 1974. The zoobenthos of lake levan, kinross and estimates of its production in the sub-littoral area during 1970 and 1971. pp. 219-239. *In:* Proceeedings of Royal Society, Edinburgh.

Needham, J.G. and Needham, P.R. 1963. A Guide to the study of Freshwater Biology. 5th edition, Liolden-day Inc., San Francisco, 106 p.

Rahman, M.S., Shahjahan, M., Haque, M.M. and Khan, S. 2012. Control of euglenophyte bloom and fish production enhancement using duckweed and lime. *Iranian J. Fish. Sci.* 11: 358-371.

Saha, G.N., Sehag, K.L., Mitra, E. and Nandy, A.C. 2006. Studies on the seasonal and diurnal variations in physico-chemical and biological conditions of a perennial freshwater pond. *J. Inland Fish. Soc. India.* 3: 79-102.

Sattar, M.A. and Rahman, M.M. 1987. Techniques of Soil Analysis, Published by M.A. Sattar, BAU, Mymensingh. 54 p.

Siddika, F., Shahjahan, M. and Rahman, M.S. 2012. Abundance of plankton population densities in relation to bottom soil textural types in aquaculture ponds. *Int. J. Agril. Res. Innov. & Tech.* 2 (1): 56-62.

Sonsupap, N., Meksumpun, C. and Meksumpun, S. 2007. Comparative study on species composition, density and distribution of benthic fauna for assessment of aquatic enrichment status in Vajiralongkom and Srinakarin Reservoir, Kanchanaburi Province. pp. 310-317. *In:* Proceedings of the 45th Kasetart University Annual Conference, Kasetart, 30 January-2 February, 2007.

Talukdar, M.Z.H., Shahjahan, M. and Rahman, M.S. 2012. Suitability of duckweed (*Lemna minor*) as feed for fish in polyculture system. *Int. J. Agril. Res. Innov. & Tech.* 2 (1): 42-46.

Uddin, M.N., Rahman, M.S. and Shahjahan, M. 2007. Effects of duckweed (*Lemna minor*) as supplementary feed on monoculture of GIFT strain of tilapia (*Oreochromis niloticus*). *Prog. Agric.* 18: 183-188.

Verneaux, V., Verneaux, J., Schmitt, A. and Lambert, J.C. 2004. Relationships of macrobenthos with dissolved oxygen and organic matter at the sediment-water interface in ten French lakes. *Archiv-fur-Hydrobiologie.* 160: 247-259.

Wetzel, R.G. and Lickens, E.E. 1979. Limnological Analysis. W.B. Saunder, 175 p.

EVALUATION OF RICE GERMPLASM UNDER SALT STRESS AT THE SEEDLING STAGE THROUGH SSR MARKERS

M. Al-Amin[1], M.M. Islam[2], S.N. Begum[3], M.S. Alam[1], M. Moniruzzaman[1], M.A.K. Patwary[1]

Abstract

Twenty eight rice germplasms were used for identification of salt tolerant rice genotypes at the seedling stage at the experimental farm and Biotechnology laboratory of the Bangladesh Institute of Nuclear Agriculture (BINA), Mymensingh during February 2009 to October 2009. Phenotyping for salinity screening of the rice genotypes was done using salinized (EC level 12 dS m^{-1}) nutrient solution in hydroponic system. Genotypes were evaluated for salinity tolerance on 1-9 scale based on seedling growth parameters following modified Standard Evaluation Scoring (SES) of IRRI. Phenotypically, on the basis of SES and % total dry matter (TDM) reduction of the genotypes viz. PBSAL-614, PBSAL-613, PBSAL-730, Horkuch, S-478/3 Pokkali and PBSAL (STL)-15 were found to be salt tolerant; on the other hand Iratom-24, S-653/32, S-612/32, S-604/32, S-633/32, Charnock (DA6), BINA Dhan-6 and S-608/32 were identified as salt susceptible. For genotyping, ten SSR markers were used for polymorphism, where 3 primers (RM127, RM443 and RM140) were selected for evaluation of salt tolerance. In respect of Primer RM127, 7 lines were found salt tolerant and 11 lines were moderately tolerant and 10 lines were susceptible. Nine tolerant, 9 moderately tolerant and 10 susceptible lines were found when the primer RM140 was used and primer RM443 identified 8 lines as tolerant, 9 lines as moderately tolerant and 11 lines as susceptible. Thus, the salt tolerant lines can be used in further evaluation for salinity tolerance and the SSR markers used in this study are proving valuable for identifying salt tolerant genes in marker assisted breeding.

Keywords: Rice, Salinity Tolerance, SSR Markers, Seedling Stage, Polymorphism

[1]Department of Genetics and Plant Breeding, Bangladesh Agricultural University, Mymensingh 2202, Bangladesh
[2]Biotechnology Division, Bangladesh Institute of Nuclear Agriculture (BINA), BAU Campus, Mymensingh 2202
[3]Plant Breeding Division, Bangladesh Institute of Nuclear Agriculture (BINA), BAU Campus, Mymensingh- 2202
*Corresponding author's email: asarowar04bau@gmail.com (M.S. Alam)

Introduction

Rice (*Oryza sativa* L.) belongs to the family Gramineae. It is the staple food crop in Bangladesh and one of the most important cereal crops throughout the world. This staple food ranked first position by production (28931 thousand metric tons) during the year 2007-08 among all cereals in Bangladesh (BBS, 2009). Total rice cultivated area 10575 thousand hectares. One of such efforts is to cultivate rice with elevated level of salt tolerance on salt affected marginal lands. Over 30 percent of the net cultivable area of Bangladesh lies in the coastal zone. Out of 2.85 million hectares of coastal and offshore land, about 1.5 million hectares are affected by varying degrees of salinity. The coastal saline soils are distributed unevenly in 64 Upazila of 13 District, covering portions of eight agro-ecological zones (AEZ) of the country (Seraj and Salam, 2000). Salinity affects plants at all stages of development, but sensitivity sometimes varies from growth stage to the next. Several studies indicated that rice is tolerant during germination, becomes very sensitive during early seedling stage (2-3 leaf stage), gains tolerance during vegetative growth stage becomes sensitive during pollination and fertilization and then become increasingly more tolerant at maturity (Pearson *et al.,* 1966; IRRI, 1967). However, some studies reported that at fertilization, rice is not sensitive to salinity (Kaddah *et al.,* 1975). Hence, early seedling stages were used to know the response of the rice plant to salinity. To screen the salt tolerant variety, we need reliable technique. IRRI standard protocol for salinity screening is such type of technique (Gregorio, 1997). Conventional breeding is time consuming and depends on environmental conditions. Molecular marker technology offers a possibility by adopting a wide range of novel approaches to improve the selection strategies in rice breeding. SSR or microsatellite markers are proved to be ideal for making genetic maps (Islam, 2004; Niones, 2004), assisting selection (Bhuiyan, 2005) and studying genetic diversity in germplasm.

Microsatellite marker analysis is promising to identify major gene locus for salt tolerance that can be helpful for plant breeders to develop new cultivars. From these points of view the present study were undertaken for determine the phenotypic performance of rice germplasm under salinized conditions at the seedling stage and to identify salt tolerant rice lines from twenty eight rice germplasm using SSR markers.

Materials and Methods

Plant materials

The experiment was carried out at the experimental farm and Biotechnology Laboratory of the Bangladesh Institute of Nuclear Agriculture (BINA), Mymensingh from February 2009 to October 2009. In this study twenty eight rice germplasm were used for salinity screening at the seedling stage following International Rice Research Institute (IRRI) standard protocol (Gregorio *et al.,* 1997). These germplasms were collected from different sources like, International Rice Gene bank (IRG) of International Rice Research Institute (IRRI), Philippines, Bangladesh Rice Research Institute (BRRI), Bangladesh Institute of Nuclear Agriculture (BINA), Mymensingh. Twenty eight rice germplasm (Horkuch, Charnock, Keshrail, Iratom-24, BRRI dhan-40, Binadhan-6, PBSAL(STL)-15, PBSAL-613, PBSAL-614, PBSAL-730, PBSAL-656, STM-1, PBRC(STL)-20, S-530/32, S-603/32, S-604/32, S-445/32, S-611/32, S-612/32, S-546/32, S-653/32, S-635/32, S-375/32, S-731/32, S-478/32, S-608/32, S633/32 and Pokkali) were evaluated for salinity screening.

Phenotypic study of salinity tolerance at seedling stage

The screening for salt tolerance were done using modified standard evaluation score (SES) of visual salt injury of the seedlings (Table 1).

Hydroponic system with IRRI standard protocol (Gregorio *et al.*, 1997) was used at the glasshouse to evaluate salt tolerance of rice germplasm. Nutrient solution (Yoshida *et al.,* 1976) was used in hydroponic system for screening salinity tolerance at the seedling stage. The modified standard evaluation score (SES) of IRRI was used to assess the visual symptoms of salt toxicity (Table 1). This scoring discriminates the tolerant, moderately tolerant and susceptible rice lines.

Table 1. Modified standard evaluation score (SES) of visual salt injury at seedling stage (Method adapted from Gregorio *et al.* (1997)

Score	Observation	Response category
1	Normal growth, no leaf symptoms	Highly tolerant
3	Nearly normal growth, but leaf tips of few leaves whitish and rolled	Tolerant
5	Growth severely retarded, most leaves rolled; only a few are elongating	Moderately tolerant
7	Complete cessation of growth; most leaves dry; some plants dying	Susceptible
9	Almost all plants dead or dying	Highly susceptible

CTAB mini preparation DNA extraction

The modified Cetyl Trimethyl Ammonium Bromide (CTAB) mini-prep method was used to extract DNA from the plant leaves. The following steps were followed in PCR-based DNA marker analysis. Healthy leaf samples were collected from 25-day old seedlings of each germplasm for isolation of genomic DNA. Leaf sample was washed in distilled water and ethanol. The collected leaf samples were then put into polythene bags with tags, placed on ice and stored in a −80oC freezer. The leaf samples were ground with pestle and mortar to collect DNA. Strict hygiene was maintained during the DNA extraction by autoclaving all glassware, micropipette tips, PCR tubes, distilled water, reagents and buffer solutions.

Isolated genomic DNA contains a large amount of RNA and pigments, which cause over estimation of DNA concentration during spectrophotometer reading. Therefore, the DNA samples were evaluated qualitatively using agarose gel electrophoresis. For checking of DNA concentration, two methods can be applied: 1) λ DNA and 2) Spectrophotometry. For this study only λ DNA was used. After electrophoresis, the gel was taken out carefully from the gel chamber and transferred in a prepared ethidium bromide solution for staining. Staining was done for 20 minutes and then placed on the UV transilluminator in the dark chamber of the Image Documentation System.

Amplification of microsatellite markers and evaluation of genotypes

Ten primers of random sequence were selected for amplification of the DNA sequences. Primers were evaluated based on intensity of bands, consistency within individual, presence of smearing and potential for population discrimination. In this experiment, OSR14, OSR17 RM152, RM443, RM127, RM140, RM9, RM169, RM18 and RM21 primers were used for polymorphism. Out of ten primers, three polymorphic SSR markers viz., RM127, RM443 and RM140 were selected to evaluate 28 rice germplasm for salt tolerance.

Table 2. The sequence and size of the microsatellite markers used for the study

Primer	PCR product size (bp)	Sequence		Annealing temp. (°C)
		Forward	Reverse	
RM443	118-124	GATGGTTTTCATCGGCTACG	AGTCCCAGAATGTCGCTTCG	55
RM127	209-226	GTGGGATAGCTGCGTCG	AGGCCAGGGTGTTGGCATGCTG	55
RM140	259-263	TGCCTCTTCCCTGGCTCCCCTG	GGCATGCCGAATGAAATGCATG	55

The PCR cocktail had total volume of 15.0 µl reaction mixture including 2 µl DNA based on salinity protocol, was placed in the PCR tubes and run in the DNA thermal cycler. The PCR tubes were set on the wells of the thermocycler plate and the machine was run according to the programme.

The germplasms having similar banding pattern to Pokkali were considered as tolerant, banding pattern having similar to Iratom-24 were considered as salt susceptible and having banding pattern at the medium position between Pokkali and Iratom-24 were considered as medium tolerant In each marker, allelic bands were designated as T for tolerant, S for susceptible and MT for medium tolerant.

Results and Discussion

Phenotypic variation of rice genotypes under salinized and non-salinized conditions at the seedling stage

Twenty eight genotypes of rice seedlings were used for screening salinity tolerance. After two or three days of salinization, salt stress symptoms were started. Several salt injuries observed in the salinized conditions which were leaves became yellowish, drying of leaves, shoot growth reduction, root growth reduction, stem became thin and week, stunted growth of seedlings and seedling drying occurred. These symptoms were also observed by several researchers (Bhuiyan, 2005; Islam, 2004; Niones, 2004; Bonilla *et al.*, 2002). The seedlings in the non-salinized condition showed normal growth over the salinized condition. Salt tolerant seedlings were distinguished from the sensitive seedlings when grown in salinized condition.

The salinity tolerant lines showed no symptom of salt injury. In the salinized setup, twenty eight germplasms showed wide variation in phenotypes under salt stress. Among the 28 germplasms, seven were found salt tolerant viz. Pokkali, Horkuch, PBSAL (STL)-15, PBSAL-613, PBSAL-730, PBSAL-656 and S-478/32 (Table.3), ten were moderately tolerant and eleven were susceptible. According to the SES of IRRI (Gregorio *et al.*, 1997) a score from 1-9 was used for grading the genotypes based on visual salt injury where, 1= highly tolerant and 9= highly susceptible. This scoring differentiated the susceptible genotypes from the tolerant and moderately tolerant genotypes. Islam *et al.* (2007) also observed wide variation in phenotypes from tolerant (score 3) to highly susceptible (score 9) using modified SES of IRRI standard protocol.

Table 3. Performance of the germplasms under salinized condition (EC 12 dS m^{-1}) at the seedling stage

Sl. No.	Name of the Variety	SES Scoring	Tolerance
1	S-611/32	5	MT
2	Keshrail	5	MT
3	S-612/32	7	S
4	S-546/32	5	MT
5	S-653/32	7	S
6	BRRI dhan 40	7	S
7	PBRC(STL)-20	5	MT
8	S-604/32	7	S
9	S-530/32	5	MT
10	S-603/32	5	MT
11	S-445/32	5	MT
12	S-635/32	7	S
13	S-375/32	5	MT
14	S-731/32	7	S
15	PBRC(STL)-15	3	T
16	Horkuch	3	T
17	PBSAL-613	3	T
18	S-478/32	3	T
19	Charnock (DA-6)	7	S
20	S-608/32	7	S
21	PBSAL-656	3	T
22	S-633/32	7	S

23	Binadhan-6	7	S
24	PBSAL-614	5	MT
25	STM-1	5	MT
26	PBSAL-730	3	T
27	Pokkali	3	T
28	Iratom-24	7	S

T= tolerant, MT= moderately tolerant, S= susceptible
1-9 scale, where 1= highly tolerant and 9= highly susceptible

Rice plant showed various degrees of responses to the salinity. The agronomic traits of tolerant lines shown lesser reduction than susceptible lines. The plant height, root length and total dry matter are greatly affected by the salinity. Iratom-28 (60.5), S-653/32(60%), Binadhan-6 (59.23%), (54.23%), and BRRIdhan 40(53.43%) showed greater plant height reduction under the salinity stress (at 12 dS m^{-1}) (Table 4).

Table 4. Performance of plant height, root length and total dry matter at the seedling stage (30DAS) under salinized and non-salinized conditions

Sl. No.	Genotypes	Plant height (cm)			Root length (cm)			Total dry matter (g)		
		Non-salinized	Salinized	% Reduction	Non-salinized	Salinized	% Reduction	Non-salinized	Salinized	% Reduction
1	S-611/32	51.2	27.8	45.70	18.2	10.2	43.95	26.2	5	80.91
2	Keshrail	56.2	43.8	22.06	18.6	12.8	31.18	25.4	14.2	44.09
3	S-612/32	44.4	24.6	44.59	15.8	12.4	21.51	10.6	4.4	58.49
4	S-546/32	49	23.6	51.83	17	14.8	12.94	15.6	5.6	64.10
5	S-653/32	52	20.8	60	16.6	11.6	30.12	27.4	4.2	84.67
6	BRRI dhan 40	40.8	19	53.43	16.4	10.8	34.14	9.6	4.4	54.16
7	PBRC(STL) -20	43.4	21.8	49.76	14.2	10.6	25.35	11.4	4.6	59.64
8	S-604/32	46.8	21.8	53.41	14.4	9.8	31.94	15.6	3.2	79.48
9	S-530/32	46.8	27.6	41.02	14.2	12	15.49	12	6.4	46.66
10	S-603/32	39.2	22.6	42.34	11.2	8	28.57	11	5.4	50.90
11	S-445/32	45.4	25.6	43.61	13.2	10.8	18.18	13.8	5.2	62.31
12	S-635/32	41	18.6	54.63	13.2	8	39.39	9.4	3.6	61.70
13	S-375/32	40.8	26.2	35.78	16.4	11.4	30.48	10.4	6.6	36.53
14	S-731/32	45.2	27	40.26	14.6	9.8	32.87	15.6	6.6	57.69
15	PBRC(STL) -15	43.8	28.2	35.61	12.8	11	14.06	14.6	7.4	49.31
16	Horkuch	58.4	45.8	21.57	14.4	12.4	13.88	17.2	12.8	25.58
17	PBSAL-613	42.8	33.4	21.96	13	12	7.69	17	13.4	21.17
18	S-478/32	53	38.8	26.79	12.4	11.1	10.48	16.6	11.6	30.12
19	Charnock (DA-6)	45.2	27	40.26	16.2	11	32.09	11.4	4.6	59.64
20	S-608/32	45.6	24.5	46.27	13.2	9.4	28.78	14.8	5	66.21
21	PBSAL-656	44.4	27.2	38.73	13.4	10	25.37	19	8.6	54.73
22	S-633/32	52	23.8	54.23	11.8	9.6	18.64	16	4.2	73.75
23	Binadhan-6	36.8	15	59.23	13.6	8.4	38.23	7.8	2.2	71.79
24	PBSAL-614	47	35.8	23.82	12.6	11	12.69	15.4	12.8	16.88
25	STM-1	45	23.8	47.11	13.4	11.4	14.92	11.2	5.2	53.57
26	PBSAL-730	46.4	33.8	27.15	16.2	14	13.58	15	11.2	25.33
27	Pokkali	58.2	45.4	22	14.2	13	8.45	17.11	13.80	23.5
28	Iratom-24	44.6	23.6	47.08	17	12.4	27.05	20	2.8	86
%CV		11.65	27.78	28.31	13.32	14.88	43.11	32.70	51.56	35.41
Range		58.4-39.2	45.8-15	60-21.57	18.6-11.2	14.8-8	43.95-8.45	27.4-9.4	14.2-2.2	86-16.88

Highest plant height reduction (60.5 %) was observed in the variety Iratom-24. On the other hand, lowest plant height reduction was observed in Horkuch (21.57 %) followed by PBSAL-613 (21.96%), Keshrail (22.06%), Pokkali (22%) and PBSAL-614 (23.82%). These results indicated that plant height reduced under the salinity stress. Such types of results were also found by several researchers (Javed *et al.*, 2006; Maiti *et al.*, 2006). They showed that the increase in salinity level reduced the seedling height. Munns and Tester (2008) also reported that salinity might directly or indirectly inhibit cell division, consequences leaves and stems of the affected plants become stunted.

Root growth also retarded by the salinity stress. At this study, several genotypes showed higher root length reduction. S-611/32 showed highest root length reduction (43.95%) followed by Iratom-24 (43.80), S-635/32 (39.39%), Binadhan-6 (38.23%) and BRRIdhan 40 (34.14%) (Table 8).On the other hand PBSAL-613 showed lowest root length reduction (7.69%). Some rice genotypes also showed lower root length reduction viz. Pokkali (8.45%), S-478/32 (10.48%), PBSAL-614 (12.69%), and Horkuch (13.88%). The result showed that root length decreased under salinity stress. Maiti *et al.* (2006) found a considerable amount of genetic variation under saline condition at the seedling stage with reference to the variables such as root length, shoot and root dry weight. Roy *et al.* (2002) reported that number of roots plant^{-1} decreased with the increasing levels of salinity. Rodrigues *et al.* (2002) reported that root length reduced due to the effect of salinity, which coincided with the present study.

Total dry matter (TDM) is an important trait to evaluate the performance of rice genotypes for salt tolerance. The highest TDM reduction found in Iratom-24 (86%) followed by S-653/32 (84.67%) and S-611/32 (80.91%). The lowest TDM reduction was found in PBSAL-614 (16.88%), followed by PBSAL-613 (21.17%), Pokkali (23.5), PBSAL-730 (25.33%) and Horkuch (25.58%). Therefore, the genotypes PBSAL-614, Pokkali PBSAL-613, PBSAL-730, Horkuch, and S-478/3 were found as tolerant based on lower percentages reduction of TDM at 12 dS m^{-1} EC level of salt stress. Comparing with modified SES system, seven genotypes viz. Pokkali, Horkuch, PBSAL (STL)-15, PBSAL-656, PBSAL-730, PBSAL-656 and S-478/32 were also found tolerant under salt stress at the same EC level (Table 3).

From the above results, it can be said that TMD reduction increased under salt stress. Javed *et al.* (2006) reported that increase in salinity level reduced the fresh biomass and dry biomass. Roy *et al.* (2002) reported that reduction of dry biomass (% TDM) increased with the increased of salinity level.

Genotyping of salinity tolerance

Identification of molecular markers tightly linked to salt tolerant genes can serve as landmarks for the physical localization of such genes facilitating marker assisted selection (MAS).

Screening twenty eight rice germplasmss for salt tolerance using SSR markers

In this experiment RM152, RM443, RM127, RM140, RM9, RM169, RM18 and RM21 primers were used for polymorphism survey of twenty eight rice germplasms. Out of these primers, three polymorphic SSR markers viz., RM127, RM443 and RM140 were shown highly polymorphic and these three markers were selected to evaluate 28 rice germplasms for salt tolerance. According to the phenotypic performance, Pokkali was considered as tolerant and Iratom-24 was considered as susceptible. The lines having similar banding pattern to Pokkali were considered as tolerant and similar to Iratom-24 were considered as salt susceptible and having both banding pattern (Pokkali and Iratom-24) were considered as moderately tolerant (Table 5).

In respect of primer RM127, 7 lines were found tolerant at salt stress. In the same reaction, 11 lines were moderately tolerant and 10 lines were susceptible with compared to the tolerant variety Pokkali and susceptible variety Iratom-24 (Fig. 1).

Nine lines were detected as tolerant, 10 susceptible and 9 moderately tolerant lines were found with compared to the tolerant variety Pokkali and susceptible variety Iratom-24 with the reaction with RM140 (Fig. 2).

With RM443, 8 tolerant lines were identified. In the same reaction, 11 saline susceptible lines and 9 moderately tolerant lines were also found with compared to the tolerant variety Pokkali and susceptible variety Iratom-24 (Fig. 3).

From the summarized Table 5 it is shown that seven lines viz. Horkuch, PBSAL-613, S-478/32, PBSAL-656, Pokkali, PBSAL (STL)-15 and PBSAL-730 were found as tolerant, five lines viz. Keshrail, S-546/32, S-530/32, S-603/32 and S-445/32 were found moderately tolerant and nine lines viz. Iratom-24, S-653/32, S-731/32 BRRIdhan 40, S-604/32, S-635/32, S-612/32, Charnock (DA-6) and Binadhan-6 were found as susceptible against all the tested markers. Three lines viz. S-611/32, PBRC (STL)-15 and STM-1 were found tolerant only with RM140 and one line PBSAL-614 was found tolerant with RM443. Five lines viz. S-611/32, S-375/32, S-608/32, PBSAL-614 and STM-1 with RM127, four lines viz. PBRC (STL)-20, S-375/32, S-633/32 and PBSAL-614 with RM140 and three lines viz. STM-1, PBRC(STL)-15 and S-611/32 with RM443 were found as moderately tolerant. PBRC (STL)-20 was found moderately tolerant with RM127, RM140 &RM443 and S-608/32 was found susceptible with both RM140 and RM443.

These three primers i.e. RM127, RM140 and RM443 showed polymorphisms in the studied twenty eight rice germplasms because they showed different banding patterns and discriminate tolerant lines from moderately tolerant and susceptible with relation to the tolerant variety Pokkali and susceptible variety Iratom-24. The SSR markers were also reported as highly polymorphic in IR29 × Pokkali for tagging salt tolerant genes (Islam, 2004; Niones, 2004). El-Refaee *et al.* (2006) also reported that 80% of the tested SSR primers showed polymorphic pattern in rice while they studied 272 SSR primers on nine rice genotypes. Sultana *et al.* (2007) found RD-2586, TBDB-100, PNR-519 and Dhol Kochuri as tolerant; BRRIdhan 40 and Y-1281 as moderately tolerant regarding the marker RM115 and concluded that genotypic screening for salt tolerance using molecular marker is more trustworthy and effective. Islam *et al.* (2007) also identified the genotypes viz. Pokkali, Jamainaru, horkoz and RT-STL33 as salt tolerant using the markers RM231 and RM24.

From the phenotypic data of twenty eight rice genotypes, based on SES and percentage TDM reduction PBSAL-614, PBSAL-613, PBSAL-730, Horkuch, S-478/3 and Pokkali were identified as salt tolerant rice genotypes whereas Iratom-24, S-653/32, S-611/32, S-604/32, S-633/32, Binadhan-6 and S-608/32 were identified as salt susceptible rice genotypes. Keshrail, S-611/32, S-546/32, S-530/32, S-603/32, S-445/32, S-375/32, PBSAL-614, STM-1 were identified as moderately salt tolerant rice genotypes.

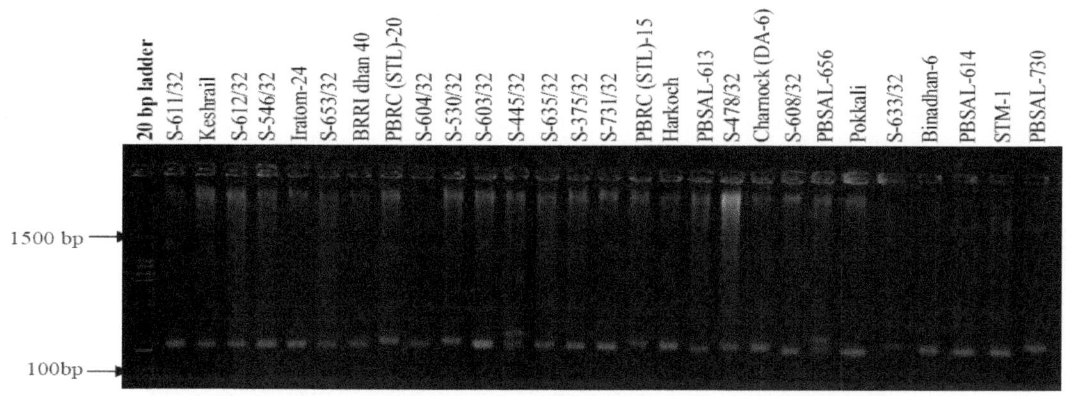

Fig. 1. Banding profiles of 28 rice germplasms using primer RM127

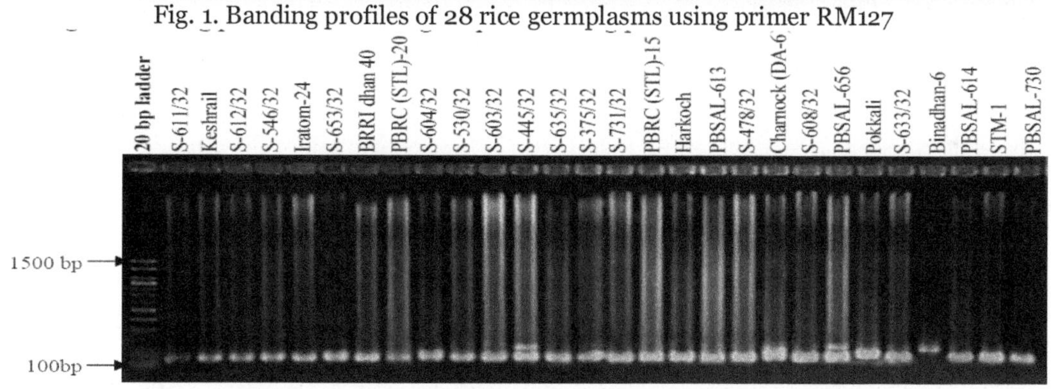

Fig. 2. Banding profiles of 28 rice germplasms using primer RM140

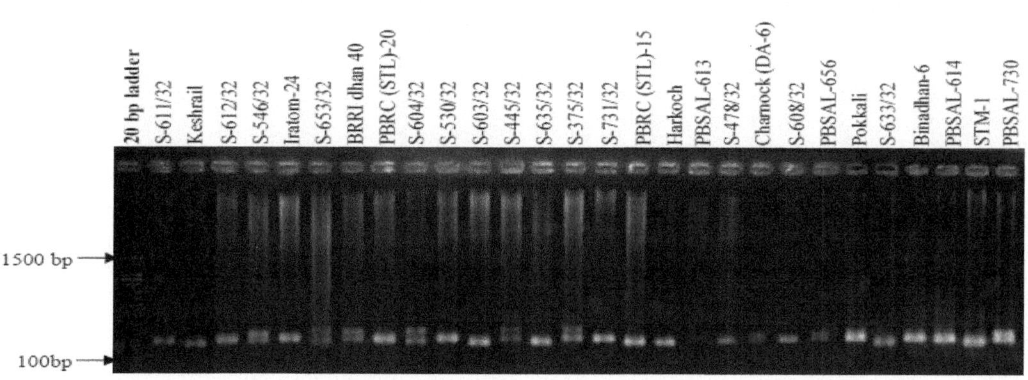

Fig. 3. Banding profiles of 28 rice germplasms using primer RM443

Table 5. Genotypic performance of twenty eight rice germplasm using SSR markers

Sl. No.	Variety/Line	Salt tolerance with SSR markers		
		RM127	RM140	RM443
1	S-611/32	MT	T	MT
2	Keshrail	MT	MT	MT
3	S-612/32	S	S	S
4	S-546/32	MT	MT	MT
5	S-653/32	S	S	S
6	BRRIdhan 40	S	S	S
7	PBRC(STL)-20	MT	MT	MT
8	S-604/32	S	S	S
9	S-530/32	MT	MT	MT
10	S-603/32	MT	MT	MT
11	S-445/32	MT	MT	MT
12	S-635/32	S	S	S
13	S-375/32	MT	MT	T
14	S-731/32	S	S	S
15	PBRC(STL)-15	T	T	MT
16	Horkuch	T	T	T
17	PBSAL-613	T	T	T
18	S-478/32	T	T	T
19	Charnock (DA-6)	S	S	S
20	S-608/32	MT	S	S
21	PBSAL-656	T	T	T
22	S-633/32	S	MT	S
23	Binadhan-6	S	S	S
24	PBSAL-614	MT	MT	T
25	STM-1	MT	T	MT
26	PBSAL-730	T	T	T
27	Iratom-24	S	S	S
28	Pokkali	T	T	T

T = tolerant, MT =Moderately Tolerant and S = susceptible

Finally, salt tolerant rice lines were identified both phenotypically (salinity screening at the seedling stage) and genotypically using twenty eight rice germplasms. The tested markers RM127, RM140 and RM443 could be efficiently used to identify salt tolerant lines in rice and can also be used in marker-assisted selection (MAS) for breeding, quantitative trait loci (QTL) mapping, studying genetic diversity in germplasms and gene pyramiding in rice salinity breeding. Microsatellite marker analysis is promising to identify major gene locus for plant breeders to develop new cultivars. The selected salt tolerant rice lines could be further tested in saline areas to observe yield potentiality for developing high yielding salt tolerant varieties suitable for saline areas.

References

BBS. 2009. Crop Statistics (Major Crops). Agriculture Wing. Bangladesh Bureau of Statistics, Ministry of planning, Government of the People's Republic of Bangladesh, Dhaka. 54 p.

Bhuiyan, M.A.R. 2005. Efficiency in evaluating salt tolerance in rice using phenotypic and marker assisted selection. M.S. Thesis, Department of Genetics and Plant Breeding, Bangladesh Agricultural University, Mymensingh, Bangladesh. 96 p.

Bonilla, P., Dvorak, J., Mackill, D., Deal, K. and Gregorio, G. 2002. RFLP and SSLP mapping of salinity tolerance genes in chromosome 1 of rice (*Oryza sativa* L.) using recombinant inbred lines. *Philippine J. Agril. Sci.* 85: 68-76.

El-Refaee, Y.Z, Atallah, M., Savo Sardaro, M.L. and Porceddu, E. 2006. Fine QTLs mapping for salinity tolerance in rice. Proceedings of the 50th Italian society of Agricultural Genetics Annual Congress. Ischia, Italy. ISBN 88-900622-7-4. pp. 10-14. Sep. 2006.

Gregorio, G.B. 1997. Tagging salinity tolerant genes in rice using amplified fragment length polymorphism (AFLP). Ph. D. dissertation. University of the Philippines Los Baños College, Laguna, Philippines. 118 p.

Gregorio, G.B., Senadhira, D. and Mendoza, R.D. 1997. Screening rice for salt tolerance. IRRI discussion paper series no. 22. Manila (Philippines). pp. 1-30.

IRRI (International Rice Research Institute). 1967. Annual Report. Los Banos. Laguna, Philippines. 308 p.

Islam, M.M. 2004. Mapping salinity tolerance genes in rice (*Oryza sativa* L.) at reproductive stage. Ph.D. Dissertation. University of the Philippines Los Baños, College, Laguna, Philippines. 149 p.

Islam, M.M., Mondol, M.N.H., Emon, R.M., Begum, S.N., Bhowmik, S.K. and Hasan, A.K. 2007. Screening of salt tolerant rice genotypes using SSR markers at seedling stage. *Bangladesh J. Prog. Sci. Tech.* 5 (1): 45-48.

Javed, M.A., Ishii, T., Kamijima, O. and Misoo, S. 2006. Discrepancy of two ecotypes of *Oryza sativa* L. to salinity at germination and seedling stages. *Ann. Biol.* 22 (2): 201-211.

Kaddah, M.T., Lehman, W.F., Meek, B.D. and Robinson, F. E. 1975. Salinity effects on rice after boot stage. *Agron. J.* 67: 436-439.

Maiti, R.K., Vidyasagar, P. and Banerjee, P.P. 2006. Salinity tolerance in rice (*Oryza sativa* L.) hybrids and their parents at emergence and seedling stage. *Crop Res. Hisar.* 31 (3): 427-433

Munns, R. and Tester, M. 2008. Mechanisms of salinity tolerance. *Ann. Rev. Plant Biol.* 59: 651- 681.

Niones, J.M. 2004. Fine mapping of the salinity tolerance gene on chromosome1 of rice (*Oryza sativa* L.) using near-isogenic lines. M.S. Dissertation. University of the Philippines, Los Baños, College, Laguna, Philippines. 78 p.

Pearson, G.A., Ayers, S.D. and Eberhard, D.L. 1966. Relative salt tolerance of rice during germination and early seedling development. *Soil Sci.* 102: 151-165.

Rodrigues, L.N., Fernandes, P.D., Gheyi, H.R. and Viana, S.B.A. 2002. Germination and formation of rice seedlings under saline stress. *Esc. Agrot. Fed. de Castanhal.* 6 (3): 397-403.

Roy, S.K., Patra, S.K. and Sarkar, K.K. 2002. Studies on the effect of salinity stress on rice (*Oryza sativa* L.) at seedling stage. *J. Interacademicia.* 6 (3): 254-259.

Seraj, Z.I. and Salam, M.A. 2000. Growing rice in saline soils. : Biotechnological approaches for Bangladesh. *Asia-Pacific Tech. Monitor.* 17 (6): 55-59.

Sultana, S., Begum, S.N., Islam, M.M, Siddika, A., Bhowmik, S.K. and Emon, R.M. 2007. Identification of salt tolerant rice genotypes via microsatellite markers at the seedling Stage. *Bangladesh J. Seed Sci. Tech.* 11 (1&2): 41-47.

Yoshida, S., Forno, D.A., Cook, J.H. and Gomez, K.A. 1976. Laboratory manual for physiological studies of rice. International Rice Research Institute, Los Baños, Laguna, Philippines. pp. 61-66.

ECONOMIC ANALYSIS OF LOCUST BEANS PROCESSING AND MARKETING IN ILORIN, KWARA STATE, NIGERIA

C.O. Farayola[1*], V. Okpodu[2] and O.O. Oni[3]

Abstract

This study was designed to estimate the economic analysis of locust bean processing and marketing in Ilorin, Kwara State, Nigeria. Primary data was used and purposive sampling technique was adopted to select the respondents used for the study. A total number of 60 respondents were interviewed. The data collected were analyzed using inferential statistical tool such as regression analysis. Budgetary analysis technique was also used to analyze the profitability of locust bean processing and marketing in the study area. Majority of the processors and marketers are making profits; 68.3% operate above breakeven point while 26.7% operate at breakeven point and the rest 5% was below the breakeven point, this indicates that they neither profit nor lost. The regression analysis result shows that quantity processed, family size and years of experience in processing are significant at 1%, 5% and 10% respectively while education level and stall rent is negative and significant at 1% and 5% respectively. F- Test also explained that independent variables are jointly significant at 1% probability level with an adjusted R^2 of 78.9%. The overall rate of return on investment indicates that average rate of return is 0.5 (50%), which is positive. It is therefore concluded that profit made by the processors and marketers can be improved on by increasing the quantity of locust bean being processed through adoption of newly discovered method of processing and improved method of preservation, packaging and marketing of the product to international standard by reducing the odour of the product without the loss of essential nutrients and palability in order to generate foreign exchange. Also, rules and regulations against cutting of economic trees for alternative uses should be enforced to maximize their values.

Keywords: Locust bean, Condiment, Income, Processing and Marketing

[1]Agricultural Development Management Department, General Management Division, ARMTI, Ilorin, Km 18, Ilorin Ajase-Ipo Highway, P.M.B. 1343, Ilorin, Kwara State, Nigeria.
[2]Rural Development and Gender Department, Entrepreneurial Development and Management Division, ARMTI, Ilorin, Km 18, Ilorin Ajase-Ipo Highway, P.M.B. 1343, Ilorin, Kwara State, Nigeria.
[3]Agricultural Development Management Department, Agricultural Finance and Rural Credit Division, ARMTI, Ilorin, Km 18, Ilorin Ajase-Ipo Highway, P.M.B. 1343, Ilorin, Kwara State, Nigeria.
*Corresponding author's email: walecharless@yahoo.com (C.O. Farayola)

Introduction

African locust bean tree "*Parkia bioglobosa*" belongs to the family of *Fabaceae* sub family "*Mimosoidea*" and genus "*Parkia*". The plant was named after Mungo Park by Robert Brown in 1826. The tree is widely recognized in West Africa as an important multipurpose tree of West Africa Savannah land. The most important part of the tree is found in its seeds and processed seed is used as condiment for soup. It is a source of natural nutritious condiment which features frequently in the traditional diet of the people (Fagbemi, 1989). Apart from the flavouring attribute of the processed locust bean "Iru", it also contributes significantly to the intake protein, carbohydrate, calcium, phosphate, iron content and essentials fatty acids, particularly vitamin B, riboflavin and vitamin A (Aju *et al.*, 2008; Oduro *et al.*, 2007; Popoola and Galaudu, 2000; Beaumont, 2002; Oladele *et al.*, 1995; Musa, 1991; Odunfa, 1985). Table 1 shows the nutrition composition of locust bean seed. It is also called by different local names in different localities; for instance, it is referred to as "kinda" in Serria Leone, "kpalugu" among the inhabitants of Nothern Ghana, "nere" in Burkina Faso. "Igba" in Yorubaland and "worku" in Ghana (Diawara *et al.,* 2000; Odunfa, 1985). The tree produces many benefits; it produces fruits which are enclosed in numerous large pods, and it tolerates a wide range of alluvial, sandy and lateritic soil, and also resists pest and diseases, survives fires and strives in full sun and tropical heat. Within Africa, clearly its prospects are greatest in West Africa, where the trees are planted and its products are known and loved. The fruits are brown in colour when ripe; it contains numerous black seeds embedded in yellowish sweet tasting pulp. This yellowish pulp can be made into colourful and refreshing drinks. Also in the dry area, locust bean trees serve as potential sources of food, edible oil, fodder lumber, fire wood and green manure. It was estimated that about 200,000 tons of Africa locust beans seeds are gathered each year in Nigeria alone, as well as large quantities are produced in the savannah region of Oyo, Osun and Kwara States of Nigeria (Onnyi *et al.*, 2004). The seeds are the most valued product of the tree. In Nigeria, it serves as food buffer during lean period while reliable income accurse to the farmers involved in harvesting and marketing of the product. This array of multipurpose uses have been reported to include food, medicine, manure, gum, shade, wind break, bee food, stabilization of degraded environmental, livestock seed, fuel, fibre fish poison among several others (Campbell-Platt, 1986; and Keay, 1989).

Table 1. Nutrition composition of locust bean seed

Component	Amount (%)
Crude protein	40.0 – 47.4
Fat	31.4 – 42.9
Carbohydrate	15.0
Crude fiber	3.1 – 5.6
Ash	3.3 – 5.6
Calcium (mg/100g)	309.0 – 880.0
Iron (mg/100g)	480.0 – 546.0
Phosphorous (mg/100g)	517.0 -584.0
Metabolic energy (Kcal/100g)	480.0 – 546.0
Gross energy (Kcal/100g)	517.0 – 618.0

Source: Odunfa, 1985

It generates reliable and dependable income for the farmers and women who involved in its processing and marketing. The processed bean known as "Iru" is a popular food condiment common in West African countries. It adds flavor and taste to most dishes, soup, sauces and stew made to accompany porridge, rice and cooked yam and cassava in the region. This means that the fermented food condiment is a crucial and widely consumed food of high nutritional value to millions of Nigeria both in rural and urban areas. It also serves as a source of protein supplements in the diet and the flavourant contribute some nutrients notable minerals and vitamins to nutritional requirements of a poor family especially in rural areas (Fagbemi, 2002; Oyerinde and Daramola, 2004; and Diawara *et al.*, 2000).

It is unfortunate that locust bean is fast losing its popularity to some other flavoring agent like maggi, royco, knorr, etc. which its nutritive value cannot be compared with that of dawadawa. The Federal Office of Statistics (now National Bureau of Statistics) in 2003 reported that about $200 million is spent annually on imported food flavor in Nigeria with a projection of 15% annual future increase despite the fact that, traditionally produced flavoring products have lower caloric value and higher dietary protein content than imported flavor. This situation may be a result of the product's odour and product quality due to the poor manufacturing practice. Locust bean processing and marketing has been facing a lot of challenges despite the dawn of science and technology. Processing is still largely in home (family yard) in a crude way and handled by local women; the production has not increased substantially due to problems associated with production process; there is no standard measurement of grading of the condiment for sale so as to know whether the condiments are properly utilized for profit maximization. It was also hypothesized that no significant relationship exists within all the explanatory variables and the net revenue. It is therefore necessary to examine the factors affecting profitability level of locust bean in the study area.

Processing and preservation of fermented locust bean

All food processes are made up of a series of steps (sometimes called "unit operations") which have to be followed in a particular sequence in order to make the food. If the steps are changed, or even if their sequence is changed, the process will produce a different product. The production of locust bean condiment (*Iru*), is essentially a traditional family yard practice done in rural cottage industry mainly by woman. The harvested bean seeds mainly sold in markets to women who processed them. Many scientists had researched into the traditional method of processing locust bean; among which are- Odunfa, 1981; Campbel-platt, 1986; Oni, 1997. Oni (1997) describes the seven stages of processing operation which are: Shelling, Pre-drying, Pounding, Winnowing or Sieving, Washing, Drying and Visual sorting. Odunfa (1985) presented a flow chart (Fig. 1), showing the stages in traditional method of processing locust bean.

In the studies conducted on modern method of processing locust bean, it was noted that the boring routine practice in cooking time was reduce by use of pressure cooker which reduced the rigor 12 hours of boiling to 2 hours. Dehuller and separator – dual purpose equipment has drastically reduced the traditional method of production of between 4 days – 6 days to 4 hours; having production capacity of 1500 kg (Audu *et al.*, 2004). Table 2 shows the comparative proximal chemical composition of iru produced locally and sample from Federal Institute of Industrial Research Oshodi, (FIIRO) Lagos, Nigeria.

Africa locust bean seed
↓
Boil in water for 12 hrs
↓
Cooling and washing
↓
Boil cotyledons in water
↓
Drain through raffia sieve
↓
Spread while hot in wide calabash tray
↓
Stack strays together, wrap with jute bag
↓
Ferment for 36 hrs
↓
Add salt and mould into desire shape
↓
Sun dry

Fig. 1. Flow chart for traditional processing of locust bean fruits to food condiment (Odunfa *et al.*, 1985)

Table 2. Comparative proximal chemical composition of iru produced locally and sample from FIIRO

Components (%)	Unfermented locust bean	Fermented locust bean	FIIRO
Crude protein	30.0	44.2	45.3
Fat	15.0	28.4	31.2
Carbohydrate	4.9	16.4	15.9
Crude fiber	3.1	6.9	4.6
Ash	2.9	4.1	3.0

Source: Onnyi *et al.* (2004)

Fermentation is a process whereby beneficial bacteria are encouraged to grow. These bacteria increase the acidity or alcohol content of a food and therefore prevent the growth of spoilage and food poisoning bacteria (Traditional foods processing for profit). The use of an incubating material set up at optimum temperature of relatively $350^{o}C$ has been discovered to reduce the time and also have better sensory properties than the one fermented locally (Adewumi and Igbeke, 1992). The physical properties of the incubating material used to determine the rate of fermentation. The higher the porosity, the higher the rate of fermented organ and faster the rate of fermentation (Onnyi *et al.*, 2004).

Preservation is an important purpose in processing. Spoilage is caused by three factors: micro-organisms, chemical reaction and enzymes. Appropriate preservative technique is necessary for different product. Iru may be preserved by grinding in a thick paste before mould into ball or circular platelet followed by sun dry. Spices, additives such as salt are incorporated before molding the final product while sun drying is undertaken to facilitate its stabilization. Also, humectants, sugar, salt create unfavourable condition for micro organism and they retain nutrients to a desirable extent. Iru is very rich in protein and used as a meat substitute in many poor home of West Africa, since it proved meaty flavor in addition to stew and soup (Cambell-platt, 1986). The average daily intake in northern Nigeria is about 1.4% of the daily calories and 5% of the total protein and is the single largest source of protein in the average diet of inhabitant of Zaria in the Northern Nigeria (Odunfa, 1985); its decline in popularity especially among the growing urban population has led to rapid increase in the import of foreign cube flavor and its efficient preservation methods has led to the necessity to modernize production technique and optimize preservation method by addition of preservatives.

Materials and Methods

The study area

This study was carried out in Kwara state, which is one of the six states in north central Nigeria. The state lies in the north central zone and covers an area of 74256 sq km or of the total area of Nigeria. Kwara state is bounded in the north by Niger state, in the south by Oyo, Osun and Ekiti states, in the east by Kogi state and in the west by Benin Republic. Because of its unique geographical position, the state is referred to as the "gateway" between the north and the south of the country. The state is divided into sixteen (16) local government areas. Kwara state was selected for the study because, in Nigeria the locust bean tree plants thrives well within the Guinea and Sudan Savannah areas as well as the lower Sahel regions of the country with a rainfall requirement of 600–1500 mm annually. Agriculture is the main stay of the economy with prevailing agricultural system combined with bush fallow and mixed cropping with emphasis on subsistent farming, while some engage in craft activities such as weaving, blacksmithing, bricklaying, carpentry, welding, etc. Fishing is also prominent along the lower River Niger Basin. Kwara state is a heterogeneous state attracting different ethnic groups including the Yorubas, Nupes, Barubas, Fulanis and Hausas. The major ethnic groups in the state are the Yorubas and their language is widely spoken across the state.

Sampling procedure and sample size

A purposive sampling technique was used to select 60 respondents for the study from Zone C of Kwara state agricultural zones (comprises of 5 Local Government areas: Ilorin west, east and south, as well as Moro, and Asa LGAs) were chosen out of the other zone in the state; because they are predominant area of production and marketing activities of locust bean.

Research instrument

Majority of the respondents (processors and marketers) are illiterates; hence structural interview schedule was used to collect the information on personal and socio-economic characteristics, processing and marketing as well as cost and returns.

Data analysis

Regression analysis was used to estimate the relationship between selected socio-economic variables and the net revenue. This is an economic tool for predicting the value of dependent variable given the values of the independents variables and measures the degree of association between two or more variables; hence, the coefficient of determination (R^2) shows the level of variation in dependent variable (Y), which is explained by variation in X(s). The model for this analysis is given below as follow:

$$Y = f(X_1, X_2, X_3, X_4, X_5, X_6, X_7, X_8)$$

where Y = Estimated profit, X_1 = Age (yrs), X_2 = Marital status, X_3 = Stall rent, X_4 = Quantity processed, X_5 = Years of experience, X_6 = Transportation cost, and X_7 = Educational level, and X_8 = Family size.

Also, enterprise budgetary analytical approach was used to estimate cost and return in locust bean processing and marketing so as to be able to know the net profit of the processors/ marketers. According to Adegeye and Dittoh (1985), profit is defined as the net flow of income. Thus, how profit is measured depends on what measure chosen to be used; in essence, profit indicates whether a business is worthwhile or not. Benefit Cost Ratio and Rate of Return on Investment were used to measure the profitability and determine the worth-whileness of locust bean enterprises.

Profit = Total Revenue – Total Variable Cost
Mathematically, $\pi = TR - TFC + TVC$

Where π denote Profit; TR is Total Revenue (amount realized from the enterprise); TFC is Total Fixed Cost (expenditure incurred on fixed assets used in processing and marketing e.g. calabash, pot-sieve e.t.c) and TVC is the Total Variable Cost (cost incurred processing and marketing locust bean e. g. cost of raw materials, transportation cost).

The Gross Margin (GM) equation is given as: $GM = TR - TVC = P \times Q - TVC$, where: GM = Gross Margin (in Naira), Q = Quantity of locust beans processed (₦ / 1.7 kg "congos"), P = Price of processed locust bean (in Naira).

Benefit cost ratio (BCR): is another measure of profitability. It was used to elevate and confirm the profitability of each respondent of locust bean processors and marketers. The formula is stated as:

BCR = Total Revenue (Benefit) ÷ Total Cost.

Investment criteria require that BCR should be greater than one {BCR > 1} before a business can be termed profitable (Adegeye and Dittoh, 1985).

Rate of return on investment (ROI): is another measure used to determine the worth-whileness of a business. Rate of Return = (TR – TC) / TC; hence, the higher the rate of return, the more profitable the business is.

Results and Discussion

Demographic characteristics of the respondents

Table 3 reveals that 78.3% of the respondents are within the age range of 21-40 years who are considered to the young agile and active in production and marketing of locust beans therefore capable of the tasks involved in the enterprise. On the average the respondent were 30.6 years old. It also shows that most of the respondents were female (88.3%) while others were male (11.7%). The table further revealed that all the respondents had education at various levels, 43.3% had primary education, 25% had secondary education while 31.7% had Arabic education, and it shows that low level of education is responsible to some of challenges, innovation and other technology ought to put in place, which results to level of their productivity.

Table 3. Frequency and percentages distribution of respondents according to their socioeconomic characteristics

Socioeconomic characteristics	Frequency	Percentage
Age group (Years)		
<20	05	8.3
21-40	47	78.3
41-60	08	13.3
Sex		
Male	07	11.7
Female	53	88.3
Education level		
Primary	26	43.3
Secondary	15	25.0
Arabic school	19	31.7
Marital status		
Single	10	16.7
Married	45	75.0
Divorced	05	8.3
Family size		
1-5	32	53.3
6-10	26	43.3
11-15	02	3.3
Year of experience		
< 10	13	21.7
11-20	23	38.3
21-30	17	28.3
31-40	07	11.7
Total	60	100.0

Source: Field survey, 2011

It also showed that 75.0% of the respondents were married, 16.7% are single and 8.3% are divorced. In addition, the table revealed that 53.3% of the respondent had 1-5 dependents; about 43.3% of them had 6-10 dependents while 3.3% respondents had 11-15 dependents. It is obscured that larger percentages of them have reasonable family size. The table further shows that majority of the respondents 66.6% have spent between 11-30 years in production and marketing of locust bean while 11.7% spent above 30 years and the remaining 21.7% accounted for those respondents that have spent less than 10 years experience in the business.

Regression analysis and socio-economic characteristics of the respondents with their profit

Regression analysis result as presented in Table 4 shows that, the coefficients of quantity processed, family size and years of experience in the enterprises are positive and significant at 1%, 5% and 10% level of significance, respectively while the coefficient of education level and market stall rent is negative and significant at 1% and 5%, respectively. The implication of all these is that , the higher the quantity processed, the larger the net revenue obtained; then, the more the years processor/producers spend in business, the more the net revenue. Stall rent and education level is significant but has an inverse relationship with the net revenue as expected; all these are in conformity with a-priori expectation. The coefficient determination of 0.798 implies that the independent variables explained 78.9% of the total variation in the dependent variable which is the net revenue, while the remaining 21.1% is attributed to other factors not included in the model i.e. error term. The F-value was found to be significant at 1%; this implies that all the explanatory variables taking together have significant effect on the net revenue. Therefore, the null hypothesis is hereby rejected.

Table 4. Regression analysis result

Model	Variable	Beta	t-Test	Significance
	Constant	4.788	5.039	0.00
X_1	Age	0.248	1.039	3.03
X_2	Marital status	3.208	0.402	0.689
X_3	Stall rent	-4.6	-2.065**	0.509
X_4	Quantity processed	0.957	11.376***	0.000
X_5	Years of experience	0.174	1.761*	0.099
X_6	Transportation cost	-6.35	-0.067	0.34
X_7	Education level	-0.408	-1.720***	0.285
X_8	Family size	0.876	2.646**	0.332

a. Dependant Variable: Profit

n = 60
$R^2 = 0.803 = 80.3\%$
Adjusted $R^2 = 0.789 = 78.9\%$
F-value = 37.742
* -Statistically significant at 10% probability level
** -Statistically significant at 5% probability level
* ** -Statistically significant at 1% probability level

Cost and return analysis

The profitability of locust bean processing and marketing enterprise was assessed using cost and return analysis. Total cost of each respondent was computed and Total revenue was estimated to determine the profitability of the processed locust bean enterprise. From the result of the analysis, the annual total cost of 60 respondents of the locust bean processor and marketers was calculated (using straight-line depreciation method to compute the total fixed cost) and the total revenue was estimated to be:

π = TR – (TFC +TVC) = TR – TC = (4,534,180 - 2,468,026: 00) = ₦ 2,066,154: 00
GM = TR – TVC = (4,534,180 - 2,405,620: 10) = ₦ 2,128,559: 90

Investment analysis

In order to further test for the profitability of the marketers and processors in the enterprises, the benefit cost ratio analysis and rate of return to naira analysis were computed (Table 5). It could be observed from table that 68.3% of the processors and marketers operate above the breakeven point while only 5.0% operates below breakeven point and 26.7% operates at breakeven point, that is, they neither gain nor lost. This shows that the business is profitable since majority of the respondents are making profit.

Table 5. Benefit cost ratio

BCR	Frequency	Percentage
< 1	3	5.0
1	16	26.7
> 1	42	68.3
Total	60	100.0

Source: Field survey, 2011

The result in table 6 below shows that, larger percentage (48.3) of the respondents realized average profit of between 0.1 k – 0.5 k as profit on every ₦1 invested in processing and marketing and 31.7% realizes between ₦0.6 k – ₦1 on every ₦1 invested, while 15% realize more than ₦1 on every ₦1 invested and only 5% run the business at lost. This shows that locust bean enterprise is a worth-while business.

Table 6. Rate of return on investment

Return on investment	Frequency	Percentage
< 0	3	5.0
0.1 – 0.5	29	48.3
0.6 – 1.0	19	31.7
> 1	9	15.0
Total	60	100.0

Source: Field survey, 2011

Conclusion

Locust bean is discovered from the researches carried out on it to be very nutritious, because of its high protein content in it (Alabi *et al.*, 2005; Oladele *et al.*, 1995). The processing and marketing of locust bean forms an essential compound of the rural cottage industry as the processors and marketers consider the net revenue derived from the enterprise as being sustainable to cater for their immediate needs; because low investment brings about low returns. The production is low due to use of rudiment equipment, time consuming, laborious, packaging and high wood consumption. Hence, the result of the research revealed that locust beans processing and marketing enterprise in Ilorin, Kwara state, Nigeria is a small scale enterprise which contributes and provided reasonable income and employment to many household members in the study area. For instance, the annual income accruing to the people in the enterprise was ₦2, 066, 154: 00 (USD 12,913 equivalent) is not compared favourably with the National Minimum wage of N18, 000 (USD 113 equivalent). The result did not agrees with that of Breman and Kessler (1995); Tee *et al.* (2009), well it may due to inflation rate. Fundamentally, the profit from the enterprise activities revealed that the people in the business were living far below the poverty line of $11,130 per year for one person or $14, 218 per year for a couple and $17, 374 per year for a family of three (Christian Boyle and Larry Mcshame, 2011) compare with $ 370 per year in 1990's (World Bank, 1990; Hauser and Pilgram, 1999).

Recommendations

Based on the findings of the study, the following recommendations were made:

1. Make the newly discovered method of processing (de-hullers, separators and press cookers) machine developed known and adopt by the processors at affordable and subsidized price to increase production and eliminate stress associated with the crude method.
2. There is need for improved method of preservation, packaging and marketing for the product to be acceptable in international markets by reducing the odour of the product without the loss of essential nutrients and palability in order to generate foreign exchange.
3. The tree with its economic value is threatened in its alternate uses as fuel-wood and charcoal; hence this should be guide against with rules and regulations.
4. Effort should be directed towards growing more of the tree to increase production of the seeds.

References

Adegeye, A.J. and Dittoh, J.S. 1985. Essential of Agricultural Economics. Impact publisher, Nigeria Limited. pp. 210 -226.

Adewumi, B.A. and Igbeke, J. 1992. The effect of steaming on the physical and dehulling characteristic of locust bean (*Parkia biglobosa*). *J. Tropical Agric.* 70: 380-382

Aju, P.C., Iwuanyanwu, U.P., Popoola, L.A. and Uwalaka, R.E. 2008. An assessment of nutrition and commercial values of *Gnetum africana* in Imo state, Nigeria. *In:* J.C. Onyekwelu Adekunle and D.O. Oke (Edn) Proceedings of the First National Conference of the Forest and Forestry Products Society. 16th-18th April, At the Federal University of Technology, Akure. pp. 18-22.

Alabi, D.A., Akinsulire, O.R. and Sanyanolu, M.A. 2005. Qualitative determination of chemical and nutritional composition of *Parkia biglobosa* (Jacq.) Benth. August. *African J. Biotech.* 4 (8): 812-815.

Audu, I., Oloso, A.O. and Umar, B. 2004. Development of concentric cylinder locust bean dehuller. *Agril. Eng. Int. J. Sci. Res. and Dev.* 5: 201-205.

Beaumont, M. 2002. Flavoring composition prepared by fermentation with bacillus spices. *Int. J. Microbiol.* 75: 189-196.

Breman, H. and Kessler, J.J. 1995. Woody plants in agro-ecosystems of semi-arid regions: with an emphasis on the Sahelian Countries. Advanced Series in Agriculture 23 Springer Verlag. p. 342.

Cambell-platt, G., 1986. African locust bean (*Parkia* spp.) and its' west African fermented food product, 'dawadawa'. *Ecology Food Nutrition J.* 9: 123-132.

Christian Boyle and Larry Mcshame. 2011 "One in every six Americans living below poverty line, U.S. Census Bureau announces" (articles.nydailynews.com/2011-09-13).

Diawara, B., Sawadogo, L, Jacobson, M. and Awug, W.K. 2000. HACCP-System of traditional fermented food (sombala) capacity building for research and quality assurance and food fermentation technology for Africa fermented foods. *WAIRTO Journal.* 26: 11-62.

Fagbemi, T. 1989. Agro forestry potentials of *Parkia bioglobosa* (Jacq) in the savanna zone of Nigeria. 'Trees for development in Sub-Saharan Africa'. Proceeding of a regional seminar held by the International Foundation for Science (IFS), ICRAF House, Nairobi, Kenya. pp. 20-25.

Fagbemi, T. 2002. Investment opportunities in renewable resources Industry-forestry. 1st Edn. Belodan, Nigeria.

Hauser, F. and Pilgrim, K. 1999. Agriculture and poverty reduction: The Benin example. *Agricultural and Rural Development*. 2: 57-60.

Keay, R.W.J. 1989. Trees of Nigeria. Oxford University Press, New-York. pp. 476.

Musa, H.L. 1991. Ginger and locust bean tree: History, growth, use and potentials. Paper presented at Tuk Ham Symposium, Kurmin Musa, March 29.

National Bureau of Statistics. 2003. The Federal Republic of Nigeria, Anbuja, Nigeria, (http://www.nigerianstat.gov.ng/)

Odunfa, S.A. and Adewuyi, E. 1985. Optimization of process conditions for fermentation of Africa locust bean, effect of time, temperature humidity. *Food Chem. Microbiol.* 9: 118-121.

Odunfa, S.A. and Oyewole, O.B. 1981. Identification of *Bacillus* species from "Iru" a fermented African locust bean product. *J. Basic Microbiol.* 26: 101-108.

Oduro, I., Ellis, W.O. and Narh, S.T. 2007. Expanding breadfruit Utilization and its potentials for Pasta Production. *Discovery and Innovation.* 19: 243-247.

Oladele, F.A., Fawole, M.O. and Bhat, R.B. 1995. Leaf anatomy of *Parkia clappertonaire* Keay (Mimosaceae). *Korean J. Biol.* 28: 21-28.

Oni, P.I. 1997. *Parkia biglobosa* (Jacq Benth) in Nigeria. "A resource assessment". An unpublished Ph.D thesis; University of North Wales, Bangor, United Kingdom, 107 p.

Onnyi, O., Odediran, O.F. and Ajuebor, N. 2004. Federal Institute of Industrial Research, Oshodi (FIIRO), Lagos. *Nigeria Food Journal.* 22: 203-208.

Oyerinde, O.V. and Daramola, A.G. 2004. Socio-economic characteristic of African locust bean South-western Nigeria. *Pakistan J. Social Sci.* 2: 291-294.

Popoola, L. and Galaudu, M.S. 2000. Prioritization of indigenous spice-species for agroforestry in the semi-arid zone of Nigeria. *Bioprospector.* 2: 103-116.

Tee, T.N., Ogwuche, J.A. and Ikyaagba, E.T. 2009. The role of locust bean and ironwood trees in human nutrition and income in Nigeria. *Pakistan J. Nutrition.* 8(8): 1172-1177.

CTA's Publication. 1997. Traditional foods processing for profit edited by Peter Fellows. Intermediate Technology Publications Ltd, 103-105 Southamptom Row, London WC1B 4HH, UK.

World Bank, 1990. World Tables. World Bank, Washington DC, USA.

28

PERFORMANCE EVALUATION OF LASER GUIDED LEVELER

M.A. Hoque[1*] and M.A. Hannan[2]

Abstract

The study was conducted at Bangladesh Agricultural Research Institute (BARI) farm on clay loam soil during Rabi season of 2010-2011. The treatments consisted of laser land leveling (T_1) and control (non-leveled) (T_2). A preliminary field survey was done using staff gage. Initially a base station was established to dispense laser ray uniformly. The laser ray erected from base station guided the sensor of the stuff gage and the leveler. Elevation data was collected from the different points of the field and made an average. The maximum gage reading were 247.0 cm and the minimum gage reading was 219.2 cm. Average gage readings of the laser leveled plot was 235.66 cm that was settled for auto adjustment. Therefore, huge amount of soils (16.46 cm high) was cut from the highest point and subsequently had to fill to the low points. Finally, an equal gage reading of 235.66 cm was observed after leveling the plot. The laser leveler (Leica MLS700) was used hitching with a TAFE tractor. The field was leveled with manual control initially and finally it was operated with auto adjustment. Two operators, 25 litter diesels and total 6 hours time were required during this leveling. Wheat was cultivated in leveled land (T_1) and non-leveled land (T_2). Laser leveling was insured for improvement in nutrient use efficiencies, option for precision farming, reduces weed problems, and improves uniformity of crop maturity. There was better distribution of water in leveled plot, which helped to reduce irrigation application time 1 hour. Due to uniformity of moisture content improved germination and crop establishment was found which reflected in higher plant population (239 m^{-2}). Maximum yield (3.41 t ha^{-1}) was obtained in T_1 due to longer panicle (10.89 cm), more grain per plant (27.47) and 1000 grain weight (47.38 g) compared to yield of T_2 (2.62 t ha^{-1}).

Keywords: Performance, Laser Guided Leveler, Bangladesh

[1]Senior Scientific Officer, F M P Engineering Division, Bangladesh Agricultural Research Institute, Gazipur-1701, Bangladesh
[2]Senior Scientific Officer, Seed Technology Division, Bangladesh Agricultural Research Institute, Gazipur-1701, Bangladesh

*Corresponding author's email: arshadulfmpe@gmail.com, arshadul@bari.gov.bd (M.A. Hoque)

Introduction

The land leveling provides smoother soil surface, reduction in time and water required to irrigate the field, more uniform distribution of water in the field, more uniform moisture environment for crops, more uniform germination and growth of crops, reduction in seed weight, fertilizer, chemicals and fuel used in cultivation, and improved field traffic ability (for subsequent operations). Limitations of laser leveling include high cost of the equipment/laser instrument, the need for a skilled operator to set/adjust laser settings and operate the tractor, and restriction to regularly shaped fields. Usually laser leveling is done for better distribution of water, water savings, improvement in nutrient use efficiencies, option for precision farming, higher crop productivity, saves 25-30% of water, improves crop establishment and improves yield, reduces weed problems, improves uniformity of crop maturity, decreases the time to complete tasks, reduces the amount of water required for land preparation.

In 1980s, leveling of surface of the irrigated blocks became major practice in Soviet Central Asian states. The land leveling was applied as water saving method and have shown promising results, water use have been reduced by 1,500 m^3 ha^{-1} in leveled fields (Ahmedjanov, 1984; Ahmedjanov *et al.*, 1988; Balabanov, 1984). Review of existing literature on land leveling indicated positive impact on water saving, crop and farm productivity (Jonish, 1991; Clemmens *et al.*, 1995; Ren *et al.*, 2003; Mallappa and Radder, 1993; Cheema and Zulfiqar, 1998). It it is applied effectively, and leveling increases crop germination and yields and improves water distribution (Rickman, 2002).

The use of laser technology in the precision land leveling not only minimizes the cost of leveling but also ensures the desired degree of precision. Precision land leveling has been a significant component of on farm water management since 1977. Land leveling of farmer's field is an important process in the preparation of land. It enables efficient utilization of scarce water resources through elimination of unnecessary depressions and elevated contours (Muhammad Asif *et al.,* 2003). It has been noted that poor farm design and uneven fields are responsible for 30% water losses. About 18 million-acre feet (MAF) of water is lost to irrigate uneven fields in Pakistan (Gill, 1998). Salinity patches in the elevated parts and leaching down of soil nutrients from the root zone in lower spots of unleveled fields can attribute towards low crop production (Muhammad Asif *et al.,* 2003). Precision land leveling (PLL) facilitated application efficiency through even distribution of water and increased water-use efficiency that resulted in uniform seed germination, better crop growth and higher crop yield (Nazir, 1994). The scarcity of canal water supplies coupled with unfit underground water has compelled the farmers to utilize available water resources more wisely and efficiently. Under these circumstances, PLL can help the farmers to utilize the scarce land and water resource more effectively and efficiently towards increased crop production.

Conventional land leveling includes surveying of the field, staking and designing the field, calculation of cut and fills areas and then use of a scraper and land planer. Despite all these labor-intensive efforts, desired accuracy is not achieved. By contrast, laser leveling involves the use of a laser (transmitter), that emits a rapidly rotating beam parallel to the required field plane, which is picked up by a senor (receiving unit) fitted to a tractor towards scraper unit. The signal received is converted into cut and fill level adjustment and the corresponding changes in scraper level are carried out automatically by a hydraulic control system. The scraper guidance is fully automatic; the elements of operator error are removed allowing consistently accurate land leveling. The laser leveling makes the flow uniform and the advance flow is not much hindered because of less irregularity in the field's micro-topography. In a word, the laser controlled land leveling facilitates advance phase and consequently more uniformity is achieved.

Laser technology can ensure very accurate and precision land leveling to the extent of ± 2 cm (Waker, 1989). However, necessary data to support its effects on crop yield and water use are scarce. It was therefore, felt imperative to evaluate the performance of the laser guided leveler and its impact on yield of wheat.

Materials and Methods

The study was conducted at block no 6 of Bangladesh Agricultural Research Institute (BARI) field on clay loam soil during Rabi season of 2010-2011. The treatments consisted of Laser land leveling (T_1) and Control (Non-leveled) (T_2). In treatments T_1 and T_2, leveling of experimental field was done as per treatment and information on the topography of each experimental unit was compiled. The net plot size of each treatment measured was 60 X 22 m^2. A preliminary field survey was done using staff gage. Initially a base station was established to dispense laser ray uniformly as shown in Fig 1. It was a battery operated device, which covers a command area with 500 m radius. The laser ray erected from base station guides the sensor of the stuff gage and the leveler (Fig. 2). Elevation data was collected from the different points of the field as located in Fig. 3. From the elevation data, an average elevation was fixed for the leveler to operate automatically.

Fig. 1. Laser base station

Fig. 2. Sensors in stuff gage and the leveler

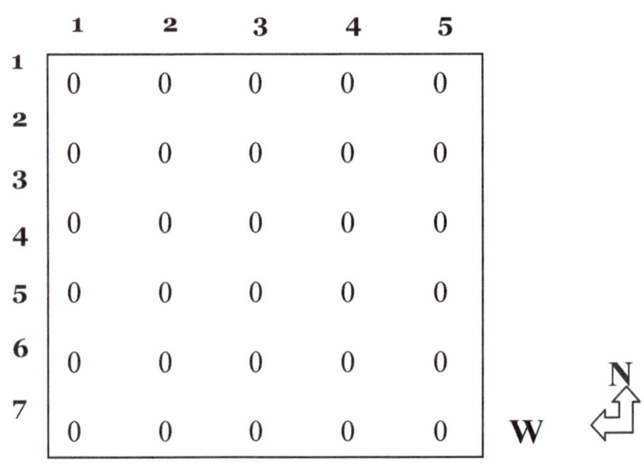

Fig. 3. Layout of elevation data collection points

The main source of irrigation was canal water, which was supplemented with tube well water as and when needed to meet the crop water requirements.

Wheat variety Bijoy was sown on December 22 by BARI Inclined plate seeder. A basal dose of NPK in the form of Urea and TSP @ 123-98-0 kg ha^{-1} were applied. Three irrigations were given till the maturity of the crop. The crop was kept free from weeds. Observations on the desired parameters were recorded using the standard procedures. The discharge available at outlet was measured every time. The time of irrigation application for each treatment was noted.

Results and Discussion

Laser leveler was used to level the field. Preliminary contour surveys were done and make an average to level. Fig. 4 shows the contour map and laser leveled line of the field from west to east respectively. Maximum gage reading indicate the lowest point of the field and minimum gage reading indicate the highest point of the field. From Fig. 4, it was observed that the maximum gage reading was 247 cm and the minimum gage reading was 219.2 cm. If whole plot would be leveled then the average gage reading should be 233.41 cm to set for auto adjustment. However, average gage reading of the laser leveled plot was 235.66 cm that was settled for auto adjustment. Therefore, huge amount soils (16.46 cm high) were cut from the highest point and subsequently have to fill to the low points. Finally, an equal gage reading of 235.66 cm was observed after leveling the plot.

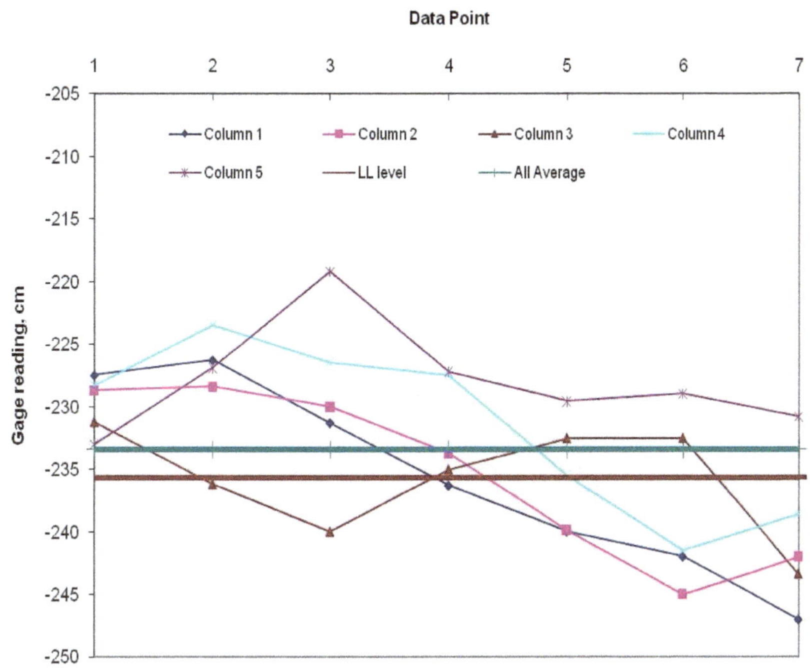

Fig. 4. The contour map and laser leveled line of the field.

The laser leveler was used hitching with a TAFE tractor. The field was leveled with manual control initially and finally it was operated with auto adjustment. Some operational parameters have shown in Table 1. For this experiment, it required 2 operators, 25 litter diesels and total 6 hours time. However, for others it will vary for the variation in land topography.

Table 1. Operational parameters of the test used in the experiment

Parameters	Values
Plot size (m^2)	1320
Time required for adjustment (hr)	1
Time required for leveling (hr)	5
Fuel consumption ($l\ hr^{-1}$)	5
No. of operators (no.)	2
Brand of the tractor	TAFE
Brand of the laser leveler	Leica MLS700
Length of the cutting unit (cm)	172
Forward speed ($km\ hr^{-1}$)	3.1

Laser leveling was insured improvement in nutrient use efficiencies, option for precision farming, reduces weed problems, improves uniformity of crop maturity, and decreased the time to complete tasks. Yield parameters and irrigating time are shown in Table 2. In treatment T_1, there was better distribution of water, which helps to reduce 1 hour irrigation application time. Due to uniformity of moisture content improved germination and crop establishment was found which reflected in higher plant population (239 m^{-2}). Longer panicle (10.89 cm), more grain per plant (27.47) and 1000-grain weight (47.38 g) insured maximum yield (3.41 t ha^{1}) in T_1 whereas in T_2 yield was 2.62 t ha^{-1}.

Table 2. Yield parameters and irrigating time

Treat ments	Plant height (cm)	Plant Population m^{-2}	Length of panicle (cm)	No. of grain/ Panicle	1000 grain wt (g)	Yield ($t\ ha^{-1}$)	Irrigation time (hr)
T_1	74.5	239	10.89	27.47	47.38	3.41	7
T_2	65.87	208.2	9.93	25.93	46.73	2.62	8

Conclusion

The maximum gage reading was 247 cm and the minimum gage reading was 219.2 cm. Average gages reading of the laser leveled plot was 235.66 cm that was settled for auto adjustment. So, a huge amount soil (16.46 cm high) was cut from the highest point and subsequently was to fill to the low points. Finally, an equal gage reading of 235.66 cm was observed after leveling the plot. Laser leveling was insured improvement in nutrient use efficiencies, option for precision farming, reduces weed problems, improves uniformity of crop maturity, and decreases the time to complete tasks. There was better distribution of water in leveled plot, which helps to reduce 1 hour irrigation application time. Due to uniformity of moisture content improved germination and crop establishment was found which reflected in higher plant population (239 m^{-2}). Longer panicle (10.89 cm), more grain per plant (27.47) and 1000 grain weight (47.38 g) insured maximum yield (3.41 t ha^{-1}) in T_1 whereas in T_2 yield was 2.62 t ha^{-1}.

References

Ahmedjanov, M. 1984. Application and investigation of the laser leveling in the field level. Fan, Tashkent, p. 144 (in Russian).

Ahmedjanov, M., Manabaev, A.N.T. and Teshabaev, A.Z. 1988. Application of laser for land leveling. *J. Mechanized Cotton Prod.* 12: 4-6 (in Russian).

Balabanov, S.V. 1984. The application of laser equipment for land leveling in the irrigated agriculture. Geodesy and photogram in the ameliorative construction. Novocherkassk. pp. 54–68 (in Russian).

Cheema, S. and Zulfiqar, M. 1998. The impact of laser land leveling on crop production in district Faisalabad. *Pakistan J. WM.* 1: 13-26.

Clemmens, A.J., Dedrick, A.R., Pereira, L.S., Sousa, P.L. and Pereira, L.S. 1995. Effect of furrow elevation differences on level-basin performance. *J. American Soc. Agric. Eng.* 38 (1): 153-158.

Gill, M. 1998. Water for life. Proceeding of seminar on world food day. Directorate of Agriculture Information. Dept. Agric. Govt. Punjab, India. pp. 9-13.

Jonish, J.E. 1991. Benefits and costs of laser land leveling Egypt. *In:* Bishay A, Dregne H (eds) Desert development Part 1: desert agriculture, ecology and biology: Proceedings of the Second International Desert Development Conference held on 25-31 January 1987 in Cairo, Egypt. vol. 4. pp. 6-9.

Mallappa, M., Radder, G.D. 1993. Evaluation of different proportion of leveling in Zingg conservation terraces on soil moisture conservation, crop growth and yield of green gram. *Mysore J. Agric. Sci.* 6 (3): 241-246.

Muhammad Asif, Manzoor Ahmed, Abdul Gafoor and Zhahoor Aslam. 2003. Wheat productivity, land and water use efficiency by traditional and laser land –leveling techniques. *Online J. Biol. Sci.* 3 (2): 141-146.

Nazir, M.S. 1994. Crop production. Crop water requirement and irrigation system. National Book Foundation. pp. 63-67.

Ren, W.T., Hu, Z.F., Cui, H.G., Yang, C.T., Liu, Y., Wang, Y.J., Zhang, Z.Y.and Li, B.F. 2003. Effect of laser-controlled land leveling and baby rice seedling direct planting on saving water. *Trans. China Soc. Agric. Eng.* 19 (3): 72-75.

Rickman, J. 2002. Land leveling. International Rice Research Institute. Philipine. p. 29.

Waker, W.R. 1989. Guidelines for design and evaluation surface irrigation system. FAO, Irigation and Drainage Paper no. 45 FAO, Rome, Italy.

STUDY ON EXISTING NUTRITIONAL CONDITION OF PERI-URBAN DAIRIES AT CHITTAGONG, BANGLADESH

M. Hasanuzzaman[1]*, M. Ahaduzzaman[2] and M.R. Begum[3]

Abstract

The study was aimed to observe the nutritive value of concentrate used in different dairy farms at peri-urban areas of Chittagong, Bangladesh. Ten (10) different concentrate feeds were randomly selected from ten different farms which analyzed for moisture, dry matter (DM), crude protein (CP), crude fiber (CF), nitrogen free extracts (NFE), ether extracts (EE), total ash (TA) and acid insoluble ash (AIA) and requirements of concentrate for unit production of milk was calculated. It was found that the nutritive value of individual feed concentrate of each farm varied significantly. The observed proportion of using feed ingredients was also significantly different among the farms. Finally, it may be summarized that the concentrate mixture used by the different farmers of peri-urban areas of Chittagong was different in regards to chemical composition, ingredients and amount required for milk production.

Keywords: Peri-urban Dairies, Concentrate Feed, Proximate, Milk Production

[1]Department of Animal Science and Nutrition, Chittagong Veterinary and Animal Sciences University, Chittagong, Bangladesh
[2]Department of Medicine and Surgery, Chittagong Veterinary and Animal Sciences University, Chittagong, Bangladesh
[3]Department of Agricultural Economics and Social Science, Chittagong Veterinary and Animal Sciences University, Chittagong, Bangladesh

*Corresponding author's email: anikzaman@yahoo.com (M. Hasanuzzaman)

Introduction

In Bangladesh, more than 80 per cent of people are involved in rearing nondescript indigenous Zebu-type animals (Khan, 2006) with an average live weight of 150 kg for cows which is 25-30 per cent less than that of Indian Zebu cattle (Jackson, 1981). This poor physical condition and low reproductive performance are mostly due to consumption of insufficient and imbalanced feed (Alam *et al.*, 2009) as feed provides the basic nutrients required for production (FAO, 1983). The concentrate mixture used in Bangladesh mainly composed of wheat bran, rice polish, split red lentil bran, split green gram bran, gram chuni, oil cakes, soybean meal, wheat flour etc. Maize is now somewhat available due to an increase in production in recent years. (Khaleduzzaman and Khandaker, 2009). Therefore, farmers can supply dairy cows with crushed maize in addition to rice polish, wheat bran and oil cakes (Kamal *et al.*, 2009). A method for the quantitative analysis of macronutrients is the Weende or proximate analysis, based on the Weende analysis that was developed in 1860 by Henneberg and Stohmann in Germany (Henneberg and Stohman, 1860). Some researchers estimated the different feed ingredients and evaluated the effect of concentrates requirement on milk yield in cows. Sen and Ray (1971) found 10.55, 1.90, 4.39, 80.66 and 1.94 per cent for crude protein (CP), crude fiber (CF), ether extract (EE), nitrogen free extract (NFE) and total ash (TA) in maize respectively. Shama (2011) reported the CP, CF, EE and TA to be 38.50, 5.40, 2.10 and 7.54 per cent, respectively in soya flakes. Chinedu and Nwinyi (2012) estimated that moisture, protein, fat, ash, crude fiber and total carbohydrates of Bambara groundnut (*Voadzeia subterranean*) were 2.86±0.02%, 32.40±0.02%, 7.35± 0.02%, 5.78±0.02%, 2.68±0.02% and 51.78±0.02% where its amount in African yam beans (*Sphenostylis stenocarpa*) were 1.96± 0.02%, 37.21±0.02%, 9.49±0.02%, 5.35±0.02%, 3.55±0.02% and 44.4±0.02%, respectively. The effect of feeding maize-based concentrates on milk yield in cows with cost-benefit analysis was done by Kamal *et al.* (2009) at smallholder farms in four districts of northern Bangladesh.

There is scant information regarding the proportion of ingredients in concentrate feeds and nutritional demand of cows for the milk production at peri-urban level. Many feed manufacturing companies are preparing concentrate feed mixtures and supplying large dairy farmers, but smallholders are unable to buy because of its high prices (Kamal *et al.*, 2009). So, it is very important to follow the proper feedingmanagement of dairy cows including the use of ingredients of optimum nutritive value and standard proportion for better outcome. Therefore, the present study was carried out to determine nutritive value of concentrate feed using in different farms and amount of concentrate feed offered in relation to milk production.

Materials and Methods

***Selection of farms*:** Ten dairy farms were selected from peri-urban areas of Chittagong district based on good transportation facilities from where ten concentrate mixture samples, one from each farm, were collected.

***Sample collection*:** Approximately 500 grams of sample was collected from each farm, wrapped up by polythene bag and preserved in the laboratory for chemical analysis. Samples were subjected to grind to make it homogenous powder. Later on, it was mixed properly and exposed to shade to cool down for better sampling.

***Sample analysis*:** The experimental samples were subjected to proximate analysis for moisture, dry matter (DM), crude protein (CP), crude fiber (CF), ether extracts (EE), total ash (TA), nitrogen free extract (NFE) and acid insoluble ash (AIA) following the method described by the Association of Official Analytical Chemist (AOAC, 2005) where all analysis were done in triplicates.

***Data analysis*:** Descriptive statistics were used to determine different concentration of feed ratio of chemical composition in different farms after that one sample t-test was used to determine any significant difference in different concentration of feed of chemical analysis by using in different farms. Percentage distribution was used to determine ingredients used in different farms and χ^2 test were used to test of significance of using observed proportion of different ingredients. Statistical packages SPSS 16.0, R 2.14.0 was used during analysis and 5% level of significance was considered.

Results and Discussion

Chemical composition of different dairy farms concentration of Mean±SE was given in Table 1. It was observed that maximum 8.70% moisture was found in F2 followed by F4 makes up 8.50% whereas minimum 7.20% in F6, and dry matter ranges from 91.30% in F2 to 93.40% in F7. In case of crude protein observed highest 14.70% in F6 followed by 14.35% in F1 whereas only 7.18% in F3 and crude fiber of F2 was maximum 18.0% and minimum 9.50% in F10. The highest and lowest percent values of EE, TA, NFE and AIA were 7.10 and 2.00 in F10 and F1, 10.92 and 4.23 in F8 and F1, 67.87 and 48.65 in F3 and F6, and 7.93 and 2.63 in F7 and F3, respectively. This varying quality may be due to difference in production, preservation and storage of the ingredients. A graphical plot was given in Fig. 1 regarding to different nutritive value ration.

Table 1. Chemical composition of different dairy farms concentrate feed rations (%DMB)

Farm	Moisture	DM	CP	CF	EE	TA	NFE	AIA
F1	7.20±0.04	92.80±0.16	14.35±0.15	17.5±0.05	2.00±0.57	4.23±0.01	61.92±1.70	3.31±0.07
F2	8.70±0.32	91.30±0.17	13.82±0.23	18.0±0.57	5.00±0.32	5.95±0.09	57.23±0.99	5.20±0.05
F3	7.00±0.33	93.00±0.07	7.18±0.58	11.5±0.05	4.90±0.66	8.55±0.23	67.87±0.54	2.63±0.90
F4	8.50±0.60	91.50±0.00	11.72±0.33	13.5±0.05	2.90±0.59	8.78±0.75	63.10±1.11	4.05±0.55
F5	7.50±0.04	92.50±0.01	11.90±0.58	16.0±0.73	5.00±1.34	9.89±0.88	57.21±0.32	2.75±0.00
F6	7.20±0.04	92.80±0.23	14.70±0.84	14.5±0.48	6.70±0.97	15.45±1.98	48.65±2.02	3.37±0.32
F7	6.60±0.37	93.40±0.32	12.25±1.34	12.0±0.10	4.10±0.02	10.68±0.77	60.97±0.72	7.93±0.98
F8	7.40±0.19	92.60±0.34	12.30±0.67	13.0±2.66	5.10±0.67	10.92±0.03	58.68±1.03	4.51±0.56
F9	7.40±1.65	92.60±0.54	11.38±0.65	12.5±0.10	6.90±0.08	6.51±0.65	62.71±0.04	5.17±0.74
F10	6.80±0.05	93.20±0.78	12.60±0.76	9.50±0.76	7.10±0.77	6.77±0.44	64.03±0.07	5.56±0.41

Table 2 represents the single mean t test of significance of chemical composition of different dairy farms where every concentration mean of nutritive value was significantly different from zero in every dairy farm. Anderson-Darling (AD) for normality test proved that the data were sampled was normally distributed (all p-values<0.05) and graphs of normal probability plot was shown in Fig. 2, in every graph the points roughly follow the fitted line. An index plot was portrayed in Fig.3, all graphs didn't exhibit obvious trend or pattern indicate that the samples were independent.

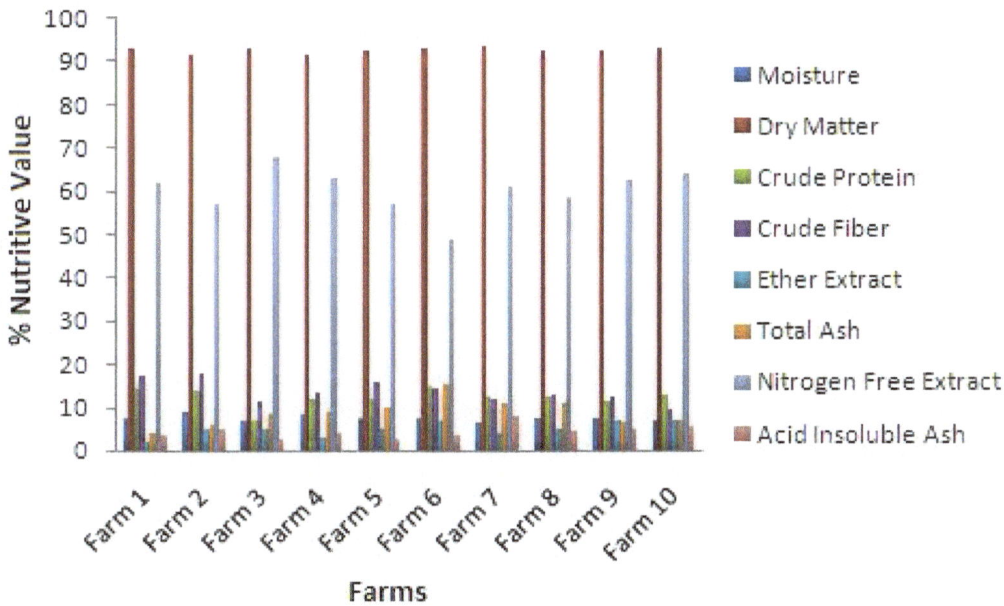

Fig 1. Percent of nutritive value of concentration using in different farms

Table 1. Test of significance of single mean of chemical composition of different dairy farms

Concentration	Mean	SD	SE	95% CI	T	P	AD (P-value)
Moisture	7.43	0.68	0.21	(6.94, 7.92)	34.64	<0.01	0.086
Dry matter	92.57	0.68	0.21	(92.08, 93.06)	431.56	<0.01	0.086
Crude Protein	12.22	2.10	0.67	(10.71, 13.73)	18.36	<0.01	0.079
Crude Fiber	13.8	2.71	0.86	(11.86, 15.74)	16.1	<0.01	0.789
Ether Extract	4.97	1.67	0.53	(3.77, 6.17)	9.4	<0.01	0.353
Nitrogen Free Extract	60.24	5.22	1.65	(56.50, 63.97)	36.49	<0.01	0.336
Total Ash	8.77	3.19	1.01	(6.49, 11.06)	8.68	<0.01	0.670
Acid Insoluble Ash	4.45	1.61	0.51	(3.30, 5.60)	8.75	<0.01	0.417

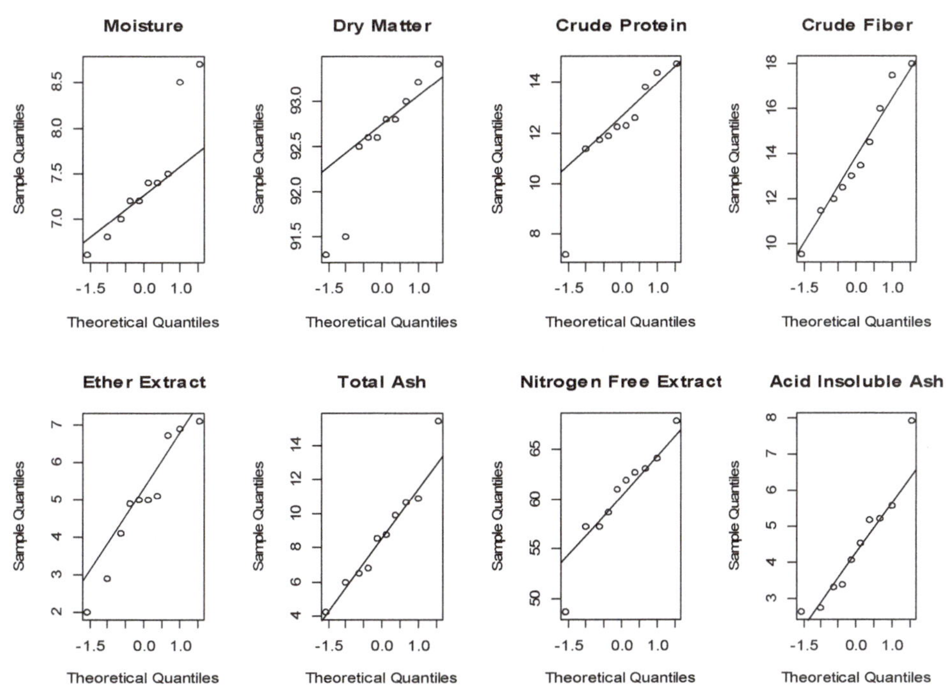

Fig. 2. Normal Probability Plot (Q-Q plot) of different concentration in different dairy farms

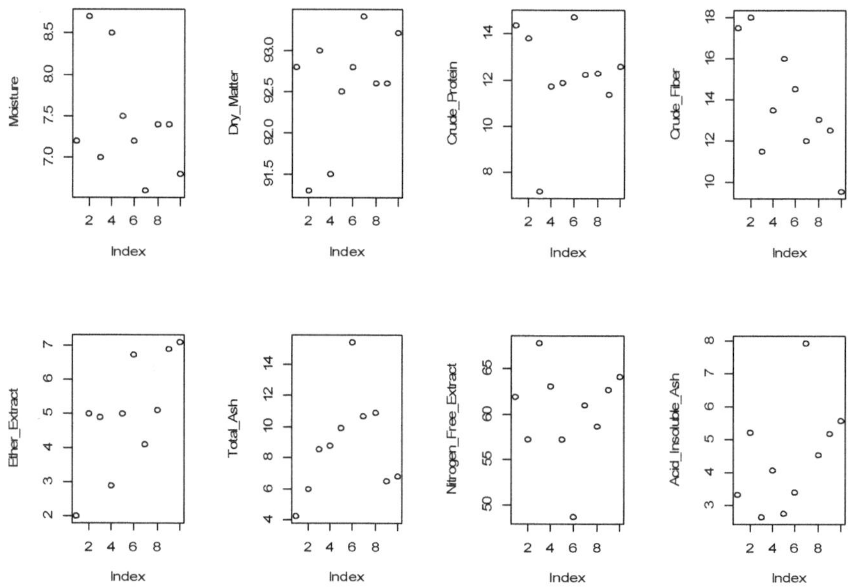

Fig. 3. Time series plot of different concentration of different farms against index

Table 3 represents distribution of dairy farms based on using different ingredients in the feed where 100, 90, 80, 60, 40, 20 and 10 percent farm used wheat bran; rice polish and mug powder; mosur hulls and maize; pea bran and oil cake; chirakura and soyabean; gram chuni, broken pea and vitamin mineral premix (VMP); and jaur, respectively. A chi-square (χ^2) test was used to examine the equality of observed proportions for using number ingredients in different farms. The result observed that there were significant difference among the observed proportion for using ingredients (Chi-square, χ^2 = 58.78; p-value<0.01). The difference is due to the availability, economic and production status of the farm.

Table 3. Percentage distribution of farms regarding to use ingredients

Ingredients	Wheat bran	Pea bran	Rice polish	Jaur	Mosur hulls	Gram chuni	Broken maize	Broken pea	Chira Kura	Mug powder	Oil cake	Flour	Soybean	V.M.P.
Used	10	6	9	1	8	2	8	2	4	9	6	5	4	2
Not used	0	4	1	9	2	8	2	8	6	1	4	5	6	8
Percent	100	60	90	10	80	20	80	20	40	90	60	50	40	20

Table 4 provides the amount of concentrate feed offered in individual farm concerning milk production, which was similar in F1, F2, F5, F9 and F10 as they offered 1.0 kg concentrate for 1kg milk. Farm F6 provided 1.0 kg concentrate feed for 1.3 kg of milk production whereas for 1.0 kg concentrates the output was 1.4.1.8 and 2.5 kg for F7 and F8, F3 and F4, respectively. The amount of feed offered in relation to milk production varied may be due to difference in nutritional values of feed ingredients in ration.

Table 4. Amount of concentrate feed offered in individual farm in regard to milk production

Farms	F1	F2	F3	F4	F5	F6	F7	F8	F9	F10
Milk : Concentrate feed	1:1	1:1	1.8:1	2.5:1	1:1	1.3:1	1.4:1	1.4:1	1:1	1:1

Conclusion

From this study, it was evident that nutritive status of different farms in the study area varied in composition, ingredients of feed and their amount required for unit milk production, which hampered the productive performances of cows.

Acknowledgement

The authors heartily acknowledge the positive help and endless support of CVASU.

References

AOAC. 2005. Official Methods of Analysis. 18th edition, Association of Official Analytical Chemist, Washington DC, USA. pp: 868-278.

Alam, M.G.S., Rahman, M.A., Khatun, M. and Ahmed, J.U. 2009. Feed supplementation and weight change, milk yield and post-partum estrus in Desi cows. *The Bangladesh Veterinarian.* 26 (2): 39-47.

Chinedu, S.N. and Nwinyi, C.O. 2012. Proximate analysis of *Sphenostylis stenocarpa* and *Voadzeia subterranean* consumed in South-Eastern Nigeria. *J. Agril. Ext. & Rural Dev.* 4 (3): 57-62.

FAO. 1983. Interactions between Livestock Production Systems and the Environment - Impact*(www.fao.org/wairdocs/lead/x6123e/x6123e04.htm).*

Henneberg, W. and Stohman, F. 1860. Begründung einer rationellen Ftterung der Wiederkäuer. Vol. 1. Schwetschtke u. Sohn, Braunschweig, Germany.

Jackson, M.G. 1981. Evolving a strategy for maximum livestock production on minimum lands. p. 15. In: Proc. of Seminar held in Bangladesh Agricultural University, Mymensingh, February 22-24, Bangladesh.

Kamal, M.M., Iqbal, D.M.H. and Khaleduzzaman, A.B.M. 2009. Supplementation of maize-based concentrates and milk production in indigenous cows. *The Bangladesh Veterinarian.* 26 (2): 48-53.

Khaleduzzaman, A.B.M. and Khandaker, Z.H. 2009. Commercial feed production and quality control: Present status and future prospects in Bangladesh. pp. 12-15. *In:* Proc. of the 6th International Poultry Show and Seminar, WPSA-BB, Dhaka, Bangladesh.

Khan, M.S. 2006. Community Extension Workers Training Guide, Participatory Livestock Development Project (PLDP), Department of Livestock Services, Dhaka, Bangladesh. pp. 11-12.

Sen, K.C. and Ray, S.N. 1971. Nutritive values of Indian cattle feeds and the feeding of farm animals. ICAR, New Delhi. pp. 67-69.

Shama, S. 2011. Effects of probiotics suppleme.ntation on the utilization of Seabuckthorn cake in broiler production. MVSc Thesis. Department of Animal Nutrition, CSK Himachal Pradesh Krishi Vishvavidyalaya, Palampur, HP, India. pp. 15-22.

AGED LEAVES EFFECT ON ESSENTIAL COMPONENTS IN GREEN AND OOLONG TEA

M.M. Rahman[1*], M.A. Kalam[1], M.A. Salam[2] and M.R. Rana[1]

Abstract

Tea leaf encompasses essential components like caffeine, polyphenol, lipid etc. The study was conducted to find out the essential constituents in green and oolong tea from aged tea leaves during the year of 2012. Tea leaves with different plucked aged were collected from Lackatoorah Tea Estate at sylhet district in Bangladesh. Collected leaves were processesed to make desirable samples for biochemical analysis in the laboratory of Department of Food Enginnering and Tea Technology, SUST. Results revealed that the essential constituents like moisture, caffeine, polyphenol, lipid, protein, ash, ascorbic acid, acidity and pH value in green tea made from different aged leaves were found slightly higher than oolong tea, ranged from 6.38±1.06 to 3.49±0.59%; 4.91±0.82 to 1.49±0.24%; 30.88±5.15 to 18.23±3.04%; 7.50±1.25 to 9.58±1.59%; 13.15±2.19 to 17.33±2.88%; 3.87±1.65 to 7.86±1.31%; 48.4±8.05 to 21.3±3.55(mg); 2.13±0.68 to 1.18±0.19% and 5.52±0.11 to 5.97±0.18%, respectively. Similarly, the moisture, caffeine, polyphenol, lipid, protein, ash, ascorbic acid, acidity and pH value from different aged leaves were found in oolong tea from 6.19±1.04 to 2.98±0.49%; 4.68±0.78 to 1.11±0.19%; 20.89±3.48 to 8.23±1.37%; 6.40±1.07 to 9.13±1.52%; 13.03±2.17 to 17.19±2.86%; 3.44±0.58 to 7.57±1.27% 6.44±1.08 to 0.98±0.17 (mg); 2.02±0.34to 1.02±0.17% and 5.53±0.11 to 5.97±0.20%, respectively. Therefore, young tea leaves (i.e. 5 to 8 days tea leaves) should be plucked for considering useful constituents in processed green tea and oolong tea.

Keywords: Tea Leaves, Components, UV-Spectrophotometer, Green Tea, Oolong Tea

[1]Department of Food Engineering and Tea Technology, Shahjalal University of Science and Technology, Sylhet-3114, Bangladesh
[2]Department of Chemical Engineering and Polymer Science, Shahjalal University of Science and Technology, Sylhet-3114, Bangladesh

*Corresponding author's email: bau_mintu@yahoo.com (M.M. Rahman)

Introduction

Tea (*Camellia sinensis* L.) is the oldest and cheapest health beverage in the world next to water which is mostly grown in the sub-tropical regions in Asia and occupies about 2.7 million hectares of cultivable area in the world (Chaudhury, 1989; Mandel and Youdim, 2004). In presence, there are almost 30 countries producing 70% black tea, 25% green tea and the remaining 5% consisted of oolong tea (FAO, 2010). Like black tea, green and oolong tea are widely made in Asian countries to consume as a beverage and has been familiar in China and Japan from centuries (Zaveri, 2006). Green tea, called non-fermented tea has a more subtle, delicate flavour, and far less caffeine than fermented tea, is medicinally beneficial because the non-fermented leaves retain a higher concentration of natural vitamins and polyphenols than fermented counterparts. Oolong tea is a semi-fermented tea especially good for digestion, is advised to take after a large meal. The chemical compositions counted in semi-fermented oolong tea are in the ranges of non-fermented green tea to fully fermented black tea. The optimal matured tea leaves are plucked for processing into green and oolong tea at an interval in order to maintain better quality and essential constituents in tea.

The chemical constituents in green and oolong tea differ due to plucking of different days tea leaves and leaves processing system. Better plucked leaves provide desirable nutrient profiles like caffeine, polyphenol, lipid, protein etc., that are responsible for maintaining better human health. The chemical composition in tea deals with the changes of biochemical occurs during the seed germination, growth of the plants, manufacturing of tea and ends up with the pharmaceutical uses for drinking tea as a beverage (Choudhury, 2010). Tea caffeine is a crucial factor contains 1 to 5 % of its dry weight (Amra *et al.*, 2006) depending on types, brand (Bennett and Bonnie, 2001) and brewing method (Hicks *et al.*, 1996). Tea leaves belong polyphenol group figured out for 25 to 35% on a dry weight basis (Balentine, 1997; Hara *et al.*, 1995). Anesini *et al.* (2008) reported the total ployphenol content in green tea that varies from 14.32±0.45

to 21.02±1.54%. The value of lipid content in tea differs due to the variety and climatic condition. Lipid presents in tea up to 7-10 percent by weight may or not have any significant role on the quality in the manufacturing process (TRIC, 1948). Green tea leaves on dry weight basis contains 15-20% protein (Cabrera *et al.*, 2006) whereas crude protein (CP) is counted 18.15% (on the dry air basis) in the tested green tea leaves. The moisture content is a vital factor, which should be 3 to 7 percent in processed tea (Harler, 1964). The mean moisture content in all tea samples was 6.8%, ranged from 5.6-7.5% with Mambila and Srilanka tea having the lowest contents of 5.6 and 6.5%, respectively (Mohammed and Suliman, 2009). The ash content analysis has ranged from 4.9% - 7.2% in Kenya and Mambila samples contained the high ash content of 7.2% while Srilanka had the lowest of 4.9%. This high ash content suggests that it is a good source of minerals. Ascorbic acid is greater in the younger shoots than mature tea leaves but lesser in shaded shoots in comparison with that of un-shaded ones. Black and oolong teas have lesser amount of ascorbic acid than green tea (Takayanagi, 1977). pH is the most responsible factor for the production of pigmented compounds- theaflavins and thearubigins having specific optimum pH at 5.0 and 6.0 respectively (Nestle, 1966). Choudhury (2001) reported that green tea leaves having a pH value of 5.06-5.80, moisture content 78-80 and 20-25 percents of dry matter content respectively.

The quality of tea depends on leaf processing and manufacturing methods. Tea component may vary in the different aged plucked tea leaves. Normally according to the leaf plucking cycle, leaves are plucked after an interval to retain essential nutrient profiles in made tea. In the tea growing developing countries like Bangladesh does not follow up this properly due to labor shortages and for this reason leaf gets more time to become mature and provides poor nutrient contents in processed tea. As there is no absolute measurement, yet how nutrient profile changes in oolong and green tea from aged tea leaves. In view of the above discussions, the experiment has been carried out to analyze the effect of aged tea leaves on essential components in green and oolong tea in Bangladesh.

Materials and Methods

Samples collection and preparation

The different aged tea leaves *(Camellia sinensis)* were collected from Lackatoorah Tea Estate, Sylhet, Bangladesh to analyze the biochemical compounds in the Department of Food Engineering and Tea Technology, Shahjalal University of Science and Technology (SUST), Sylhet, Bangladesh during the period of May-July, 2012. Tea leaves were divided into two parts in order for producing the green tea and oolong tea. The processing system of green tea and oolong tea were outlined below:

a) Manufacturing process of green tea

Different aged plucked tea leaves were manually steamed to retain greenish colour for 45-60 second and then dried in sunlight, when leaves were fully dried, crushed or grinded to make a powder form for further analysis in the lab. Then each sample was preserved separately in a small plastic pot with the air tight condition to protect this from micro-organisms. As it is processed without fermentation, the sample colour was almost green; no typical aroma was developed and liquor has not trace of red or brown coloration.

b) Manufacturing process of oolong tea

After plucking, aged tea leaves were spread thinly on the flat bamboo basket for withering under direct sun for 30-60 minutes. The withering process time varies depending on the availability of sunlight. After then, tea leaves were transferred to a floor with the shady condition for further withering for 6-8 hours and agitated by hand one hour interval. In this process, the border of tea leaves became red and a strong aroma evaporates with the decreasing of moisture content gradually. The withered leaves were then dried to inactivate the enzymes for fermentation. The partially dried leaves were crushed and re-dried under sunlight to receive oolong tea having moisture content below 7 percent and stored in an air-tight plastic pot for further biochemical analysis.

Determination of essential components in green and oolong tea

Moisture content for each tea sample was determined using an oven set at 102° C and dried to constant weight (AOAC, 1984). The caffeine was determined by separating from the sample by refluxed and filtered. The filtered is then extracted with chloroform as solvent which removed by evaporation and weigh the caffeine (AACC, 2000). The total polyphenol content (TPC) was determined by spectrophotometer (T-60UV-Spectrophotometer) using Gallic acid as standard, according to the method described by the International Organization for Standardization (ISO, 2005). Lipid content in the samples was determined using soxhlet extraction method (AOAC, 2000). Protein content was analyzed using Kjeldhal digestion method followed by distillation and titration (AOAC, 2000). Total ash was determined using Muffle furnace set at 550°C for 3 hrs (AACC, 2000). pH of the tea infusion was determined by using portable pH meter according to method AOAC (1990). The amount of ascorbic acid was measured by preparing iodine solution and then titrating. Total acidity of made tea sample was determined by titrating samples against 0.1N sodium hydroxide solution to persistent pink color following protocols of AOAC (1990).

Results and Discussion

Aged tea leaves effect on essential components in green tea and oolong tea

Results revealed the effect of aged tea leaves on essential components in green and oolong tea, which are presented in Table 1 and Table 2. The values of moisture content decreased gradually from 5 to 12 days tea leaves (i.e., young to mature leaves) which had in the range of 6.38±1.06 to 3.49±0.59% at green tea and 6.19±1.04 to 2.98±0.49% at oolong tea, is similar with the finding of Harler (1964). The variation in the moisture may be attributed to the degree of drying type and nature of tea involved (Kumar *et al.*, 2005).

The caffeine content in green and oolong tea samples analyzed varied from 4.91±0.82 to 1.49±0.24% and 4.68±0.78 to 1.11±0.19% at 5 days to 12 days tea leaves, which is similar with the reporting of Amra *et al.* (2006) who found 1 to 5% of its dry weight caffeine contain in various tea. The result of polyphenol content in Table 1 showed that green tea from 5-12 days tea leaves had the highest polyphenol content of 30.88±5.15% and 18.23±3.04% while oolong tea was the lowest of 20.89±3.48% and 8.23±1.37% in Table 2, is compared with the finding of Anesini *et al.* (2008) and Balentine *et al.* (1997). Polyphenol content is always lower in the first leaf harvested after pruning than that of subsequent flushes and applicable for tea grown under shade (Eden, 1960).

The values of lipid content were gradually increased in green and oolong tea from 5-12 days tea leaves which ranged from 7.50±1.25 to 9.58±1.59% and 6.40±1.07 to 9.13±1.52%, which is significant with the result of Hara *et al.* (1995) found 7-10% lipid in various tea. Protein contents from 5-12 days tea samples are below as shown in Table 1 and 2, range of 13.15±2.19 to 17.33±2.88% in green tea and 13.03±2.17 to 17.19±2.86% in oolong tea, which is within the acceptable range of Cabrera *et al.* (2006). The ash content in tea is an indication of its mineral content; it does not necessarily indicate high quality except when there is a favorable balance of the essential minerals (Ahmed *et al.*, 1989). The results of the ash content analysis in green tea at 5-12 days tea leaves are outlined in Table 1 with ranged from 3.87±1.65 to 7.86±1.31% and in oolong tea from 3.44±0.58 to 7.57±1.27% in Table 2. These results were within the limit recommend for tea sample (Ahmed *et al.*, 1989).

Table 1. Aged leaves effects on essential components in green tea*

Sample Types	Essential components in green tea								
	Moisture%	Caffeine%	Polyphenol%	Lipid%	Protein%	Ash%	Ascorbic Acid, mg	Acidity%	PH %
5 days	6.38±1.06	4.91±0.82	30.88±5.15	7.50±1.25	13.15±2.19	3.87±1.65	48.4±8.05	2.13±0.68	5.52±0.11
6 days	5.98±0.82	4.22±0.70	28.08±4.68	7.95±1.33	13.65±2.29	4.54±0.76	41.7±6.95	2.01±0.34	5.55±0.12
7 days	5.23.±0.88	3.82±0.63	27.79±4.63	8.21±1.36	14.37±2.39	5.11.±0.86	34.7±5.75	1.87±0.32	5.61±0.21
8 days	4.89±0.82	3.34±0.55	26.02±4.33	8.75±1.45	15.05±2.50	5.58±0.93	32.6±5.45	1.82±0.30	5.72±0.23
9 days	4.47±0.75	2.83±0.47	24.11±4.02	8.99±1.22	15.34±2.55	6.02±1.00	30.2±5.05	1.67±0.28	5.78±0.09
10 days	4.14±0.56	2.34±0.39	22.50±3.75	9.11±1.52	15.77±2.62	6.43±1.08	27.7±4.65	1.53±0.26	5.86±0.14
11 days	3.82±0.64	1.88±0.31	20.44±3.40	9.34±1.56	16.23±2.70	7.14 ±1.19	25.6±4.25	1.38±0.23	5.92±0.25
12 days	3.49±0.59	1.49±0.24	18.23±3.04	9.58±1.59	17.33±2.88	7.86±1.31	21.3±3.55	1.18±0.19	5.97±0.18

**The values indicates Mean± Standard Deviation*

Table 2. Aged leaves effects on essential components in oolong tea*

Sample Types	Essential components in oolong tea								
	Moisture%	Caffeine%	Polyphenol%	Lipid %	Protein %	Ash%	Ascorbic acid, mg	Acidity %	PH %
5 days	6.19±1.04	4.68±0.78	20.89±3.48	6.40±1.07	13.03±2.17	3.44±0.58	6.44±1.08	2.02±0.34	5.53±0.11
6 days	5.97±0.99	3.92±0.65	19.26±3.21	6.85±1.15	13.54±2.25	3.86±0.65	5.09±0.85	1.91±0.31	5.50±0.12
7 days	5.07.±0.85	3.35±0.55	18.08±3.01	7.23±1.20	12.24.±2.35	4.53.±0.76	4.76±0.79	1.78±0.29	5.60±0.11
8 days	4.39±0.74	2.93±0.48	17.05±2.84	7.52±1.26	14.97±2.49	5.09±0.82	4.31±0.72	1.65±0.28	5.62±0.13
9 days	3.98±0.66	2.43±0.40	14.55±2.43	7.92±1.32	15.17±2.52	5.83±0.98	3.86±0.65	1.43±0.24	5.76±0.19
10 days	3.53±0.59	1.99±0.34	12.79±2.13	8.31±1.38	15.65±2.60	6.22±1.04	2.34±0.40	1.27±0.22	5.83±0.15
11 days	3.14 ±0.53	1.48±0.25	9.70±1.61	8.86±1.47	16.13±2.68	7.05 ±1.18	1.32±0.22	1.13±0.18	5.93±0.17
12 days	2.98±0.49	1.11±0.19	8.23±1.37	9.13±1.52	17.19±2.86	7.57±1.27	0.98±0.17	1.02±0.17	5.97±0.20

**The values indicates Mean± Standard Deviation*

The ascorbic acid in green tea from 5-12 days tea leaves was found 48.4±8.05 to 21.3±3.55 mg while oolong tea was 6.44±1.08 to 0.98±0.17 (mg) in Table 2. Nagao *et al.* (2005) reported that green tea leaves contained 17.5 mg kg^{-1} ascorbic acid. The content of acidity in green and oolong tea samples manufactured from 5-12 days tea leaves ranged from 2.13±0.68 to 1.18±0.19% and 2.02±0.34 to 1.02±0.17%, respectively. The ranges of pH content in green tea at 5-12 days tea leaves were 5.52±0.11 to 5.97±0.18% while oolong tea had 5.53±0.11 to 5.97±0.20%, which is similar result of Choudhury (2001) that green tea leaves having pH value of 5.06-5.80%. Brunton and Hussain (2001) reported that both conventional oolong tea and herbal tea eroded dental enamel. They also stated that the pH of conventional oolong and herbal tea had been measured to be lower than the critical pH (5.5) that is necessary for the demineralization of enamel. According to their results, the erosive effect of herbal tea was found to be five times greater than that of conventional oolong tea.

Conclusion

Normally tea is collected from tea garden with maintaining proper plucking cycle retained the optimum quality in processed tea, but after over plucking, tea can get more time to make mature and show lower nutrient values like caffeine and polyphenol content in made green tea and oolong tea. The study revealed that the useful compositions found in green and oolong tea are acceptable in young tea leaves (i.e. 5 to 8 days) than mature ones (i.e.9 days to others). Based on the findings can be recommended leaf plucked from tea estate within 5 days to 8 days to maintain excellent quality in processed tea.

References

AACC. 2000. The determination of crude fibre in food. Standard using the Fibrecap 2021/2023.

Ahmed, I., Zaidi, S.S.H. and Khan, Z.A. 1989. The determination of major, minor and trace elements in tea, tea liquor, instant coffee and cocoa samples. *Pakistan J. Sci. Ind. Res.* 32 (8): 513-515.

Amra, P., Mojca, S., Zeljko, K., Bernd, W., Frank, O. and Sabine, G. 2006. Extraction of active ingredients from green tea (*Camellia sinensis*); Extraction efficiency of major catechins and caffeine. *Food Chem.* 96: 597-605.

Anesini, C., Ferraro, G.E. and Filip, R. 2008. Total polyphenol content and antioxidant capacity of commercially available tea (*Camellia sinensis*) in Argentina. *J. Agric. Food Chem.* 56: 9225–9229.

AOAC. 1984. Official Methods of Analysis of the Association of Analytical Chemists. 14th edn. Washington DC,USA.

AOAC. 1990. Official Methods of Analysisof the Association of Analytical Chemists. 15th edn. Washinton DC, USA

AOAC. 2000. Official Methods of Analysis of the Association of Analytical Chemists. 17th edn. Washinton DC, USA.

Bennett, A.W. and Bonnie, K.B. 2001. The world of caffeine: The science and culture of the World's Most Popular Drug. Rutledge. p. 228.

Brunton, P.A. and Hussain, A. 2001. The erosive effect of herbal tea on dental enamel. *J. Dentistry.* 29: 517-520.

Cabrera, C., Artacho, R. and Giménez, R. 2006. Beneficial effects of green tea: a review. *J. American College Nutr.* 25: 79-99.

Chaudhury, M.S.H. 1989. Tea Growing. 1st Ed. Ananda printer 166. Arambagh. Motijheel. Dhaka-1000. pp. 101-125.

Choudhury, M.A. 2001. Optimum Fermenting Factors in the formation of Theaflavins from Crude Catechin *in vitro* Model System and their Significance in storing. International Conference on O-Cha (tea) Culture and Science. Shizuoka. Japan. pp. 29-59.

Choudhury, M.A. 2010. Optimum quality teas processing and pharmaceutical uses.1st Ed. S. Ahmed Computer, Srimangal. pp.53-55.

Eden, T. 1960. The development of tea culture. *In:* Tea. Ed. Eden. Longman, London. pp. 1-4.

FAO. 2010. Committee on commodity problems, intergovernmental group on tea, Market development and outlook. Retrieved March 16th 2012, from http://www.fao.org/docrep /meeting/018/K7503E.pdf.

Hara, Y.L., Wickremasinghe, S.J. and Yamanishi, R.L. 1995. Special issue on tea. *Food Res. Int.* 11: 371–545.

Balentine, D.A. and Harbowy, M.E. 1997. Tea chemistry. *Crit. Res. Plant Sci.* 16: 415–481.

Harler, C.R. 1964.The Cuttivation and Marketing of Tea. 3rd edn. Oxford: Oxford University Press, London.

Hicks, M.B., Hsieh, Y.H.P. and Bell, L.N. 1996. Tea preparation and its influence on methylxanthine concentration. *Food Res. Int.* 29 (3-4): 325–330.

ISO. 2005. Determination of substances characteristic of green and black tea. Part 1: Content of total polyphenols in tea, Colorimetric method using Folin-Ciocalteu reagent. 14502-1(22).

Kumar, A., Nair, A.G.C., Reddly, A.V.R. and Garg, A.N. 2005. Availability of essential elements in India and US tea brands. *Food Chem.* 89: 441-448.

Mandel, S. and Youdim, M.B. 2004. Catechin polyphenols: neurodegeneration and neuroprotection in neurodegenerative diseases. *Free Radical Biol. Med.* 37 (3): 304–317.

Mohammed, M.I. and Suliman, M.A. 2009. Proximate, caffeine and tannin analyses in some brands of tea consumed in Kano Metropolis, Nigeria. *Bayero J. Pure & Appl. Sci.* 2 (2): 19 – 21.

Nagao, T., Komine, Y., Soga, S., Meguro, S., Hase, T., Tannaka, Y. and Tokimitsu, I. 2005. Ingestion of a tea rich in catechins leads to a reduction in body fat and malonaldialdehyde-modified LDL in man. *American J. Clinical Nutr.* 81: 122-129.

Nestle. 1966. Patent specification, Great Britain. 1,034,670.

Takayanagi, H. 1977. Relationship between the contents of manganese and vitamin C in leaves of shaded tea. Study of Tea. 52 p.

TRIC, 1948. Tea Research Institute of Ceylon, St. Coombs Estate in Talawakelle, Sri-Lanka. 50 p.

Zaveri, N.T. 2006. Green tea and its polyphenolic catechins: medicinal uses in cancer and noncancer applications. *Life Sci.* 78: 2073-2080.

AGRICULTURAL ACTIVITIES OF THE MALAYALI TRIBAL FOR SUBSISTENCE AND ECONOMIC NEEDS IN THE MID ELEVATION FOREST OF PACHAMALAI HILLS, EASTERN GHATS, TAMIL NADU

V. Anburaja* and V. Nandagopalan

Abstract

This paper is aimed to study the Agricultural activities of the Malayali Tribal for Subsistence and Economic needs in the mid elevation forest of Pachamalai Hills, Eastern Ghats, Tamil nadu. About 8 percent of the Indian population belongs to a category listed as "Scheduled Tribes" enumerated in the Schedule to Article 342 of the Constitution of India. Tribal people has been seen to be strongly associated with the forests, hills and remote areas, practicing a unique life style, having a unique set of cultural and religious beliefs. For millennia, tribal communities have lived in forests and survived on hunting and gathering. However, with growing population and resource pressure, it is now witnessing that a rise in livelihoods based on settled farming. In the study area, cereals are the major crop cultivated for their edible grains. The tribal people living in the hills tops of the study area were cultivated 11 major agricultural crops. They are, *Eleusine coracana, Panicum miliare, Oryza sativa* (verity I) (Mara Nellu), *Oryza sativa* (verity II), *Manihot esculenta, Macrotyloma uniflorum, Vigna mungo, Sesamum orientale, Paspalum* sp, *Pennisetum americanum* and *Setaria italica.* Among them, *Paspalum* sp gives the maximum yield with 655 kg $acre^{-1}$.

Keywords: Agricultural Activities, Malayali Tribal, Mid Elevation Forest, Pachamalai Hills, Tamil Nadu

PG and Research Department of Botany, National College, Tiruchirappalli – 620 001, Tamil Nadu, India.

*Corresponding author's email: vanburaja@gmail.com (V. Anburaja)

Reviewed by Md. Rahsedur Rahman, Bangladesh Agricutural University, Mymensingh, Bangladesh.

Introduction

Today, India ranks second worldwide in farm output. Agriculture and allied sectors like forestry and logging accounted for 16.6% of the GDP in 2007, employed 60% of the total workforce (CIA, 2008) and despite a steady decline of its share in the GDP, is still the largest economic sector and plays a significant role in the overall socio-economic development of India. The population of India is 1.2 billion. Simultaneously, farmland is expected to be engulfed by urbanization and environmental degradation. To maintain a satisfactory food security system, crop production should increase by 3% a year (Ravishankar and Selvam, 1996).

The tribal population in India is 84.51 million, which constitutes 8.14% of tribal population (Census, 2001). There are about 449 tribes and sub tribes in different parts of India. Half of the India's tribal people live in the forests and forest fringes and their economy is link with the forests. Most of the tribals in Tamilnadu are cultivators, cattle rearing agriculture laborers or dependent on forests of their livelihood. In Tamilnadu 36 scheduled tribe communities have been identified (Source: Tamilnadu Forest Department Web site: http://www.forests.tn.nic.in).

Tribal communities, namely Irulas, Malayalis and Muthuvans living in the state of Tamil Nadu, have been cultivating traditional cultivars of paddy, millets, pulses and vegetables. The traditional cultivars sown by them, over generations, form the principal crops of their agricultural system (Ravishankar and Selvam, 1996). The subsistence life style, local diet habits and organoleptic preferences of these tribes, as well as their dependence on monsoon rain for irrigation, have led them to cultivate and conserve local seeds for consumption and for sowing the following season. This traditional practice is a blessing in disguise and has saved many forms of specific and intra-specific varieties of millets and paddy. Historically, the economy of most tribes was subsistence agriculture or hunting and gathering. Tribal members traded with outsiders for the few necessities they lacked, such as salt and iron (Thakur and Sunil Pandey, 2009)

Their knowledge of seed selection, their traditional methods of conserving seeds and grains in eco-friendly traditional granaries and their communities' participation in maintaining germplasm and plant protection methods provide important insights to global efforts aimed at genetic conservation (Ravishankar and Selvam, 1996).

Shifting cultivation is an agricultural system in which plots of land are cultivated temporarily, and then abandoned. This system often involves clearing of a piece of land followed by several years of wood harvesting or farming, until the soil loses fertility. Once the land becomes inadequate for crop production, it is left to be reclaimed by natural vegetation, or sometimes converted to a different long-term cyclical farming practice. In addition, ecological consequences are often deleterious, but can partially mitigate if new forests are not invaded (Anderson, 1997).

Shifting cultivation has resulted in large-scale deforestation, soil and nutrient loss, and invasion by weeds and other species. The indigenous biodiversity has affected largely. Shifting cultivation is prevalent mostly in tropical countries. In India, the people of eastern and north-eastern region practice shifting cultivation on hill slopes. 85% of the total cultivation in northeast India is by shifting cultivation (Singh and Singh, 1992). Due to increasing requirement for cultivation of land, cycle of cultivation followed by leaving land fallow has reduced from 25–30 years to 2–3 years. Earlier the fallow cycle was of 20 – 30 years duration, thereby permitting the land to return to natural condition (Patro and Panda, 1994). Due to reduction of cycle to 2–3 years, the resilience of ecosystem has broken down and the land is increasingly deteriorating. The paper discusses the shifting-cultivation practices in the eastern and northeastern regions of the country, and suggests certain strategies to revive the deforested areas for achieving ecological sustainability Agricultural practices are at the cost of loss of biodiversity resources; estimates indicate that one unit of energy in agronomic production costs loss of greater energy from the forests. However, in the Central Himalayan eco-systems, where agriculture practice is more scientific compared to shifting cultivation, one unit of energy in agronomic production entails an expenditure of about 10–12 units of energy from the surrounding forests as firewood, fodder and leaf manure (Singh, 1986). Loss of energy from the forests per unit of agricultural production may be far greater in shifting-cultivation areas. Agricultural practices are at the cost of loss of biodiversity resources; estimates indicate that one unit of energy in agronomic production costs loss of greater energy from the forests. However, in the Central Himalayan eco-systems, where agriculture practice is more scientific compared to shifting cultivation, one unit of energy in agronomic production entails an expenditure of about 10–12 units of energy from the surrounding forests as firewood, fodder and leaf manure (Singh, 1986).

Agricultural practices are at the cost of loss of biodiversity resources; estimates indicate that one unit of energy in agronomic production costs loss of greater energy from the forests. However, in the Central Himalayan eco-systems, where agriculture practice is more scientific compared to shifting cultivation, one unit of energy in agronomic production entails an expenditure of about 10–12 units of energy from the surrounding forests as firewood, fodder and leaf manure (Singh, 1986). Loss of energy from the forests per unit of agricultural production may be far greater in shifting-cultivation areas.

The increase in human population, particularly in the developing countries, has put tremendous pressures on land. The extension of crop lands, for increasing food production, has been directly responsible for the reduction in areas under forests and grass lands. According to one estimate, about 40% of the land surface of the earth was converted into crop lands and permanent pastures by early 1990s. More than 6% area under tropical forests was converted to shifting cultivation between 1980 and 1990 across all tropical countries (World Resource Institute, 1996). This paper is aimed to study the Agricultural activities of the Malayali Tribal for Subsistence and Economic needs in the mid elevation forest of Pachamalai Hills, Eastern Ghats, Tamil nadu.

Materials and Methods

The data collections were done during 2006–2008 to collect information on the agricultural practices of Malayali tribes in the Pachamalai hills. The above villages lie in these hills are situated at the mid regions of Tamilnadu with Latitudes 11°09'00" to 11°27'00" N and longitudes 78°28'00" to 78°49'00" E and included to the Salem and Tiruchirappalli districts.

This study was made mainly by the interviews with the trial people of the study area. Interview was a two-way communication processes between a minimum of two persons namely a respondent and an interviewer. In the present study, interview was made between various user groups and farm hold peoples. Non-verbal responses like smile, nodding head, change in tone, eye contact were personally noted down. Such non-verbal signals have lot of hidden meaning which cannot be recorded without proper interception. To overcome this, probing process was adapted. The data collected from the tribal people of the study area were statistically calculated with Microsoft excel program 2007.

Results and Discussion

Their knowledge of seed selection, their traditional methods of conserving seeds and grains in eco-friendly traditional granaries and their communities' participation in maintaining germplasm and plant protection methods provide important insights to global efforts aimed at genetic conservation. By care full selection and conservation, these communities have enhanced the genetic potential of their seeds and have been identified from the tribal communities. The tribals prefer to continue the cultivation of their own traditional cultivars is these crops are ecologically suitable, drought resistant, pest tolerant and disease resistant.

Mostly the hills top land was used the cultivation of 11 agricultural crops. Among them, *Eleusine coracana* gave 8%, *Panicum miliare* gave 12%, *Oryza sativa* (verity I) (Mara Nellu) gave 9 % and *Oryza sativa* (verity II) gave 11%, totally 20%, the productivity of *Manihot esculenta* was 16%, *Macrotyloma uniflorum* 2%, *Vigna mungo* 3% of the total productivity of the uphill crops. *Sesamum orientale* supplies the edible oil need with the 11% of total productivity, *Paspalum* sp give maximum yield with 17% of the total productivity *i.e.*, 655 kg acre^{-1}. *Pennisetum americanum* and *Setaria italica* gave 6% and 5% of the productivity respectively (Table 1). During the agricultural process, the native flora of forest was disturbed and devasted from parental lands. The present study have also observed the low rainfall concomitantly with increasing temperature on hills tops has caused the shifting of agricultural land have lead to the damaged species distribution pattern in natural ecosystem which indirectly has affected the ecosystem functioning.

In case land, cultivated for a long time, it is highly probable that the land may turn infertile. This is due to lack of organic matter and minerals rich in nitrogen and phosphorus. The land may turn acidic and lose its percolation capacity. After some period of cultivation, the land would be left as it lacks the minerals and organic matter for the growth of the crop plants. According to the view of some environmental scientists, the system of shifting cultivation is an unproductive and wasteful process of cultivation. They also opine that shifting cultivation is the prime cause for widespread environmental degradation and destruction of tropical forests in the world. In a stable shifting cultivation system, the fallow land is used for natural vegetation for a longer period of time, to regain the fertility of the soil. If the fallow land is kept uncultivated for a longer period, it will be more fertile in nature. During this period, the soil temperature becomes lower and nutrients are extracted from the subsoil. This would also lead to the reduction in the soil acidity level.

The irrigation system of the Malayali tribes of the study area is mainly depending on the seasonal monsoon rainfall. In addition, there is no proper irrigating system in the study area except in the few villages.

Animal husbandry

Goat rearing also plays a vital role in supplying their meat and money. Cows and bullocks are maintained only for the ploughing. After the ploughing period the both the cows and bullocks are left in the village surrounding area until the next season. So the production of milk from the cows is almost nil.

Poultry - indigenous varieties of hen and cocks are maintained for egg and meat production especially in the tribal areas. Indigenous Cow and bullocks are also maintained. These are mainly for ploughing purpose. After ploughing, the cattle are simply left in the areas surrounding the tribal area until the next season.

Piggery is the most popular Animal Husbandry Venture in the tribal areas. Tribal families of the study area maintain the indigenous pigs for their meat during the festivals, marriage ceremonies and some other needed days.

Recommendations

Extension and continuous training programmes in the tribal villages to popularize Animal Husbandry as a venture for augmenting milk, meat and egg production and to increase the rural employment potential in the study area. Hybrid cows have to be supply to increase the milk production. Hybrids of goat may be given to the tribes with a training will give a additional economic support to them. Fodder seed production farm, developing grazing land, establishing Silvi Pastural System and fodder training cum extension programme as well as free supply of fodder seeds and cutting is needed in the tribal area to reduce the deforestation due to over grazing. Supplying exotic pigs to tribes and a new scheme of insurance of pigs belonging to tribals has been taken up apart from providing practical training in pig keeping.

Several factors may contribute to the persistence of their local knowledge. The lack of modern and government facilities and remote geographical features of tribal area, as well as a strong belief in folk knowledge continue the preference for traditional knowledge for their life style (Shandesh Bhattarai *et al.*, 2006).

Table 1. Check List of Plants Species with Agricultural Importance to Tribal Community

Sl. No.	Name of the crops	Type of irrigation	Yield (Kg $acre^{-1}$)
1.	*Eleusine coracana* (L.) Gaertner	Depends on rain water. (Manavari)	309 ± 73.74
2.	*Panicum miliare* Lam.	Depends on rain water	486 ± 74.68
3.	*Oryza sativa* L. (Mara Nellu)	Depends on rain water.	334 ± 70.01
4.	*Oryza sativa* L.	Stream water (only available October to February)	417 ± 79.97
5.	*Manihot esculenta* Crantz.	Depends on rain water.	635 ± 110.67
6.	*Macrotyloma uniflorum* (Lam.) Verdc.	Depends on rain water.	82 ± 25.88
7.	*Vigna mungo* (L.) Hepper.	Depends on rain water.	100 ± 22.78
8.	*Sesamum orientale* L.	Depends on rain water.	432 ± 74.58
9.	*Paspalum sp.*	Depends on rain water.	655 ± 83.16
10.	*Pennisetum americanum* (L.) Leeke	Depends on rain water.	250 ± 38.47
11.	*Setaria italic* (L.) P. Beauv.	Depends on rain water.	181 ± 52.95

Photo 1: Paddy cultivated field

Photo 2: The stream for irrigation

Photo 3: Tribal busy in agriculture activities

Photo 4: Wasted land with a herds of cows

Fig. 1. Agricultural practices in the study area

References

Anderson, A. 1997. Prehistoric Polynesian impact on the New Zealand environment: Te Whenua Hou. In Historical Ecology in the Pacific Islands: Prehistoric Environmental and Landscape Change (Eds, Kirch, P. V. and Hunt, T. L.) Yale University Press, New Haven and London. pp. 271-283.

Census. 2001. Web address: http://tribesindia.com

CIA. 2008. CIA Fact book: India. Central Intelligence Agency. https://www.cia.gov/library/publications/the-world-factbook/geos/in.html. Retrieved 2008-06-10.

Patro, S.N. and Panda, G.K. 1994. Eastern Ghats in Orissa: Environment, Resources and Development, Orissa Environment Society.

Ravishankar, T. and Selvam. V. 1996. Contributions of Tribal Communities in the Conservation of Traditional Cultivars. In: Using Diversity: Enhancing and Maintaining Genetic Resources On-farm, International Development Research Centre, New Delhi, India. pp. 78-86.

Shandesh Bhattarai, Ram P Chaudhary and Robin S.L. Taylor. 2006. Ethnomedicinal plants used by the people of Manag district, central Nepal. *J. Ethnobiology and Ethnomedicne*, 2:41 (doi: 10.1186/1746-4269-2-41).

Singh, J.S. 1986. Eco-Development Guidelines and Model of Development of the Central Himalaya, Department of Botany, Kumaun University, Nainital. p. 48.

Singh, J.S. and Singh, S.P. 1992. Forests of Himalaya, Gyanodaya Prakashan, Nainital, p. 294.

Thakur, A.P. and Sunil Pandey. 2009. 21st Century India. Global vision publishing house, New Delhi. 510 p.

Web: http://www.forests.tn.nic.in

WRI. 1996. World Resource, World Resource Institute, Oxford University Press, Oxford, 365 p.

PHEROMONE USE FOR INSECT CONTROL: PRESENT STATUS AND PROSPECT IN BANGLADESH

Md. Azharul Islam

Abstract

The insect's world is filled with many odors. Insects use these odors to cue them in a variety of complex social behaviors, including courtship, mating, and egg laying. Scientists and pest control specialists have known about these complex communication systems for decades. The main aim of this study was to visualize the availability, trends and differences in the sources of pheromone control in agricultural growth of Bangladesh. It also concerned on constrains and present use of pheromone and their possible recommendation on behalf of Bangladesh agriculture. It concentrated on the data during last three decades (1980-2010), comprising status of pheromone use in Bangladesh agriculture and its future. Review revealed that Bangladesh has been enormously successful in increasing pheromone use in agricultural production (especially for vegetables). Understanding of the nature of pheromones and their potential for pest control along with the future prospective of pheromone technique in agriculture were stated. Since the pheromone, technologies for control of major crop pests in Bangladesh are still limited. So that this review emphasized on more attention to the authority to increase the research works and project facilities related to develop and promote pheromone techniques. It is highly recommended to increase availability of pheromone in market, more investment in research and development, introduction of newly identified pheromone for specific pest, to assist government and non-government organizations to work with farmers to reduce harmful insecticide use and promote pheromone tactics as one part of integrated crop management (ICM).

Keywords: Sex pheromone, Integrated Crop Management, Monitoring, Trapping, Mating disruption

---

Assistant Professor, Graduate Training Institute, Bangladesh Agricultural University, Mymenshingh, Bangladesh

*Corresponding author's email: islam162@yahoo.com, mdaislamgtibau@gmail.com (Md. Azharul Islam)

Reviewed by Duraikannu Vasanthakumar, UPASI Tea Research Institute, Tamil Nadu, India

Introduction

Bangladesh, one of the smallest countries (area 57 K sq. miles) in South-East Asia, has a predominantly farming-based economy. A delta, historically originated through the sedimentation of the Bay of Bengal, the country is blessed with highly fertile agricultural lands. However, due to very high population, the nation has always been struggling against poverty and starvation. Bangladesh economy draws its main strength from agriculture sector. The sector contributes 19.10% (at current prices) and employs about 51% of the labor force (Mondal, 2010). Despite increase in the shares of fisheries, livestock, and forestry, crop sub-sector alone accounts for 60.83% share of agricultural GDP (BBS, 2011). Agricultural land per capita is decreasing over the years in Bangladesh. Agriculture and environment are closely interlinked. Agricultural production system depends on the environment for utilization of land, rainfall, daylight duration, insect pests and diseases. Pest problem is one of the major constraints for achieving higher production in agriculture crops. Bangladesh loses about 30% of its crops due to pests and diseases each year (BBS, 2011). So how do farmers control their insect pest problems? Farmers in Bangladesh depend on synthetic insecticides because they are readily available, highly promoted, inexpensive, easy to apply and quick acting. However, applied insecticides also kill non-target arthropods, typically insects involved in pollination and predators such as spiders and ground beetles. Insecticide residues find their way into watercourses, particularly in rice cultivation, and affect the water we drink and food we eat (Cork *et al.* 2003, 2005). Furthermore, quite often the indiscriminate and unscientific use of pesticides has led to many problems, such as pests developing resistance, resurgence of once minor pest into a major problem besides environmental and food safety hazards.

There are many alternative approaches to control insect pests. Cultural practices, including crop sanitation, use of resistant varieties and methods to promote the activities of natural enemies and predators all act to reduce the threat from potential insect pests. Thus keeping insect pests below damage thresholds, does not aim reduce populations to a level where genetic change is induced, leading to the development of resistant biotypes. Such biorational pesticides include insecticides such as neem kernel seed extract, different extracts of plant parts, fungi (*Beaveria, Metarhizium*), viruses (nuclear polyhedrosis virus), bacteria (*Bacillus thuringiensis*) and semiochemicals (Akhtar and Mandal, 2008; Islam *et al.*, 2008a; Mamun *et al.*, 2008; Islam, 2009; Islam and Becerra, 2011; Islam and Begum, 2011). Country needs to adopt the total integrated pest management system as the insecticide-based management system has failed to control many pests. The pests are becoming resistant to almost all chemical pesticides as the frequency of spraying is gradually increasing while their efficacy is gradually decreasing. Biological control involves use of a specially chosen organism to control a specific pest. This chosen organism might be a predator or a parasitoid, which attacks harmful insects. Field demonstration results prove that the use of biological agents has no adverse effect on human health and the cost effectiveness of bio-control measures is very attractive, safe and sustainable. Insects communicate by means of scents–pheromones, chemicals used for 'signaling'. With these, they both locate and identify their mates. They are natural chemicals emitted in micro quantities in the form of a vapor by virtually all known insects. Each insect species has its own unique signature scent. In fact, sex provides us with a powerful means of surveillance and control in the insect world. A female insect typically puffs out a thousand millionth of a gram of her signature several times a minute. Males of her species follow this scent to mate with the female. It follows that if you can identify and then duplicate that scent, you have the means of controlling the males of that species. This is the mysterious incidence of pheromone technology.

The existence of pheromones has been known for centuries, apparently originating in observations of mass bee stinging in response to a chemical released by the sting of a single bee. The first isolation and identification of an insect pheromone (silkworm moth) occurred in 1959 by German scientists. Since then, hundreds, perhaps thousands of insect pheromones have been identified by increasingly sophisticated equipment. Today we have a much clearer view of the limitations and possibilities associated with insect pheromones in IPM programs. The two primary uses of insect pheromones are for detection and monitoring of populations and for mating disruption. These uses take advantage of sex pheromones on which a vast majority of insect pests relies to mediate reproduction.

To date, the research works on pheromone in pest management in Bangladesh still limited. There are few scientists involved in trapping of moths using pheromone (Alam *et al.*, 2003; Cork *et al.*, 2001, 2003, 2004a, 2004b, 2005; Uddin, 2008; Mazumder and Khalequzzaman, 2010) in Bangladesh but no one works on the identification of pheromone from Bangladeshi insect species. The survey report on pheromone practice in Bangladesh agriculture is also limited (Islam, 2012). Major objective of this review article is to discuss the pheromone technology in pest management, present status of pheromone use in Bangladesh agriculture and suggest possible opportunities to address the topic that may assist the government and non-government policy makers to develop national economy.

Uses of Pheromones in Pest Management

The use of pheromone for controlling pest insects requires three items: a pheromone chemical, a trap, and a support to hang the trap in the field. Technically sex pheromones can be used in three principal ways:

Detection and Monitoring: The principle use of insect pheromones is to attract insects to traps for detection and determination of temporal distribution. In most instances, the males are responders to female-produced pheromones. Trap baits, therefore, are designed to closely reproduce the ratio of chemical components and emission rate of calling females. Trap baits of many designs have been tested over the years. Trap design is also critical to effective use of traps for monitoring insect populations. Traps vary in design and size dependent on the behavior of the target insects. The information from trap catches can be very useful for decision making on insecticide applications or other control measures. For example, trap catches may indicate a loss of effect of pheromone on mating disruption and the need to reapply a pheromone treatment. Careful monitoring and experience in interpreting collected data are important for success. Traps may also be placed with the objective of destroying males for population control.

Examples of the use of pheromones in pest management programmes for detection, monitoring and timing of pesticide spray programmes are sesiid moth, *Macroscelesia japona* , in orchards of Ibaraki prefecture, Japan (Islam *et al.*, 2007), codling moth, *Cydia pomonella,* in apple and pear orchards in

Australia (Williams, 1989), citrus leaf miner (*Phyllocnistic citrella*) in orchards in Ogasawara (Bonin) Islands and Ehime Prefecture, Japan (Vang *et al.*, 2008), *Heliothis spp.* in USA (Lopez *et al.*, 1990), nettle moth, *Parasa lepida lepida* (Limacodidae) in orchards in Gifu prefecture in Japan (Islam *et al.*, 2009), apple leaf roller, *Bonagota cranaodes* in Brazil (Kovaleski *et al.*, 1998), *Spodoptera litura* in India (Ranga Rao *et al.*, 1991), *Leucinodes orbonalis* in Bangladesh (Alam *et al.*, 2003, Uddin *et al.*, 2008, Mazumder and Khalequzzaman, 2010).

Mass trapping: Sex pheromone baited traps can capture male moths continuously, thus preventing mating and multiplication of the pest. This approach has proven to be particularly efficient and economical. *Rhynchophorus palmarum* is the primary pest of oil, coconut and palm in Central and South America. By 1994 the number of trees needing to be felled was reduced to less than 3,000 per annual demonstrating that mass trapping can be highly effective in controlling palm weevil populations (Alpizar *et al.*, 2002, Hallett *et al.*, 1999, Oehlschlager *et al.*, 2002). Highest mass trapping of males of *Macroscelesia japona* reported by lure baited with *E*2,*Z*13-18: Ald and *E*2,*Z*13-18:OH (Islam *et al.*, 2007). Examples of mass trapping are lures baited by *Z*7,9-10:OH were examined on nettle moth, *P. lepida lepida* (Islam *et al.*, 2009), cotton weevil (*Anthonomus grandis*) successfully baited with its aggregation pheromone (Cork *et al.*, 2003), Japanese strain of *Phyllocnistic citrella* trapped only the lure containing *Z*7,*Z*11-16:Ald (Vang *et al.*, 2008).

Brinjal (*Solanum melongena* L.) is an economically important crop throughout South and South East Asia. Fruit losses in excess of 50% are commonly reported due to the boring activity of larvae of the brinjal shoot and fruit borer, *Leucinodes orbonalis* (Cork *et al.*, 2005). Zhu *et al.* (1987) reported (*E*)-11-hexadecenyl acetate as the pheromone of *L. orbonalis* and traps baited with up to 500 µg attracted more male moths than six virgin females. Subsequently Attygalle *et al.* (1988) identified (*E*)-11-hexadecen-1-ol in addition to the related acetate using insects obtained from Sri Lanka. In field trails conducted in India where blends containing between 1 and 10% *E*11-16: OH caught even more male *L. orbonalis* than *E*11-16: Ac alone. At the 1000 µg dose, addition of 1% *E*11-16: OH to *E*11-16: Ac was found to be significantly more attractive to male *L. orbonalis* than either 0.1 or 10% *E*11-16:OH. Trap catch was found to be positively correlated with pheromone release rate, with the highest dose tested, 3000 µg, catching significantly more male moths than lower doses (Cork *et al.*, 2001). In order to reduce the cost of pheromone based technologies for control of *S. incertulas*, a programme of research to develop an effective mass trapping system. This proved to be highly effective using indigenous traps and lures at a density of 20 traps ha^{-1} (Cork and Krishnaiah, 2000).

Mating disruption: Sex pheromone can be used for disruption of mating, which is achieved by placing high concentrations of pheromone at regular intervals throughout the field. The high concentration of pheromone saturates the area resulting in males failing to find females, which produce very minute quantities of these chemicals, thus preventing mating and multiplication of the pest. The major pest of cotton in Egypt in the early 1980's was the pink bollworm (PBW), *Pectinophora gossypiella*. The female sex pheromone was identified by workers in the USA (Bierl *et al.*, 1974). The diversity of mating communication system in lepidopteran insects was also reported (Islam *et al.*, 2008b). The economic importance of *P. gossypiella* and the fact that its pheromone is relatively cheap and chemically stable, the decision was made to try to control it using mating disruption. It has also been identified as a pest control method in which the insect does not become resistant.

Pheromone traps

Various types of traps are available commercially, while others can be made by farmers inexpensively at home. A pheromone-baited lure inside the trap will bring male moths inside the trap. Proper trap design is critical to kill the pest once it enters the trap. The type of trap to be used depends on the behaviour of the target insect. Various research works showed the most effective traps in pest control are delta traps, winged traps and funnel traps. Different available and relatively low cost traps are shown in Fig. 1. Pheromone traps are very sensitive, meaning they attract insects present at very low densities. They are often used to detect presence of exotic pests, or for sampling, monitoring, or to determine the first appearance of a pest in an area.

Present Status on Pheromone use in Bangladesh

There is significant evidence that insecticide use in Bangladesh is increasing dramatically (Fig. 2). The country imports pesticides worth 24000 crore taka every year a huge business. Farmers are applying pesticides where they are not actually necessary, at high cost to themselves and the environment. These pesticides are becoming redundant because of the development of resistance to them and more importantly affect environmental pollution. Governments are becoming aware of the negative environmental and health aspects associated with the use of these compounds, so now there is an urgent need for the development of alternative control technologies.

Fig. 1. Different types of pheromone traps used in field; traps used in crop field (a-e), traps with trapped pest insects (f-h). **a.** one lure septum hanged with crop plant, **b.** net trap, **c.** water trap, **d.** bucket trap, **e.** bucket with window trap, **f.** delta trap, **g.** Yellow sticky trap, **h.** wing trap

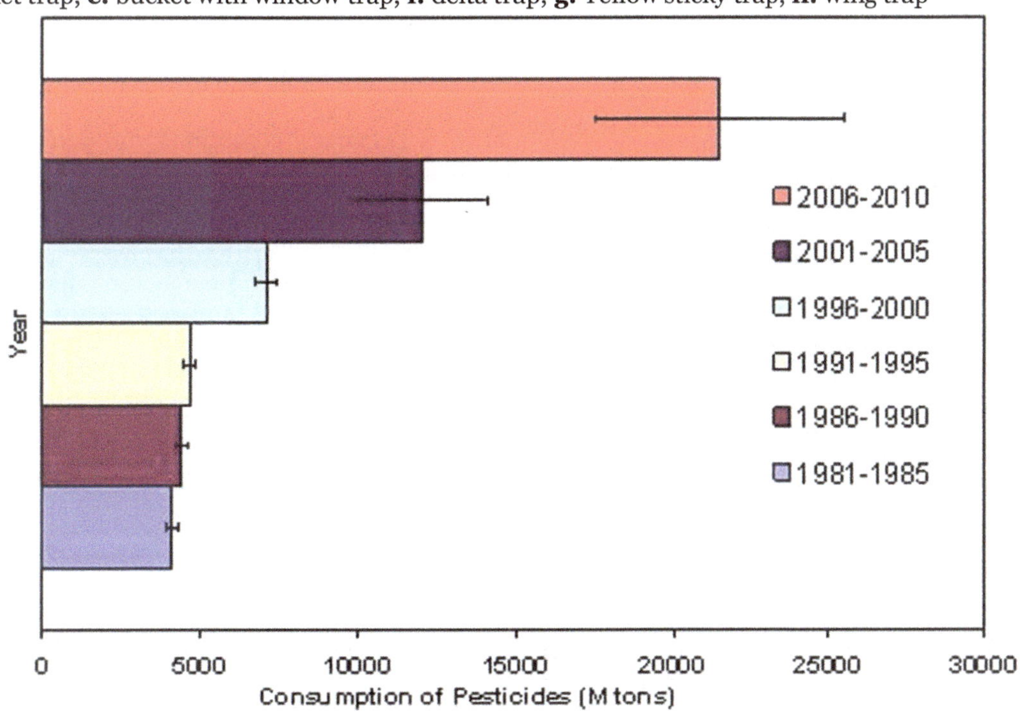

Fig. 2. Pesticide consumption in Bangladesh over 30 years (Source, BBS 2011).

In Bangladesh, about more than 100 major crops are cultivated over a year. There are more than 1000 harmful insects' causes of economic loss on the major cultivated crop. However, the availability of pheromone components of very few insects is reported and available in market. Presently, most of the pheromone components either imported from abroad or synthesized in Bangladesh. However, those chemicals are not confirmed by proper research work for the insects available in Bangladesh. The regional variation in chemical communication reported among the same insect species occurred due to their races differences (Vang *et al.*, 2008). So it should be noted that identified pheromone of an insect from another country may not match with the Bangladeshi strain. This review also emphasized one note that there are extreme limited reports available about the identification of insects' pheromone in Bangladesh.

Bangladesh Agriculture Research Institute (BARI) and Department of Agriculture Extension (DAE) provided synthetic sex pheromones and beneficial insects to a number of farmers in vegetable growing districts like Jessore, Narsingdi, Comilla, Bogra, Pabna and other 244 Upazilas experiment and find out if these inputs could replace harmful pesticides to reduce damage to public health as well as harmful pests can be controlled by using sex pheromone traps, locally known as magic trap. The magic traps are popular among farmers of limited area in Bangladesh that demand for those has increased very quickly. However, ironically the supply is so little that, farmers face problems in expanding this new device for controlling pests in a natural, environment-friendly, safe and secured way to boost crop production. The knowledge and availabilities of pheromone in Bangladesh related similar information reported by Islam (2012).

Limited organizations got permission to import pheromones from abroad. Those organizations got special permission to import several sex pheromones item from plant protection wing of DAE, MOA, GOB. There is a remarkable gap between different research institute and such company or organization engaged in commercialization of pheromones. Limited reports available about the economical benefit achieved after using pheromone technologies in Bangladesh. For example, few farmers in Sikandarpur village in Jessore are being doubly benefited. They are earning more money by producing vegetables at lesser costs while protecting the environment also. They use pheromone trap to kill insects instead of applying insecticides. Alam *et al.* (2003) reported on sex pheromone trap technology for the control of shoot and fruit borer in brinjal and cucurbits in that region of Bangladesh. As above, adult female yellow stem borers (*Scirpophaga incertulas*) attract their mates with a pheromone and this can be exploited in a pest management strategy by developing a synthetic pheromone blend and a lure and trapping system, which attract and trap male moths in Bangladesh. Uddin *et al.* (2008) also reported on such male moth trap, where emphasize on trap types and height to find effective trapping.

Building on work conducted by scientists in India, Syngenta has been collaborating with BRRI and NRI to adapt mass trapping technology for use in Bangladesh (Cork *et al.*, 2004a). According to scientists from three organizations (Natural Resources Institute, NRI; Bangladesh Rice Research Institute, BRRI and the multinational agrochemical company Syngenta) worked in collaboration to find sustainable control method, sleeve traps with a pheromone lure are both effective and cost-efficient in controlling the insect which is responsible for 70-80% of pest damage to rice crops (Anonymous, 2003). Mazumder and Khalequzzaman (2010) reported that the sex pheromones which have been extensively studied and already are in management programmes to improve their efficacy in totality in an Integrated Pest Management Programme on the basis of feedback from the extension workers, and farmers. They studied mainly on the efficiency of male moth catch of eggplant shoot and fruit borer, *Leucinodes orbonalis*.

The previous research reports supported that a wide range of trap designs, pheromone blends and concentrations were tested with farmers in their fields in Comilla and Mymensingh districts in 2001-03, along side a socio-economic study of farmer's resources, constraints and perceptions to ensure the resulting technology was appropriate for adoption (Cork *et al.*, 2004b). On farm, large-scale mass trapping trials demonstrated that 20 traps ha^{-1} were sufficient to reduce male yellow stem borer populations significantly. The trials provided good evidence that mass trapping could significantly reduce the level of mating, with consequent reductions in larval progeny (Cork *et al.*, 2004b).

It is also good news for us that the pheromone of fruit and shoot borer moth is now synthesized and produced in the factory. It is available in the market and to use the sex pheromone in the field, one needs two items: the chemical (or the pheromone) lure and a suitable trap, which are available in Bangladesh with much cost effective price. Thus pheromonal control as an IPM strategy may control eggplant shoot and fruit borer with minimal use of pesticides (Prosad *et al.*, 2005; Uddin et al., 2008; Mazumder and Khalequzzaman, 2010).

Disseminating the idea to farmers is already done in many areas of the country though it is limited. Some of the participating farmers are working in their fields to demonstrate how pheromone/lures worked in their field. It is also important that neighbouring farmers will get preparation to work together in the technology worked. Although most of the farmers in Bangladesh are illiterate but they can still learn the benefits of using the lures around the traps through demonstrations. They already known about the natural enemy of harmful insects are saved, that the danger of insecticide use is avoided, their environment is safe and that they can have a good crop. Recently, Islam (2012) reported on the knowledge and practices on pheromone used by Bangladeshi farmers, where found the present knowledge on pheromone technology and some important suggestions. Farmers now understood the money that they save in reducing their use of pesticides can be spend instead on herbicides which are less toxic and give a more effective return.

Besides, the above discussion on present status of pheromone technology in Bangladesh agriculture few broad problems are pointed out, such as:

1. Insufficient investment in pheromone research in Bangladesh,
2. Use of IPM technology is limited to rice and few vegetables.
3. Lack of proper knowledge on identification and synthesis of pheromones.
4. Lack of scopes for promotion and training of potential scientists on pheromone research.
5. Limited commercial manufacture of pheromone technology for control of insects of rice and brinjal developed.
6. Training programmes for pesticide dealers, farmers on pheromone and ICM technologies are still limited.
7. Insufficiency of materials required for pheromone technology.

Recommendations

In view of the several disadvantages /limitations associated with the unscientific use of pesticides in agriculture, there is an urgent need for minimizing the use of chemical pesticides in the management of insect pests. Growing public concern ever potential health hazards of synthetic pesticides and steep increase in cost of cultivation/low profit making by farmers has led to the exploration of eco-friendly pest management tactics such as Integrated Crop Management (ICM). Admitting the scarcity of sex pheromones and beneficial insects, researchers commented by using huge dozes of pesticides destroyed our biodiversity. The government is yet to consent to bulk import and mass use of insect pheromones. We expect the government and the private sector entrepreneurs to set up more laboratories to identify and synthesize pheromone components of the harmful insects for developing a sustainable pest management system.

To overcome the limitations related to pheromone technology for crop protection; the government, NGOs and personnel engaged in agro-research/works should be concern about the following suggested recommendations:

1. Establishment of a new institution/ foundation is recommended to meet the need for pheromone techniques and adaptation in Bangladesh agriculture.
2. Investment should be raised in pheromone research works collaborated with different agro-based research institutes and agricultural universities.
3. The fund should be raised to help generate pheromone-based technologies with climate change hazards and disseminate such technologies at farmers' level.
4. Government might encourage establishment of farmer's cooperatives to ensure the availability of the pheromones on time.
5. Researcher should be conduct to test the adoptability of identified pheromone for pest insects available in Bangladesh.
6. Researcher should be take similar attention on harmful pests of other major crops like rice and vegetables.
7. More specific pheromone component(s) for the harmful insect(s) should be developed to control the pest(s).
8. Contribution of private sector and NGOs has to be encouraged to quality pheromone production.
9. It is also necessary to expand IPM practice to other economic crops (e.g. Fruits, Pulses etc.).
10. Attempts on pheromone technologies should now be made to transfer at the field level.
11. Farmers should as well be motivated to reduce their dependence on the use of chemical pesticides to control harmful pest insects.
12. Conduction of appropriate training programs on pheromone technologies with farmers, field level agricultural officers, researchers/scientists are necessary.

It is essential to review the present activities on pheromone tactics to pest control with the relevant experts, professionals, and farmer's representatives and update it based on their comments and suggestions. Author would like to illustrate the high attention on status of pheromone use in Bangladesh agriculture and future activities is addressed by the government to ensure sustainable agriculture and food security. This issue needs to be spelled out in the present National Agricultural Policy (NAP).

Conclusion

Bangladesh has been enormously successful in increasing its agricultural production in an effort to attain self-sufficiency. This has been largely achieved through new technologies in agriculture. The farmers were immensely benefited by using a combination of pheromones and beneficial insects when they could no longer control harmful pests even after spraying insecticides. Damage to production was 40 to 50 per cent even after spraying cocktail insecticides but after using pheromones and beneficial insects, the rate of damage diminished to 10 per cent. The identification and uses of insect pheromones is an active area of research and new developments continue to be made. Potentially, pheromones may be used to trap out certain Bangladeshi harmful insects and to attract insects to insecticide or chemosterilant baits, reduce the number of insecticide applications, or to confuse insects and disrupt mating. Use of traps as a sampling tool to determine need for and timing of control measures can provide the basis of an ICM/IPM strategy for these pests.

Pheromone programs have been used for several decades around the globe and to date (2012) there is no documented public health evidence to suggest that agricultural use of synthetic pheromones is harmful to humans or to any other non-target species. However, continuing research is being conducted. In order to reduce the pesticidal load in the environment and with sustainability, certain behavioral chemicals could be harnessed. Such an endeavor is the use of sex pheromones. This dynamic and paradigm shift in management strategies satisfies all the bio-safety concern as well as playing a pivotal role in combating insect pests of high-value and damage sensitive crops. The research from socio-economists will help to guide the strategy for promotion of the pheromone technology in Bangladesh successfully. This review facilitates technical ways for identification/confirmation, understanding of the nature of pheromones, their potential for pest control and overview in respect of Bangladesh.

References

Akhtar, N. and Mandal, K.A.M.S.H. 2008. Effects of caffeine and castor oil on growth of *Tribolium castaneum* Herbst (Coleoptera: Tenebrionidae). *Bangladesh J. Entomol.,* 19 (1): 1-8.

Alam, S.N., Rashid, M.A., Rouf, F.M.A., Jhala, R.C., Patel, J.R., Satpathy, S., Shivalingaswamy, T.M., Rai, S., Wahundeniya, I., Cork, A., Ammaranan, C., Talekar, N.S. 2003. Development of an integrated pest management strategy for eggplant fruit and shoot borer in South Asia. Technical Bulletin 28, AVRDC . The World Vegetable Center, Shanhua, Taiwan, 66 p.

Alpizar, D., Fallas, M., Oehlschlager, A.C., Gonzalez, L.M., Chinchilla, C.M. and Bulgarelli, J. 2002. Pheromone mass trapping of the west Indian sugarcane weevil and the American palm weevil (Coleoptera: Cuculionidae) in palmito palm. *Florida Entomologist*, 85: 426-430.

Annonymous. 2003. Pheromones– the trappings of sustainable pest control? *New Agriculturist*. Retrieved from http://www.new-ag.info/03-4/develop/dev03.html

Attygalle, A.B., Schwarz, J. and Guanawardena, N.E. 1988. Sex pheromone of brinjal shoot and pod borer *Leucinodes orbonalis* Guenee (Lepidoptera: Pyralidae: Pyraustinae). *Z. Naturforschung*, 43: 790-792.

BBS (Bangladesh Bureau of Statistics). 2011. Monthly Statistical Bulletin, Bangladesh. (September, 2011). Bangladesh Bureau of Statistics Division, Ministry of Planning, Govt. People's Republic of Bangladesh, Dhaka. p. 142.

Bierl, B.A., Beroza, M., Staten, R.T., Sonnet, P.E. and Alder, V.E. 1974. The pink bollworm sex attractant. *J. Econ. Entomol.,* 53: 172-173.

Cork, A. and Krishnaiah, K. 2000. Pheromones for control of yellow stem borer, *Scirpophaga incertulas* (Walker) (Lepidoptera: Pyralidae) in India. *Proceedings of XXI International Congress of Entomology*, Iguassu, Brazil. pp. 132-141.

Cork, A., Alam, S.N. and Talekar, N.S. 2005. Development and commercialization of mass trapping for control of brinjal borer, *Leucinodes orbonalis* in South Asia. pp. 29-33. *In*: Proceedings of National Symposium on Recent Advances in Integrated Management of Brinjal Shoot and Fruit Borer, 3-4 October 2005. Indian Institute of Vegetable Research, Varanasi, India.

Cork, A., Alam, S.N., Das, A., Das, C.S., Ghosh, G.C., Farman, D.I., Hall, D.R., Maslen, N.R., Vedham, K., Phythiam, S.J., Rouf, F.M.A and Srinivasan, K. 2001. Female sex pheromone of brinjal fruit and shoot borer, *Leucinodes orbonalis* blend optimization. *J. Chem. Ecol.,* 27(9): 1867–1877.

Cork, A., Alam, S.N., Rahman, M., Kamal, N.Q. and Talekar, N.S. 2004a . Mass trapping for control of rice and vegetable pests in South Asia. *In*: Proceedings of the XXII International Congress of Entomology, 15-21 August 2004, Brisbane, Australia (Abstract). p. 78.

Cork, A., Kamal, N.Q., Ahmed, S., Jayanth, D.K.P. and Casagrande, E. 2004b. Commercial adoption of pheromone as a component in ICM of rice in Bangladesh. *Asia: Reports on DFID crop protection programme*, (September 2001-October 2003) pp. 165-168.

Cork, A., Kamal, Q.N., Alam, S.N., Choudhury, S.C.J. and Talekar, N.S. 2003. Pheromones and their application to insect pest control-A Review. *Bangladesh J. Entomol.*, 13: 1-13.

Hallett, R.H., Oehlschlager, A.C. and Borden, J.H. 1999. Pheromone trapping protocols for the Asian palm weevil, *Rhynchophorus ferrugineus* (Coleoptera: Curculionidae). *International J. Pest Management,* 45: 231-237.

Islam, M.A. 2009. Study on the Lepidopteran Sex Pheromones including Multiple Double Bonds. *PhD Thesis,* Graduate school of bio-applications and system engineering, Tokyo University of Agriculture and Technology, Tokyo, Japan. p. 147.

Islam, M.A. 2012. Knowledge and Practice of Pheromone Technologies: A Case Study of a Representative District in Bangladesh. *Academic Research International,* (In Press).

Islam, M.A. and Becerra, J.X. 2011. Comparative analyses of chemical composition in the leaves of three *Bursera* species and their effect on insect pest. *J. Expt. Biosci.,* 2(2): 29-34.

Islam, M.A. and Begum, S. 2011. Quantitative analysis of α–mangostin in mangosteen fruit rind extract. *Int. J. Agril. Res. Innov. & Tech.,* 1(1&2): 55-59.

Islam, M.A., Ahmed, K.H., Khan, A.B. and Uddin, M.K. 2008a. Individual and combined effects of three plant extracts and a chemical insecticide (Sevin 85SP) on pulse beetle. *Bangladesh J. Entomol.,* 19(1): 9-17.

Islam, M.A., Masanobu Yamamoto, Mieko Sugie, Hideshi Naka, Toshirou Suzuki, Tetsu Ando. 2008b. LC/MS Analyses of Dienyl Lepidopteran Sex Pheromones for Unraveling the Diversity of Mating Communication Systems. 4th Pan Pacific Conference on Pesticide Science. Honolulu, Hawaii, USA (June, 1-5). p. 96.

Islam, M.A., Yamakawa, R., Do, N.D., Numakura, N., Suzuki, T. and Ando, T. 2009. Instrumental analysis of terminal conjugated dienes for reexamination of the sex pheromone secreted by a Nettle moth, *Parasa lepida lepida. Biosci. Biotechnol. Biochem.,* 73 (5): 1156-1162.

Islam, M.A., Yamamoto, M., Sigie, M., Naka, H., Tabata, J., Arita, Y. and Ando, T. 2007. Synthesis and Characterization of 2,13- and 3,13-Octadecadienals for the identification of the sex pheromone secreted by a clearwing moth. *J. Chem. Ecol.,* 33: 1763-1773.

Kovaleski, A., Botton, M., Eiras, A.E. and Vilela, E. 1998. Lagarta enroladeira da macieira *Bonagota cranaodes* (Meyrick, 1937) (Lepidoptera: Tortricidae): Bioecologia, moitoramento e controle. EMBRAPA, CNPUV, Bento Goncalves. p. 67-87.

Lopez, J.D., Shaver, T.N. and Dickerson, W.A. 1990. Population monitoring of *Heliothis* spp. using pheromones. pp. 473-496. In: R.L. Ridgway, R.M. Silverstein, M.A. Inscoe, (eds) *Behaviour-modifying Chemicals for Insect Pest Management.* Marcel Dekker, New York.

Mamun, M.S.A., Shahajahan, M. and Ahmad, M. 2008. Laboratory evaluation of some indigenous plant extracts as repellent against red flour beetle, *Tribolium castaneum* Herbst. *Bangladesh J. Entomol.,* 18(1): 91-99.

Mazumder, F. and Khalequzzaman, M. 2010. Eggplant shoot and fruit borer *Leucinodes orbonalis* Guenee. male moth catch in sex pheromone trap with special reference of lure elevation and IPM. *J. Bio-sci., 1*8: 9-15.

Mondal, M.H. 2010. Crop Agriculture of Bangladesh: Challenges and opportunities. *Bangladesh J. Agril. Res.,* 35 (2): 235-245.

Oehlschlager, A.C., Chinchilla, C., Castillo, G., and Govzalez, L. 2002. Control of red ring disease by mass trapping of *Rhynchophorus palmarum* (Coleoptera: Curculionidae), *Florida Entomologist,* 85: 507-513.

Prasad, H., Singh, H.M., Singh, A.K. 2005. Effective range of sex pheromone of *Leucinodes orbonalis* Guen. *J. Appl. Zool. Res.* 16(1): 81-82.

Ranga Rao, G.V., Wightman, J.A. and Ranga Rao, D.V. 1991. The development of a standard pheromone trapping procedure for *Spodoptera litura* (F.) (Lepidoptera: Noctuidae) population in groundnut (*Arachis hypogaea* L.) crops. *Tropical Pest Management.* 37: 37-40.

Uddin, A.B.M.A., Alam, S.N. and Alam, M.Z. 2008. Effect of different pheromone trap designs and installation height in the brinjal field for catching brinjal shoot and fruit borer male moth. *Bangladesh J. Entomol.* 18(1): 11-21.

Vang, L.V., Islam, M.A., Do, N.D., Hai, T.V., Koyano, S., Okahana, Y., Ohbayashi, N., Yamamoto, M. and Ando, T. 2008. 7,11,13-Hexadecatrienal identified from female moths of the citrus leafminer as a new sex pheromone component: synthesis and field evaluation in Vietnam and Japan. *J. Pestic. Sci.,* 33 (2): 152-158.

Williams, D.G. 1989. Forecasting codling moth spray dates with pheromones traps and weather data. pp. 115-118. In: T.E. Bellas (ed). Application of pheromones to pest control. *Proceedings of a workshop held at CSIRO Entomology,* Canberra, CSIRO. Entomology 1989.

Zhu, P., Kong, F., Yu, S., Yu, Y., Jin, S., Hu, X. And Xu, J. 1987. Identification of the sex pheromone of eggplant borer, *Leucinodes orbonalis* Guenee (Lepidoptera: Pyralidae). *Z. Naturforschung,* 42: 1347-1348.

ALLELOPATHIC STUDIES ON MILK THISTLE (*Silybum marianum*)

Shamima Sultana[1*] and Md. Asaduzzaman[2,3]

Abstract

Declining crop yield due to weeds and their resistance to herbicides are major constraint for successful crop productions. Milk thistle (*Silybum marianum*) is common weed species in Australian cropping rotation. Allelopathic potentiality of milk thistle on different crops has been documented sporadically, but there is no literature on about ryegrass and canola. Therefore, a laboratory based allelopathic extracts bioassay was conducted. The hot water extracts was prepared from milk thistle plant parts added into water with ration of 1: 10 (plant sample: distilled water) where mixture was heated 10 minutes. After heat treatment samples was immediately sieved and centrifuged and the resulted solution was treated as 100% concentration. Separately, to get the fresh water extract plant sample was added into water (1:10) and kept 24 hours in room temperature. After 24 hours, the sample was sieved and centrifuged and collected samples result was treated 100% concentrations. To obtain 50% concentration, both hot and fresh samples were diluted with distilled water. Therefore the experiment was conducted with five different treatment concentrations (0, 50% hot water extracts, 50% fresh water extracts, 100% hot water extracts and 100% fresh water extracts). The experiment was comprised with RCBD design with three replications under control conditions. During experimental period the allelopathic effects of donor species on germination and seedling growth of ryegrass and canola was observed. Results shows, germination and seedling growth of both receiver species are inhibited by milk thistle extracts. Extracts from fresh water at 100% was more toxic to receiver species followed by 50% concentration of fresh and 100% from hot water extracts. This concentration reduced the root, shoot growth of ryegrass and canola 84.971%, 84.269% and 89.898%, 87.394%, respectively. The result also revealed that allelopathic pattern of hot water extracts was same however; it is less toxic to both receiver species.

Keywords: Allelopathy, milk thistle, ryegrass and canola

[1]College of Development Alternative (CODA), 14 A Dhanmondi, Dhaka 1205, Bangladesh
[2]Environmental and Analytical Laboratories, Faculty of Science, Charles Sturt University, Wagga Wagga, NSW 2678, Australia & [3]School of Agricultural and Wine Sciences, Faculty of Science, Charles Sturt University, Wagga Wagga, NSW 2678, Australia

*Corresponding author's email: shamima.agro@yahoo.com (Shamima Sultana)

Reviewed by Mirza Hasanuzzaman, Shere-Bangla Agricultural Univestity, Dhaka, Bangladesh

Introduction

Weeds are disliked and are not seen as of such use except for some recognition that they are part of nature (Qasem and Foy, 2008). Although, their ultimate values in natural and agro-ecosystems have not been completely discovered, but many of their negative and positive impacts on both systems are very familiar. To minimize the negative impacts, agricultural crops have increasingly relied on chemical herbicides. However, problems associated with intensive herbicide use include soil and groundwater contamination, development of herbicides resistance weeds and the escalating cost of developing of new herbicides (Worsham, 1989). Recent assessments of the allelopathic effect of crops and/or weed on weeds have the goal of using naturally produced allelochemicals to reduce reliance on herbicides (Einhelling and Leather, 1988; Putnam and Duke, 1978). In allelopathy, plants provide themselves with a competitive advantage by releasing the phytotoxins into nearly environment (Pratley, 1996). The concept of allelopathy interaction between weeds is interesting. There is evidence that certain weeds species have the potential to be used in solving problems of other weed species and represents an excellent source of natural chemicals that may be involved in developing natural herbicides (Qasem and Foy, 2008).

Milk thistle (*Silybum marianum* Gaertn.) is a winter annual or a biennial noxious weed belonging to *Asteraceae* (Young *et al.,* 1978; Austin *et al.,* 1988; Groves and Kaye, 1989). Its current distribution includes most temperate areas of the world including Australia (Chambreau and MacLaren, 2007). In Australia, it is classified as a declared plant (noxious weed) and particularly prevalent in Victoria and New South Wales, where dense stands can develop on soils of high fertility (Dodd, 1989). Research suggests that ripe fruit of milk thistle contains flavonoids that are used to prepare antihepatotoxic drugs, and have different medicinal value. Apart from the medicinal value, milk thistle may be used for the phytoremediation of polluted). It is a major weed in sugar beet wheat and canola (*Brassica napus* L.) causing large yield reductions (Omtvetd, 1984; Khan and Marwat, 2006; Shimi *et al.,* 2006). However, researches have documented the allelopathic effect of milk thistle on mustard (*Brassica juncea* L.), cucumber (*Cucumis sativus* L.), wheat, and sorghum (*Sorghum bicolour* L.) (Inam and Hussain, 1988) but no such reports exist of milk thistle caused negative effect on annual ryegrass (*Lolilum rigidum* L). The present study was conducted to evaluate and determine the germination and initial growth performance of ryegrass and its associated crop canola under different extracts concentration of milk thistle. The study also aimed to determine the phototoxicity and stability of milk thistle under heat treatment. The significant results will help to create a sustainable weed control option especially against herbicides resistance ryegrass.

Materials and Methods

The allelopathic effects of milk thistle was investigated by implying laboratory study during 10 December 2011 to 15 Decemebr, 2011 at NSW DPI, Wagga Wagga, Australia.

Extracts preparation

Above ground mature milk thistle plants were collected from Wagga Wagga campus of Charles Sturt University, Collected samples were chopped out into 2 cm pieces. 10 g chopped plant samples were added to 100 mL distilled water in a 500 mL volumetric flask and boiled 10 minutes. To get the fresh water extracts, another 10 g chopped samples was added to 100 ml distilled water in a 500 ml) volumetric flask and wrapped with aluminum foil and kept for 24 hours at room temperature (25°C). The heated and soaked extracts were filtered through two layers of cheesecloth used centrifuged for 10 minutes at 5000 rpm using an Eppendorf 5810 bench top centrifuge. It was separately treated with 100% concentration of hot water and fresh water extract, respectively. The 100 % concentration of both extracts (hot and fresh water) was diluted with distilled water to obtain 50% and 0% (control).

Allelopathy assessment

To know the allelopathic effect of milk thistle annual ryegrass and canola were used as the receiver species. Twenty seeds (non- surface sterilized) of ryegrass and canola were sown into 9 cm petri dishes lined with one layer with one layer of Whatman No. 1 filter paper. Five mililitre of each extract from different concentrated were delivered to each petri dish and distilled water (5 ml) was used as control. Each petri dish with its cover was sealed with a piece of parafilm to reduce evaporation. All dishes were maintained in a control growth room at 21/19°C with day/light for 12 h/12 h. First two days a black polythene was used to cover the all petri dishes, and after two days then cover was removed. Germinated seeds > 1 mm radical were recorded and root and shoot lengths were measured after 5 days of incubation.

Experimental design and statistical analysis

A randomized block design with three replications were used for the experiment. The experiment was repeated twice to confirm the consistent results. All experimental data were subjected to analysis using Genstat 5 (version 13) and treatments means were tested separately with least significant difference (LSD) at 5% level of probability. Percent of germination was calculated as (germinated seeds/total given seeds) X 100.

Results and Discussion

Germination percentages

In our bioassay, the milk thistle extracts significantly inhibited the germination of both species.

Annual ryegrass

Types of extracts concentration noticeably retarded the germination of ryegrass at 3 and 5 days after sowing. Germination of viable ryegrass seeds significantly reduced by fresh water extracts at 100% and 50% concentration followed by 100% extract from hot water extract. The maximum germination percentage 88.33 and 98.33 were obtained by control treatment at 3 and 5 days after sowing, respectively. Fresh water extracts were more toxic compare to hot water extracts for germination of ryegrass. The control gave the highest germination percentage, which was 88.33 and 98.33 at 3 and 5 days, respectively followed by concentration at 50 % from hot water and there was no significance than control. It might be due to the heat treatment diluted the phyto-toxicity of milk thistle and which is less toxic to ryegrass germination process.

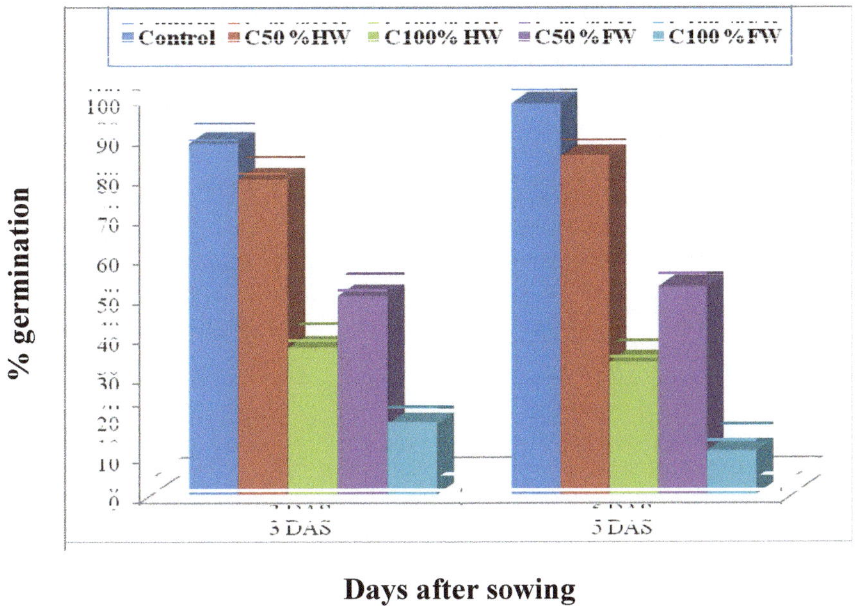

Fig. 1. Efffect of milk thistle extracts on germination of annual ryegrass at 3 and 5 days after sowing. LSD at 5% level for 3 and 5 DAS is 18 and 15

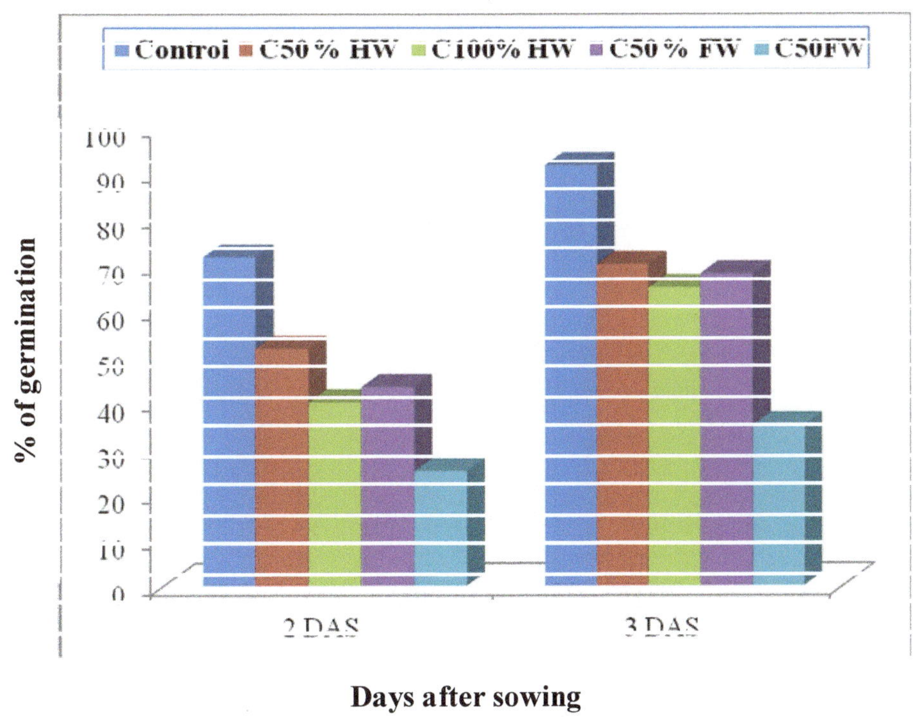

Fig. 2. Efffect of milk thistle extracts on germination of canola at 3 and 5 days after sowing. LSD at 5% level for 3 and 5 DAS is 13 and 12

Canola

In canola, the rate of germination was noticeably reduced by all milk thistle extracts at 2 as well 4 days after sowing. Statistically significant lowest percentage of germination 25 and 35 was occurred by extract from fresh water treatment at 100% at 2 and 4 days, respectively. Although at 3 days after sowing there was no statistically noticeable difference among the germination rate caused by concentration at 100%, 50% from hot and 50% from fresh water extracts, but germination rate was significantly reduced by concentration at 100% from fresh water extract. In the contrary, control gave the maximum germination percentages of canola at each observation.

Root and shoot growth

Annual ryegrass

The radical and plumule growth of ryegrass was demonstrably inhibited in both hot and fresh water extracts especially in fresh water extracts. Significantly, lowest root (12; 45 cm) and shoot length (14; 48 cm) were measured by extracts from fresh water at 100 % and 50 %, respectively (Fig. 3 and 4). These results suggest, due to 24 hours submerged of milk thistle, it released more phytotoxic substances into water that inhibited the seedling growth of ryegrass. The hot water extracts was also inhibited the root growth of ryegrass, although it was less toxic compare to fresh water extracts but root and shoot length was reduced by heat treated extracts at 100 % and 50%. Our finding also reporting that the radical of ryegrass was more sensitive to milk thistle than plumule growth. Similar research also reported separately by Wu *et al.* (1999) and Asaduzzaman *et al.* (2012). They reported that root growth of ryegrass is suppressed more than its shoot growth due to wheat and canola allelopathy.

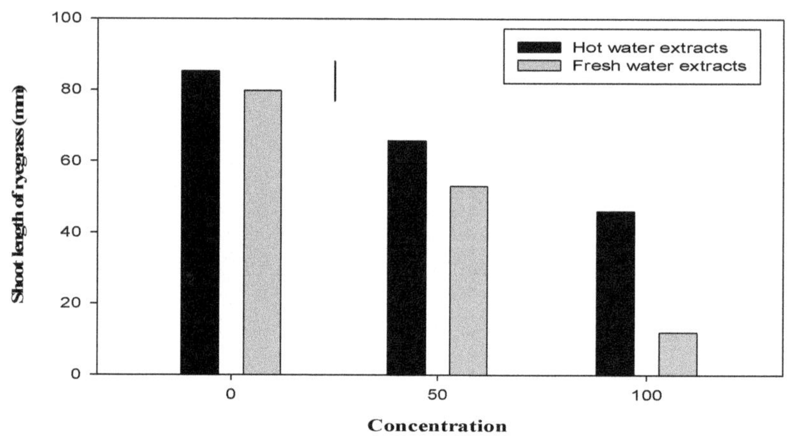

Fig. 3. Allelopathic extract effect of milk thistle on root grwoth of ryegrass. LSD at 5% level is 11

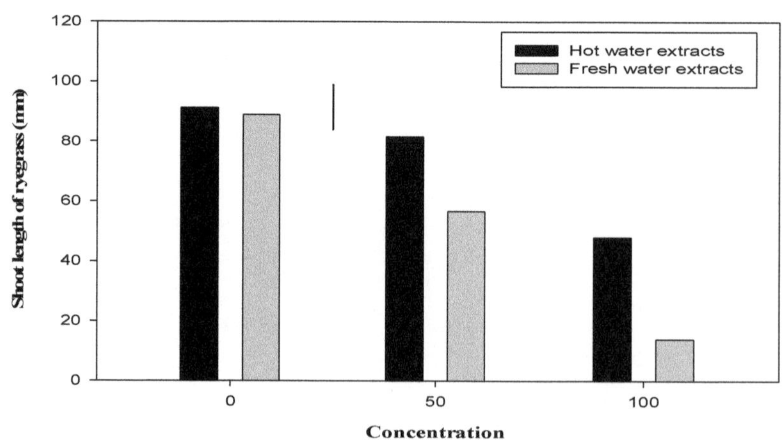

Fig. 4. Allelopathic extract effect of milk thistle on shoot grwoth of ryegrass. LSD at 5% level is 15

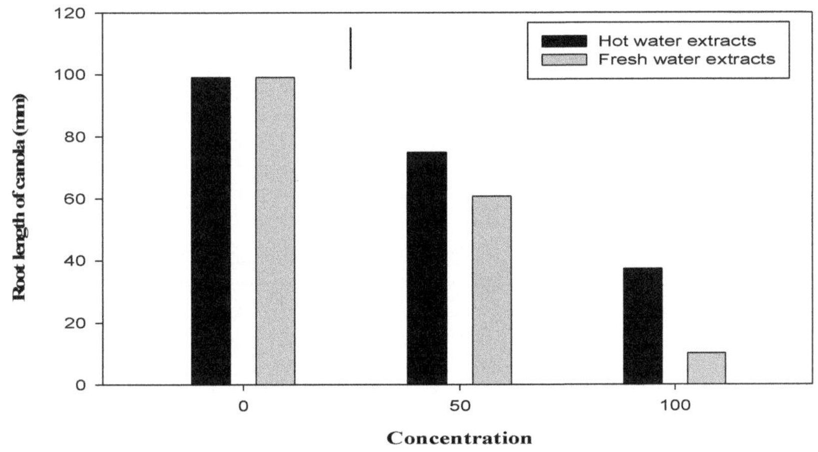

Fig. 5. Allelopathic extract effect of milk thistle on root grwoth of canola. LSD at 5% level is 13

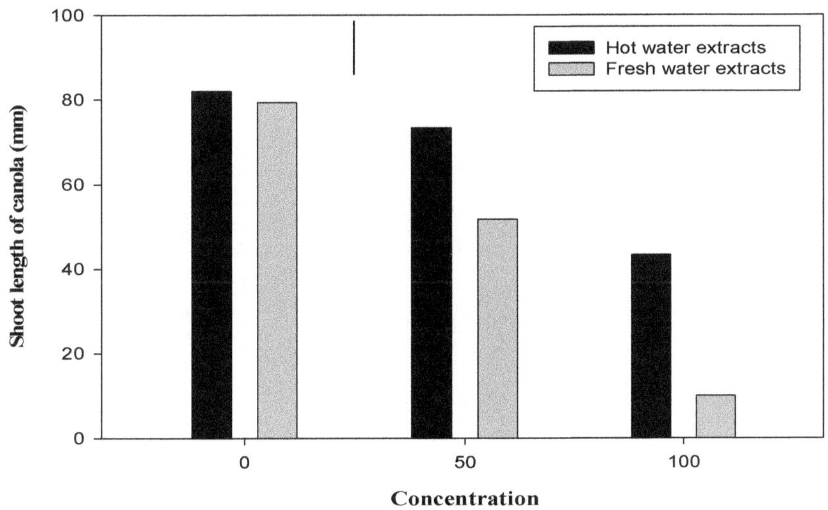

Fig. 6. Allelopathic extract effect of milk thistle on shoot grwoth of canola. LSD at 5% level is 12.6

Canola

Aqueous extracts from both fresh and hot water samples significantly reduced the root and shoot growth of canola compare to control (Fig. 5 and 6). Similarly, ryegrass, the root and shoot growth of canola was also inhibited by fresh water extracts. Concentration at 100% from fresh water restricted the 89.898% and 87.394% root and shoot length of canola, respectively. Suppression of canola seedling growth may be cause by slow germination that was affected by hydrolysis products of milk thistle into fresh water. Another explanation could be associated with a, increasing extract concentration that increasing the phytotoxicity of donor species, which leaded to, restricted the cell division and elongation of meristematic tissues of canola. This results are in close to agreement with the findings of Inam and Hussain (1988) who documented the allelopathic effect of milk thistle and reported that aqueous extracts from the leaves, stems, inflorescences, and roots decreased the germination and early growth of mustard (*Brassica juncea* L.), cucumber (*Cucumis sativus* L.), wheat, and sorghum (*Sorghum bicolour* L.). Separately the maximum root and shoot length were recorded at control followed by extracts from hot water treatment at concentrations 50% and 100%. These results might be due to that, heat treatment diluted or destroy the phytotoxicity of milky substances of donor; therefore, it was less toxic to germination and seedling growth of canola.

Conclusion

In this studies milk thistle demonstrated allelopathic effects on ryegrass and canola including reduced seed germination and reduce seedling growth. Over all the allelopathic potential of milk thistle increased with increased concentration. In addition, stability and level of phytotoxicity of milk thistle vary on methods for extract preparation and solvent or media. The fresh water extract is more toxic than the heated extract but it is clear that in both condition there is some inhibitory substances presence in milk thistle tissues causing this allelopathicity. These could be used as a potential natural herbicide resource but they must first be identified and their mode of action studies. Future plans are also need to conduct field experiment.

References

Asaduzzaman, M., An, M., Pratley, J.E., Luckett, D.J. and Lemerle, D. 2012. Allelopathic impacts of growth duration and density of canola on annual ryegrass. (Personal communication).

Austin, M.P., Fresco, L.F.M., Nicholls A.O., Groves, R.H. and Kaye, P.E. 1988. Competition and relative yield estimation and interpretation at different densities and under various nutrient concentrations using *Silybum marianum* and *cirsium vulgare*. *J. Ecol.*, 76: 157–171.

Chambreau, D. and MacLaren, P.A. 2007. Got milk thistle? An adaptive management approach to eradicating milk thistle on dairies in King county,Washington state. pp. 107-109. In: *Meeting the Challenge: Invasive Plants in Pacific Northwest Ecosystems* (ed. by Harrington T.B. and Reichard S.H.). General technical report. No. PNW-GTR-694, June. Pacific Northwest Research Station, Forest Service, United States Department of Agriculture, Portland.

Dodd, J. 1989. Phenology and seed production of variegated thistle, *Silybum marianum* (L.) in Australia in relation to mechanical and biological control. *Weed. Res.*, 29: 255–263.

Einhelling, F.A. and Leather, G.R. 1988. Potential for exploiting allelopathy to enhance crop production. *J. Chem. Eco.*, 14: 1829-1844.

Groves, R.H. and Kaye, P. E. 1989. Germination and phenology of seven introduced thistle species in Southern Australia. *Aust. J. Bot.*, 37: 351–359.

Inam, B. and Hussain, F. 1988. Allelopathic effects of *Silybum marianum* Gaertn. *Sarhad J. Agric.*, 4: 481–494.

Khan, M.A. and Marwat, K.B. 2006. Impact of crop and weed densities on competition between wheat and *Silybum marianum* Gaertn. *Pak. J. Bot.*, 38: 1205–1215.

Omtvedt, I.T. 1984. *A Descriptive Guide for Major Nebraska Thistles*. Agricultural Experiment Station, University of Nebraska–Lincoln, Lincoln, NB.).

Pratley, J.E. 1996. Allelopathy in annual ryegrass. *Plant Prot. Quar.*, 11: 213-214.

Putnam A.R. and Duke, W.B. 1978. Allelopathy in agro- ecosystem. *Ann. Rev. Phyto-path.*, 16: 431-451.

Qasem, J.R. and Foy, C.L. 2008. Weed allelopathy, Its ecological impacts and future prospects. *J. Crop Prod.*, 4(2): 43-119.

Shimi, P., Poorazar, R., Jamali, M. and Bagherani-Torshiz, N. 2006. Evaluating clopyralid as a broad leaf herbicide in canola fields of Iran. *Pak. J.Weed Sci. Res.*, 12: 307–311.

Worsham, A.D. 1989. Current and potential techniques using allelopathy as an aid in weed management. (In C. H. Chou and G. R. Waller eds.), *Phyto-chem. Eco.*, pp. 275-291.

Wu, H., Pratley, J.E., Lemlerle, D. and Haig, T. 1999. Crop cultivars with allelopathic capability. *Weed Res.*, 39: 171-180.

Young, J.A., Evans ,R.A. and Hawkes, R.B. 1978. Milk thistle (*Silybum marianum*) seed germination. *Weed Sci.*, 26: 395–398.

ASSESSMENT OF POTABLE WATER SUPPLY SOURCES IN ILORIN METROPOLIS, KWARA STATE, NIGERIA

J.J. Musa[1*] and G.A. Fumen[2]

Abstract

Water distribution system plays a vital role in preserving and providing a desirable life quality to the public. In this regards, the reliability of supply is a major component. This study was conducted using the investigative research approach method; 5000 questionnaires which were randomly distributed round the township of Ilorin and 4658 (93.16%) were returned. Among them, 2350 respondents said that they have their own wells dugged in their respective homes, 1100 respondents had their taps running at specific periods of the week, 750 individuals said that they have boreholes available, 308 people declared that they make use of the streams and rivers running around their areas while about 150 of the remaining buy water from the water vendors so they cannot tell the exact source of the water. It was discovered that majority of the residential areas in Ilorin do not depend on the tap water supply rather they depend on dugged wells and bore holes.

Keywords: Boreholes, Dam, Flow Rate, Pumps, Supply Rate, Tanks, Water, Well

[1]Department of Agricultural and Bioresources Engineering, Federal University of Technology, P M B 65 Minna, Nigeria
[2]College of Agriculture, Ahmadu Bello University, Zaria, Nigeria

*Corresponding author's email: johnmusa125@rocketmail.com (J.J. Musa)

Introduction

Water distribution system plays a vital role in preserving and providing a desirable life quality to the public, of which the reliability of supply is a major component. Reliability analysis of a water distribution system is concerned with its ability to deliver water to individual consumers in the required quantity and under a satisfactory pressure (Al-Zahrani and Syed, 2005). Understanding the causes of water poverty is a matter of fundamental importance. The insufficient distribution of safe water is observed because of the unsustainable demographic growth, particularly in developing countries. The population pressure determines increasing national water stress, defined as a ratio between the total volume of water theoretically available in a country and the total internal population, equivalent to a value below 1000 m^3/year/per capita. The macro-regionalization of water scarcity indicates that eighty countries, involving the 40% of the total world population, are currently suffering for water stress. The logical solutions sorting out from the consideration of water stress would rather being concentrated on increasing water supply, from one side, and on policies for population control, from the other (Ahmad, 2000).

Asa dam water works situated in the northwestern part of Ilorin in the north central part of Nigeria. The city has a population of over 800,000 inhabitants. Asa dam water work is in one phase, which supplies over 300,000 m^3 of water per day when the treatment plants are operational. The maximum height of the dam is about 45 m. The raw water intake from the dam reservoir is through pipe openings at different levels in circular concrete intakes tower located within the reservoir. There are eight different levels giving the opportunity to select a particular level for raw water abstraction depending on its characteristics. Due to the height of the dam, there are three broad regions in the reservoir. The top 5 m in which the effects of sunlight and air are felt (called epilimnon); the bottom which is 10 meters below the surface dose not experience any form of sunlight or air (hypolimnon) and the region between them is called metalimnon (Shinstine *et al.*, 2002). Pipelines are essential part of water distribution system, facilitating the transportation of water in varied quantities from headwork's through the treatment plants to the service reservoir and eventually to the consumers. The main trunk conveys water from source to the plant and to the service reservoir while the distribution mains from part of the distribution network and the service pipe connections made on distribution main (Lansey *et al.*, 1989).

Untreated water also called raw water may come from groundwater or surface water sources such as lakes, rivers and streams. The raw water is usually transported to a water treatment plant where it is processed to produce treated water also known as portable finished water. The degree to which the raw water was processed to achieve portability depends on the characteristics of the raw water, relevant drinking water standard treatment processes and the characteristics of the distribution system (Gleick, 1996).

Before water is distributed, the treated surface water usually enters a unit clear well service, which is of three main purposes in water treatment. This provide the contact time for disinfectant such as chlorine which is added towards the end of the treatment process during which adequate contact time is required to achieve acceptable level of disaffection. Secondly, the clear well provides storage that act as a buffer between the treatment plant and the distribution system. The distribution system naturally fluctuates between periods of high and low water usage. Thus, the clear well stores excess untreated water during periods of low demand and deliver it during period of peak demand. Thirdly, the clear well can serve also as a source for back was water for cleaning plant filters which when needed is used at a high rate for a short period of time (Lawrence *et al.*, 2002).

Transmission and distribution system of water can either be looped or branched. In the looped systems, there may be several different paths that the water can follow to get from the source to a particular consumer while in a branch system (tree system); the water has only one possible path from the source to consumers (Allen, 1998).

Planners carefully research all aspects of a water distribution system and try to determine which major capital improvement projects are necessary to ensure the quality of service for the future. The process is used to project system growth and water usage for several years to come for system and population growth. Not only can such planning be used to identify potential problem area such as future low pressure areas, water quality size, locating new transmission main pumping stations and storage facilities but can also be used to ensure that the predicted problems will not occur (Al-Zahrani and Syed, 2005).

The aim of this study is to acertain the various sources of water used for domestic services in the metropolitan city of Ilorin in the north central area of Nigeria.

Methodology

Ilorin is the state capital of Kwara located on coordinates of 8° 30'N and 4° 33'E (www.ilorin.org). This study was conducted using the Investigative Research Approach method, which deals with the distribution of questionnaires, personal interview and field observations. Five thousand (5000) questionnaires were randomly distributed round the township of Ilorin, the capital of Kwara State. The areas for which the questionnaires were distributed includes Adewole Estate, Harmony Estate, Government Reservation Area (GRA), the heart of Ilorin township, Saw-mail area, Kulende Housing Estate and its environs, the university permanent site, Tanke area and the Kwara State polytechnic area. The feedback from these questionnaires were analyzed and presented in graphical and chart forms.

Results and Discussion

Out of the 5000 questionnaires distributed, only 4658 (93.16%) were returned. Table 1 shows the detailed distribution of water sources in the township area of Ilorin. Table 1 shows that 2350 respondents said that they have their own wells dugged in their respective homes, 1100 respondents said that they have taps running at specific periods of the week whereas another 750 said that they have boreholes available and 308 people said that they make use of the streams and rivers running around their areas. About 150 of the remaining buy water from water vendors so they cannot tell the exact source of the water. Figure 1 shows the percentile distribution of water sources in Ilorin

Table 1. Detailed distribution of water sources in Ilorin

Sl. No.	Water sources	Population of people using this sources
1	Well	2350
2	Tap	1100
3	Bore Hole	750
4	Stream/River	308
5	Other sources	150
	Total	4658

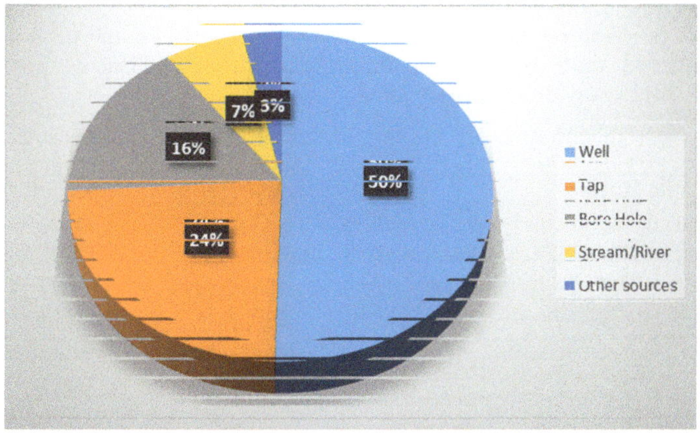

Fig. 1. Percentile distribution of water sources in Ilorin

Though most of the residents in Ilorin that have open dugged wells also have water, pipelines laid to their various houses and when asked how frequent the tap water flows they said that it runs during the rainy season only while during the dry season the taps do not run. This prompted them to get another source of water supply. This also applied to those residents with boreholes which service them during the off peak periods.

Table 2. Functionality of the various water sources in Ilorin

Sl. No.	Water Sources	Functional	Non-functional
1	Well	2350	-
2	Tap	728	372
3	Bore Hole	750	-
4	Stream/River	308	-
5	Other sources	84	66

Table 2 shows the functionality of the various sources of water throughout the year. It was discovered that only the tap water was not regular throughout the year while some of the areas where they had limited wells and boreholes complained of the slow recharge rate which was discovered was due to the rate of drawing water from the wells and boreholes was higher than the rate at which they are recharged. This implies that these various residents need to have specific times of collecting water from the wells and boreholes.

It was discovered that out of the 750 houses that had access to tap water, 180 of such houses had regular supply, which means that at any particular time of the day, water is always running from these taps. Such houses are mostly located around the government residential/reserved area of the town. One of such factor also could be the location of Agba dam, which was established to service the Government House and the principal officers working there. 230 houses visited during the course of this study indicate that the residents usually have water supply every other day. According to the respondents, the taps only start running from 8am in the morning to about 12 midnight of the same day. In other to have enough water to use, they install water tanks of various sizes to store water for future use. They also opinioned that the pressure of the water running from the taps was always very high which may be due to the closeness of these houses to Agba dam. Another 198 residents stated that their taps only run once a week for 24 hours during which they store water in storage tanks for future use and when the tap does not run, they supplement their stores with the well water. They also complained about the pressure of water running from the tap, which they said was low. Most of these houses are located far away from the dam and treatment plant with a lot of branch offs from the main distribution pipeline. Because of this, a reduction of water pressure in the tap may be one of the major reasons for the poor distribution of water in the area. They also stated that only during the dry season that the rate of flow of water at once a week is usually not consistent which means that they result into going to the dam area to collect water for drinking and other domestic uses with their cars and where water vendors are available, they buy from them.

The remaining 142 residents stated that their taps run only once in two weeks which again is not regular and the pressure are usually the pressure is very low. This prompted the people in the area to get together to construct motorized boreholes with a 500,000 m^3 overhead tank for easy distribution of water to the various houses. However, the Federal and State governments alongside some International donor agencies usually support these kinds of projects. Table 3 shows the rate of tap supply to the metropolitan city of Ilorin while Figure 2 shows the percentile rating of water supply to some residential areas of Ilorin. It was discovered that every other day water supply had the highest water supply rate of 230 houses and closely followed by the once a week supply rate of 198 houses. The lowest water supply rate was that of once in two weeks.

Table 3. Rate of tap water supply

Sl. No.	Supply Rate	No of Houses	Percentile Rating
1	Regular	180	24
2	Every other day	230	30.67
3	Once in a week	198	26.40
4	Once in 2 weeks	142	18.93
5	Others	-	-

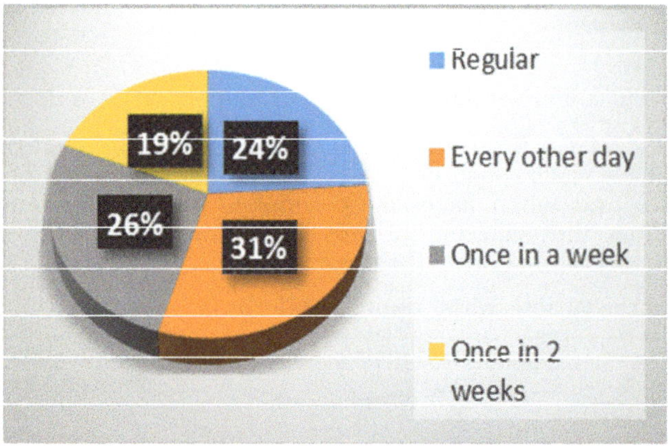

Fig. 2. The percentile rating of water supply to some residential areas of Ilorin

Most residents of Ilorin interviewed stated that the wells and boreholes were most consistent both during the dry and rainy seasons while the tap water was always regulated. The State Government through the State Ministry of Water Resources has constructed some storage tanks in some specific location of the town with no good pump stations to effectively pump water from these tanks to the public.

Conclusions

Evaluation of water distribution system in Ilorin is studied to measure the ability of meeting the consumers demand. It was discovered that majority of the residential areas in Ilorin do not depend on tap water supply rather they depend on dugged wells and bore holes. Thus, in the need to drill and dig boreholes and wells respectively resulting in almost every house. These methods of underground water extraction possess many environmental hazards such as landslides. Due to the topographical nature of Ilorin and its environment, more water storage facilities and pump stations should be provided at major areas of the town to service the residents with treated water. This will reduce the rate of drilling and digging of boreholes and wells respectively and will further enhance prompt supply of water at every other day to the residents of the town.

References

Ahmad, M. 2000. Water pricing and markets in the Near East: policy issues and options. *Water Policy*. 2 (3): 229-242.

Allen, J.A. 1998. Moving water to satisfy uneven global needs: 'trading' water as an alternative to engineering it. *ICID Journal* (International Commission on Irrigation and Drainage). 47 (2): 1-8.

Al-Zahrani, M.A. and Syed, J.L. 2005. Evaluation of municipal water distribution system reliability using minimum cut-set method. *J. King Saud Univ*. 18: 67-82.

Gleick, P.H. 1996. Basic water requirements for human activities: meeting basic needs. *Water International*. 21: 83-92.

Lansey, K.E., Duan, N. and Mays, L.W. 1989. Water distribution system design under uncertainties. *ASCE J. Water Resources Planning & Management*. 115 (5): 630-644.

Lawrence, P., Meigh, J. and Sullivan, C. 2002. The water poverty index: an international comparison, Keele Economics Research Papers, 19 p.

Shinstine, D.S., Ahmed, I. and Lansey, K.E. 2002. Reliability/Availability analysis of municipal water distribution networks: case studies. *ASCE J. Water Resources Planning & Management*. 128 (2): 140-151.

www.ilorin.org (Accessed on the November 16, 2008).

ABUNDANCE OF PLANKTON POPULATION DENSITIES IN RELATION TO BOTTOM SOIL TEXTURAL TYPES IN AQUACULTURE PONDS

F. Siddika, M. Shahjahan* and M.S. Rahman

Abstract

Plankton is an important food item of fishes and indicator for the productivity of a water body. The present study was conducted to evaluate the effects of bottom soil textural conditions on abundance of plankton in aquaculture pond. The experiment was carried out using three treatments, i.e., ponds bottom with sandy loam (T1), with loam (T2) and with clay loam (T3). The ranges of water quality parameters analyzed were suitable for the growth of plankton during the experimental period. Similarly, chemical properties of soil were also within suitable ranges and every parameter showed higher ranges in T2. A total 20 genera of phytoplankton were recorded belonged to Chlorophyceae (7), Cyanophyceae (5), Bacillariophyceae (5), Euglenophyceae (2) and Dinophyceae (1). On the other hand, total 13 genera of zooplankton were recorded belonged to Crustacea (7) and Rotifera (6). The highest ranges of phytoplankton and zooplankton densities were found in T2 where low to medium-type bloom was observed during the study period. Consequently, the mean abundance of plankton (phytoplankton and zooplankton) density was significantly highest in T2. The highest abundance of plankton in the T2 indicated that pond bottom with loamy soil is suitable for the growth and production of plankton in aquaculture ponds.

Keywords: Phytoplankton, zooplankton, aquaculture, soil textural classes, water quality

Department of Fisheries Management, Bangladesh Agricultural University, Mymensingh-2202, Bangladesh.

*Corresponding author's email: shahjahanm75@yahoo.co.uk (M. Shahjahan)

Reviewed by Abdullah-Al Mamun, Noakhali Science and Technology University, Bangladesh

Introduction

Live organisms of the water consists of three major groups namely plankton, nekton and benthos. Among these, plankton is of fundamental importance to fisheries. Plankton is also a vital factor influencing the fish production. Phytoplankton is the basic primary producers of all types of water bodies and is used as food by fish directly or indirectly. The qualitative and quantitative abundance of phytoplankton indicate the productive status of a water body, whether it is an oligotrophic or a eutrophic one. Therefore, a thorough knowledge of abundance of phytoplankton and its quality in time and space in relation to environmental conditions has become a prerequisite for fish production. Existence of zooplankton production primarily depends on the primary production. Zooplankton is a link in food chain between the primary producers and nektonic and benthonic animals in higher trophic levels. Their functions decrease phytoplankton populations through grazing (Raymont, 1963); accelerate phytoplankton growth excreting nutrient substances which are finally metabolized (Ketchum, 1962); and supply themselves as food to predators.

The nutrients status of both soil and water plays a significant role in the growth and abundance of aquatic organisms, especially plankton and benthos. The chemical properties (nutrients status) have some growth promoting effect on the various species of benthos fauna (Habib *et al.,* 1984). On the other hand, nutrient status of soil depends on the type of soil texture. In soil science, the USDA (Donahue *et al.,* 1990) defines twelve major soil textural classes. Soil textures are classified by the fractions of each soil separate (sand, silt, and clay) present in a soil. Classifications are typically named for the primary constituent particle size or a combination of the most abundant particles sizes, e.g. sandy clay or silty clay. A fourth term, loam, is used to describe a roughly equal concentration of sand, silt, and clay, and lends to the naming of even more classifications, e.g. clay loam or silty loam or sandy loam. Loam soils generally contain more nutrients and humus than sandy soils. However, so far there is no study on the effects of bottom soil textural conditions on growth and abundance of benthic fauna. Productivity of plankton and productivity of water body depends on the kind of textural class of pond bottom-soil. In the present study, an experiment was conducted to evaluate the effects of bottom soil textural conditions on abundance of plankton in aquaculture ponds.

Materials and Methods

Experimental design

The experiment was conducted in the ponds situated at the campus of Bangladesh Agricultural University, Mymensingh during August to November 2011. The experiment had three treatments with three replications, i.e., ponds bottom with sandy loam (T1), with loam (T2) and with clay loam (T3). The average depth of the experimental ponds was 106.68 cm.

Water quality parameters

Various physical and chemical water quality parameters of the ponds such as water temperature (°C), transparency (cm), dissolved oxygen (mg L^{-1}), pH, free CO_2 (mg L^{-1}), total alkalinity (mg L^{-1}), PO_4-P (mg L^{-1}) and NO_3-N (mg L^{-1}) were estimated fortnightly following the standard method.

Chemical parameters of pond bottom-soil (sediment)

Various chemical parameters of the ponds bottom soil (sediment) such as pH, available phosphorus (ppm), total nitrogen (%), organic carbon (%) and organic matter (%) were estimated fortnightly using standard method (Sattar and Rahman, 1987).

Study of plankton

Plankton population of ponds water such as phytoplankton density (cells L^{-1}) and zooplankton density (cells L^{-1}) were estimated fortnightly. The counting of plankton (both phytoplankton and zooplankton) was done with the help of Sedgwick-Rafter Counting Cell (S-R cell) under a compound binocular microscope. The plankton population was determined by using the formula of Rahman, (1992). Identification of plankton (phytoplankton and zooplankton) up to generic level were made according to Prescott (1964), Needham and Needham (1963) and Belcher and Swale (1978).

Statistical analysis

Values are expressed as means ± standard error of the mean (SEM). Data were analyzed by one-way analysis of variance (ANOVA) followed by Tukey's post hoc test to assess statistically significant differences among the different sampling days and different treatments. Statistical significance was set at $P < 0.05$. Statistical analyses were performed using SPSS Version 14.0 for Windows (SPSS Inc., Chicago, IL).

Results and Discussion

The present study was conducted to evaluate the effects of bottom soil textural conditions on the abundance of plankton in aquaculture ponds. The highest abundance of plankton found in the ponds bottom with loam indicated that loamy bottom soil is suitable for the growth and production of plankton in aquaculture ponds.

Water quality and bottom-soil (sediment) parameters of ponds

The water quality parameters of the experimental ponds were within the productive ranges for the growth of plankton and benthos and showed no abrupt changes during the tenure of experiment (Table 1). Within limit productive ranges of such water quality parameters have also observed by a number of other authors (Wahab *et al.*, 1995; Kohinoor *et al.*, 1998; Haque *et al.*, 1998; Uddin *et al.*, 2007; Chowdhury *et al.*, 2008). Similarly, the ranges of pH, organic carbon (%), organic matter (%), available phosphorus (ppm) and total nitrogen (%) of pond bottom-soil in the aquaculture ponds were within the suitable ranges and showed no abrupt changes during the experimental period in all the treatments (Table 2). These results is in agreement with Akter (2006), who observed more or less similar results during his study on effect of bottom soil properties on the abundance of benthic fauna in nursery ponds.

Table 1. Water quality parameters (Means ± SEM; n = 3) of the ponds during the experimental periods

Parameters	Treatments		
	Treatment 1	Treatment 2	Treatment 3
Water temperature (°C)	27.20 ± 3.13	27.00 ± 3.07	27.20 ± 3.20
Air temperature (°C)	27.50 ± 2.88	27.50 ± 2.88	27.50 ± 2.88
Transparency (cm)	32.30 ± 0.91*	16.00 ± 1.15	15.30 ± 1.11
Dissolved oxygen (mgL^{-1})	7.86 ± 0.24	7.21 ± 0.70	5.86 ± 0.48
Free CO_2 (mgL^{-1})	3.86 ± 0.90	3.43 ± 1.62	4.00 ± 0.82
Total alkalinity (mgL^{-1})	82.72 ± 8.28	149.14 ± 9.05*	51.00 ± 6.86
Phosphate-phosphorous (mgL^{-1})	2.20 ± 0.58	2.68 ± 0.23	1.73 ± 0.46
Nitrate-nitrogen (mgL^{-1})	3.11 ± 0.55	3.58 ± 0.19	2.78 ± 0.33

* indicates the significant difference among the treatment

Table 2. Chemical parameters of pond bottom-soil (means ± SEM; n = 3) during the experimental periods

Parameters	Treatments		
	Treatment 1	Treatment 2	Treatment 3
pH	7.03 ± 0.13	7.13 ± 0.16	7.32 ± 0.13
Organic carbon (%)	0.78 ± 0.03	0.94 ± 0.02	0.84 ± 0.03
Organic matter (%)	1.36 ± 0.01	1.64 ± 0.01	1.41 ± 0.01
Available phosphorus (ppm)	13.86 ± 0.51	25.76 ± 0.13	16.99 ± 0.38
Total nitrogen (%)	0.073 ± 0.005	0.076 ± 0.005	0.08± 0.01

Abundance of plankton

A total 20 genera of phytoplankton (Table 3) were recorded belonged to Chlorophyceae (7), Cyanophyceae (5), Bacillariophyceae (5), Euglenophyceae (2) and Dinophyceae (1) were recorded in the present study, which is more or less similar to the findings with Kohinoor, (2000) who recorded 24 genera of phytoplankton belonging to Euglenophyceae, Cyanophyceae, Bacillariophyceae and Chlorophyceae. More or less similar numbers of genera were recorded in the plankton population by a number of authors in the ponds of Bangladesh Agricultural University campus (Dewan *et al.,* 1991; Wahab *et al.,* 1995; Kohinoor *et al.,* 1998; Uddin *et al.,* 2007 and Chowdhury *et al.,* 2008).

Table 3. Generic status of phytoplankton under different major groups found in the aquaculture ponds during the experimental periods

Major groups	Generic names		
	T1	T2	T3
Chlorophyceae	*Chlorella*	*Chlorella*	*Chlorella*
	Oocystis	*Pediastrum*	*Oocystis*
	Pediastrum.	*Scenedesmus*	*Pediastrum*
	Scenedesmus	*Closterium*	*Scenedesmus*
	Ulothrix	*Actinastrum*	*Closterium*
		Oocystis	
Cyanophyceae	*Anabaena*	*Microcystis*	*Microcystis.*
	Gomphospaeria	*Anabaena*	*Anabaena*
	Microcystis	*Gomphospaeria.*	*Gomphospaeria*
		Aphanocapsa	*Oscillatoria*
		Oscillatoria	
Bacillariophyceae	*Cyclotella*	*Asterionella*	*Asterionella*
	Diatoma	*Cyclotella*	*Cyclotella*
	Asterionella	*Diatoma*	*Diatoma*
		Fragillaria	*Fragillaria*
		Tabellaria.	*Tabellaria.*
Euglenophyceae	*Euglena*	*Euglena*	*Euglena*
	Phacus	*Phacus*	*Phacus*
Dinophyceae	*Ceratium*	*Ceratium*	*Ceratium*

Table 4. Fortnightly variations in mean abundance of total phytoplankton (x10^5 cells L^{-1}) in the experimental ponds under three treatments during the study period.

Treatments	Sampling days						
	1	2	3	4	5	6	7
T1	9.07	9.16	8.30	7.85	8.00	9.40	10.24
T2	30.05	32.90	29.70	29.39	29.70	32.72	32.30
T3	19.00	16.90	15.80	19.50	15.80	21.10	19.10

The densities of phytoplankton (means ± SEM; n = 3) were ranged from 7.85 to 10.24, 29.39 to 32.90 and 15.80 to 21.10 (x 10^5) cells L^{-1} in the ponds of T1, T2 and T3, respectively (Table 4). Phytoplankton abundance in aquaculture ponds were recorded in some other studies ranged from 2.0 - 8.0 x 10^5 cells L^{-1} (Dewan *et al.,* 1991), 9.26 - 16.03 x 10^4 cells L^{-1} (Wahab *et al.,* 1991) and 10.70 - 50.65 x 10^4 cells L^{-1} (Haque *et al.,* 1998). The mean abundance of total phytoplankton (Fig. 1) was significantly higher in T2 followed by T3 and T1 where light to medium-type bloom was observed during the study period. Similar observations were noted by Mathias (1991), Chowdhury and Sultana (1989) in various habitats.

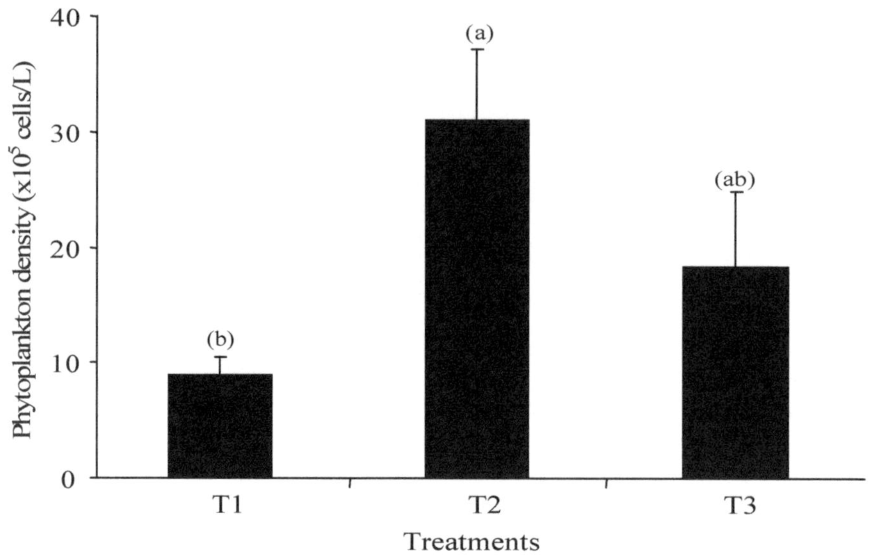

Fig. 1. Cell densities (means ± SEM; n = 3) of total phytoplankton in different treatments during the study period. Values accompanied by different letters are statistically significantly different ($p < 0.01$).

Total 13 genera of zooplankton (Table 5) were recorded belongs to Crustacea (7) and Rotifera (6). Kiran *et al.* (2007) found in their study that the population of zooplankton consisted of cladocerans (4), copepods (2) and rotifers (7), which are more or less similar to the present study. Masud *et al.* (1996) recorded 11 genera of zooplankton belonging to crustacean (7) and rotifera (4) which are lower than those of the present study.

Table 5. Generic status of zooplankton under different major groups found in the aquaculture ponds during the experimental periods

Major groups		Generic names		
		T1	T2	T3
Crustacea	Cladocera	*Daphnia* *Diaphanosoma* *Ceriodaphnia*	*Daphnia* *Diaphanosoma* *Ceriodaphnia* *Moina*	*Daphnia* *Diaphanosoma* *Ceriodaphnia* *Moina*
	Copepoda	*Cyclops* *Diaptomus*	*Cyclops* *Diaptomus*	*Cyclops* *Diaptomus*
	Crustacean larva	Nauplius	Nauplius	Nauplius
Rotifera		*Brachionus* *Filinia* *Keratella* *Hexarthra* *Polyarthra.*	*Asplanchna* *Brachionus* *Filinia* *Keratella* *Hexarthra* *Polyarthra*	*Asplanchna* *Brachionus* *Filinia* *Keratella* *Hexarthra* *Polyarthra*

Table 6. Fortnightly variations in mean abundance of total zooplankton (x10^3 cells L^{-1}) in the experimental ponds under three treatments during the study period.

Treatments	Sampling days						
	1	2	3	4	5	6	7
T1	25.00	25.40	25.50	24.10	22.70	22.70	26.50
T2	78.50	81.30	84.10	86.30	74.40	83.40	93.80
T3	66.90	61.00	67.00	67.90	60.40	55.90	76.00

The densities of zooplankton (means ± SEM; n = 3) were ranged from 22.7 ± 2.32 to 26.50 ± 3.32, 74.40 ± 3.50 to 93.8 ± 4.56 and 55.90 ± 3.65 to 76.00 ± 3.85 (x 10^3) cells L^{-1} in the ponds of T1, T2 and T3, respectively (Table 6). The mean abundance of total zooplankton (Fig. 2) was significantly higher in T2 followed by T3 and T1. Dewan *et al.* (1991) found zooplankton density ranged between 2.0 $\times 10^5$ cells L^{-1} and 2.0 $\times 10^5$ cells L^{-1} which are more or less similar to that of the present study.

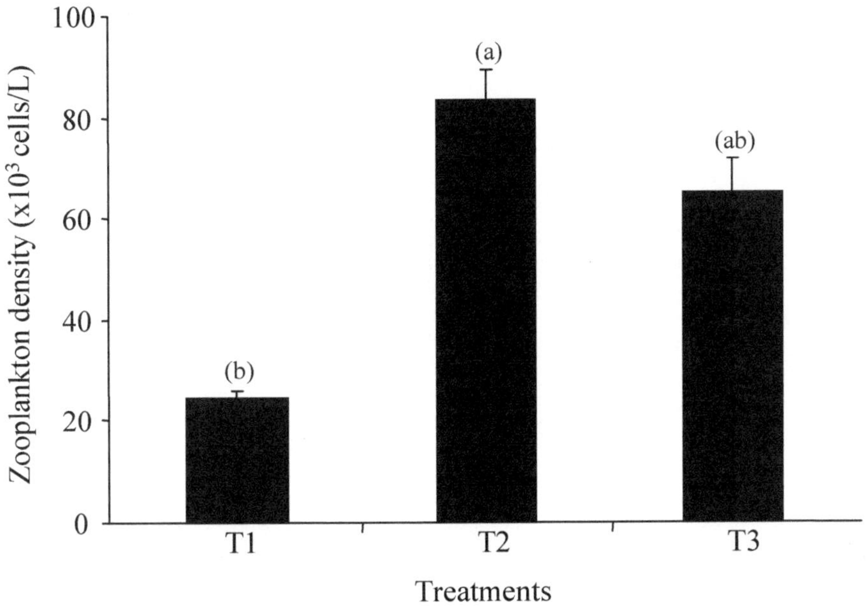

Fig. 2. Cell densities (means ± SEM; n = 3) of total zooplankton in different treatments during the study period. Values accompanied by different letters are statistically significantly different ($p < 0.01$).

However, in the present study, significantly highest mean abundance of phytoplankton and zooplankton were recorded in T2 (Fig. 1 & 2). The occurrence of these groups of plankton might be due to the suitable ecological conditions of the ponds that favoured the growth of these groups. Hulyal and Kaliwal (2008) found in an experiment that the distribution and population density of zooplankton depends upon the physico-chemical factors of the environment. On the other hand, chemical properties of soil were comparatively higher in T2 (Table 2) may be an important cause to the higher abundance of macro-benthos population in the present study. This argument also supported by Ali *et al.* (1987), Verneax *et al.* (2004) and Kailasam and Sivakami (2004) who found the significant effect of chemical properties on the plankton growth and production. Moreover, loam soils generally contain more nutrients and humus. Humus is a temporary intermediate product left after considerable decomposition of dead plants and animals, which might be support food for macro-benthos. This result indicated that pond bottom with loamy soil is suitable for growth and production of macro-benthos in aquaculture ponds.

In conclusion, suitability of bottom soil textural conditions on abundance of plankton population density was analyzed in aquaculture ponds. Most of the water quality parameters of the ponds were more or less similar and within productive limit, and chemical properties of soil were also within suitable ranges. The mean abundance of plankton density was significantly highest in T2 indicated that loamy soil bottom is suitable for the growth and production of plankton in aquaculture ponds.

References

Akter, N. 2006. Effect of bottom soil properties on the abundance of benthic fauna in nursery ponds. MS thesis, Department of Aquaculture, Bangladesh Agricultural University, Mymensingh, Bangladesh. p. 74.

Ali, M.M., Rahmatullah, S.M. and Habib, M.A.B. 1987. Abundance of benthic fauna in relation to metrological and physical-chemical factors of water. *Bangladesh J. Agril. Sci.,* 12: 239-247.

Belcher, H. and Swale E. 1978. A Beginner's Guide to Freshwater Algae. HMSO, London. p. 47.

Chowdhury, A.N.B. and Sutltana, N. 1989. Occurrence and seasonal variation of zooplankton in a fish pond in relation to some physicochemical factors. *Bangladesh J. Zool.,* 17: 101-106.

Chowdhury, M.M.R., Shahjahan M., Rahman, M.S. and Sadiqul Islam M. 2008. Duckweed (*Lemna minor*) as supplementary feed in monoculture of nile tilapia, *Oreochromis niloticus*. *J. Fish. and Aquatic Sci.,* 3: 54-59.

Dewan, S., Wahab, M.A., Beveridge, M.C.M., Rahman, M.H. and Sarker, B.K. 1991. Food selection, electivity and dietary overlap among planktivorous Chinese and Indian Major Carps fingerlings grown in extensively managed, rain-fed ponds in Bangladesh. *J. Aqua. and Fish. Manag.,* 22: 277-294.

Donahue, R.L., Miller, R.W. and Shickluna, J.C. 1990. Soils-An Introduction to Soils and Plant Growth. Prentice-Hall of India Private Ltd, New Delhi. 110001. 667 p.

Habib, M.A.B., Haque, A.K.M. and Islam M.A. 1984. Status of macro-benthos in two ponds in Bangladesh Agricultural University Campus, Mymensingh. *Bangladesh J. Agril. Sci.,* 11: 39-46.

Haque, M.S., Wahab, M.A., Wahid, M.I. and Haq, M.S. 1998. Impacts of Thai silver barb (*Puntius gonionotus* Bleeker) inclusion in the polyculture of carps. *Bangladesh J. Fish. Res.,* 2: 15-22.

Hulyal, S.B. and Kaliwal, B.B. 2008. Water quality assessment of Almatti Reservoir of Bijapur (Karnataka State, India) with special reference to zooplankton. *Environment Monitoring Assessment,* 139 (1): 299-306.

Kailasam, M. and Sivakami, S. 2004. Effect of thermal effluent discharge on benthic fauna of Tuticorin Bay, South East coast of India. *Indian J. Marine Sci.,* 33: 194-201.

Ketchum, C. 1962. On the study of primary produce in the sea by soviet scientist. *Int. Rev. Gel. Hydrobiol,* 45: 107-111.

Kiran, B.R., Puttaia, E.T. and Devidas-Kamath. 2007. Diversity and seasonal fluctuation of zooplankton in fish pond of Bhadra fish farm, Karnataka. *Zoos Print Journal,* 22(12): 2935-2936.

Kohinoor, A.H.M. 2000. Development of culture technology of three small indigenous fish mola (*Amblypharyngodon mola*), punti (*Puntius sophare*) and chela (*Chela cucius*) with notes on some aspects of their biology. Ph.D. Thesis, Department of Fisheries Management, BAU, Mymensingh. 263 p.

Kohinoor, A.H.M., Islam, M.L., Wahab, M.A. and Thilsted, S.H. 1998. Effect of mola (*Amblypharyngodon mola* Ham.) on the growth and production of carps in polyculture. *Bangldesh J. Fish. Res.,* 2: 119-126.

Masud, A.K.M.S., Chowdhury, H.A., Rahmatullah, S.M., Salam, M.A. and Islam, A.K.M.S. 1996. Comparative study on the zooplankton population of Derelict, Extensive and Semi-intensive Fish Ponds. *Bangladesh J. Aquaculture,* 18: 33-37.

Mathias, P.M. 1991. Seasonal trends in the fluctuation of phytoplankton and physico-chemical factors in a tropical lake (Govindgarh lake, Maddhya Pradesh, India) and their interrelationship. *Indian J. Inland Fish. Soc.,* 17 (1-2): 11-24.

Needham, J.G. and Needham, P.R. 1963. A Guide to Study of Freshwater Biology. 5th Edn., Holden-Day, Inc., San Francisco. 106 p.

Prescott, G.W. 1964. Algae of Western Great Lakes area. Wm. C. Brown Co. Dubuque, IOWA, 946 p.

Rahman, M.S. 1992. Water Quality Management: Aquaculture. BRAC Prokashana, Mohakhali, Dhaka, Bangladesh. 84p.

Raymont, J.E.G. 1963. Plankton and productivity in the oceans. Macmillan, New York, USA. 60 p.

Sattar, M.A. and Rahman, M.M. 1987. Techniques of Soil Analysis, Published by A. Sattar, BAU, Mymensingh. 54 p.

Uddin, M.N., Rahman, M.S. and Shahjahan M. 2007. Effects of duckweed (*Lemna minor*) as supplementary feed on monoculture of GIFT strain of tilapia (*Oreochromis niloticus*). *Progres. Agric.,* 18: 183-188.

Verneaux, V., Verneaux, J., Schmitt, A. and Lambert, J.C. 2004. Relationships of macrobenthos with dissolved oxygen and organic matter at the sediment-water interface in Ten French Lakes. *Archiv-fur-Hydrobiologie,* 160(2): 247-259.

Wahab, M.A., Ahmed, Z.F., Islam, M.A. and Rahmatullah, S.M. 1995. Effect of introduction of common carp, *Cyprinus carpio* (L.) on the pond ecology and growth of fish in polyculture. *Aquaculture Research,* 26: 619-628.

Wahab, M.A., Begum, M.A. and Ahmed, Z.F. 1991. The effects of silver carp introduction in the polyculture of major Indian carps. *BAURes Progrssive,* 5: 429-437.

MYCOBIAL CONTAMINATION AND MYCOTOXINOGENESIS OF *Tinospora cordifolia*: AN IMPORTANT MEDICINAL PLANT OF INDIA

S. Sharma, G. Sumbali* and V. Sharma

Abstract

An investigation of mycoflora and associated mycotoxins was carried out from dried market samples of stem portions of *Tinospora cordifolia,* an important medicinal plant of India. These samples were collected from various wholesale and retail shops of eight districts of Jammu and Kashmir state viz., Kathua, Jammu, Udhampur, Rajouri, Poonch, Doda, Srinagar and Leh. A total of 39 fungal species representing 18 genera were recovered by using surface washing technique. Assessment of mycobial load of *T. cordifolia* showed the presence of many such fungal species that are widely acknowledged as the most important mycotoxin producers. Analyses of samples for mycotoxin contamination was done by multi-mycotoxin detection method. The dried samples of *Tinospora cordifolia* were detected to be contaminated with aflatoxin B_1, aflatoxin B_2, ochratoxin A, patulin and citrinin. Among the various mycotoxins detected, aflatoxins were present in maximum number of samples, which is probably because *Aspergillus flavus* was recovered from all the investigated samples. However, fusarial species and their toxins were not detected from the investigated samples.

Keywords: Medicinal Plant, Mycoflora, Mycotoxins, Stem Portions, *Tinospora cordifolia*

Department of Botany, University of Jammu, B.R. Ambedkar Road, Jammu-180006, India

*Corresponding author's email: geetasumbalippl@yahoo.co.in (G. Sumbali)

Introduction

Herbal medicines are used for primary health care by approximately 80 percent of the population in the developing countries (Akerele, 1993). These medicines have always remained the first choice of the people of developing countries because of low cost. Moreover, these herbal medicines have better cultural acceptability, better compatibility with human body and lesser side effects. People of the developed countries had more inclination towards chemical preparations, but during the last few years, there has been a considerable change in the medicinal system of the developed world also. Widespread toxicity and harmful after-effects associated with long-term use of synthetic drugs have forced the developed countries to think for some safe alternative healing methods (Gijtenbeek *et al.,* 1999). They have realized that 'natural is better' and thus there is an increasing demand for 'green medicine' which is now commonly sold in health stores.

The state of Jammu and Kashmir is the sixth largest state (area wise) of India. It is situated in the north and surrounded by North-Western Himalayas and has a rich heritage of over 300 medicinal plants (Kaul, 2010). These plants are frequently harvested, dried and used either in the system of Ayurveda or in local health care traditions practiced by hakeems of Kashmir, vaidyas of Jammu and amchis of Ladakh. However, due to lack of proper knowledge and use of unscientific methods of collection, drying and storage (Kaul, 1997), these dried medicinal plants may become prone to contamination with fungal spores. Another effect is that these medicinal plants also undergo deterioration even before they are used in making drugs. Realizing the fact that mycoflora and mycotoxin contamination of dried medicinal plants has not received the attention that the magnitude of the problem warrants and since no such work has been attempted from Jammu and Kashmir State, which is a large reservoir of medicinal plants, an investigation was undertaken on one such important medicinal plant, *Tinospora cordifolia* Miers.

Tinospora cordifolia, which is known by the common name Guduchi, is an herbaceous vine (Fig. 1) of the family Menispermaceae (Hooker, 1897). It is used in the treatment of diabetes, high cholesterol, allergic rhinitis (hay fever), lymphoma and other cancers, hepatitis, fever, gonorrhea, syphilis, and to boost the immune system. Research has demonstrated that a combination of *T. cordifolia* extract and turmeric extract is effective in reducing the hepatotoxicity (Adhvaryu *et al.,* 2008).

Alcoholic extract of the stem shows activity against *Escherichia coli* (Nagaprashanthi *et al.*, 2012). A number of products of *T. cordifolia* are now available in the market, which are being used for curing different ailments (Table 1).

In spite of all these curative properties, it is not possible to capitalize the full medicinal wealth of *T. cordifolia* because of the associated fungal contamination of its market samples, which leads to altered efficacy of its drug products. Thus, an investigation was undertaken to isolate, purify and identify the mycoflora associated with the market samples of this plant and analyze the market samples for the presence of natural mycotoxin contaminants like aflatoxins (B_1 and B_2), ochratoxin A, patulin, citrinin, zearalenone, zearalenol and deoxynivalenol.

Fig. 1 Habit of *Tinospora cordifolia*

Fig. 2 Dried stem portions of *Tinospora cordifolia*

Materials and Methods

(i) Isolation of mycoflora and assessment of frequency (%)

Dried market samples of stem portions of *Tinospora cordifolia* were collected in pre-sterilized polythene bags from various wholesale and retail shops of eight districts of Jammu and Kashmir viz., Kathua, Jammu, Udhampur, Rajouri, Poonch, Doda, Srinagar and Leh. Sample bags were brought to the laboratory and surface mycoflora associated with the market samples of dried stem portions of *T. cordifolia* was determined by using surface washing technique (Singh and Kainsa, 1983). For isolating maximum number of fungal propagules from the surface of each sample, three different media–Dichloran Rose Bengal Chloramphenicol agar (DRBC), Dichloran 18% Glycerol agar (DG 18) and Malt Salt agar (MSA) were used. The petriplates were incubated for 7 days at 28 ± 2°C till the proper growth of the fungal colonies was obtained. The recovered fungal species were identified by studying their cultural and morphological characters and by using various keys and relevant literature.

Percent frequency of each fungal species was calculated by using the formula given below:

$$\text{Frequency (\%)} = \frac{\text{Number of samples from which an organism was isolated}}{\text{Total number of samples tested}} \times 100$$

(ii) Extraction of mycotoxins from dried samples

Dried stem portions of *T. cordifolia* (Fig. 2) obtained from different markets were analysed for mycotoxin contamination by multi-mycotoxin detection method developed by Stoloff *et al.* (1971). In this method, 25 g of finely ground sample was taken in 250 ml Erlenmeyer flask containing 100 ml mixture of acetonitrile and 4% potassium chloride (90:10v/v). This solution was put on a mechanical shaker for 30 minutes and then filtered through Whatman no. 41 filter paper. 50 ml of this filtrate was taken in 250 ml separating funnel, defatted and extracted twice with 50 ml of iso-octane. The upper iso-octane layer was discarded. To the lower acetonitrile layer, 12.5 ml of water was added, shaken and extracted thrice with 20 ml of chloroform each time. The chloroform acetonitrile layer was filtered through Whatman no. 41 filter paper having a bed of anhydrous sodium sulphate. The extract was collected in a beaker and then evaporated to dryness on a water bath. The residue was dissolved in 1 ml of benzene: acetonitrile (98:2 v/v) solution and stored in a clean screw cap vial for thin layer chromatography and HPLC analysis.

(iii) Qualitative and quantitative estimation of mycotoxins

For qualitative estimation, known amount of sample extracts and standards of investigated mycotoxins obtained from Sigma Aldrich Co. were spotted on activated TLC (Thin Layer Chromatographic) plates and developed with different solvent systems (Table 2). Quantitative estimation of detected mycotoxins was done through high performance liquid chromatography (Table 2). The analytical equipment of HPLC (CLASS-LC10 SHIMADZU) consisted of a liquid chromatographic pump LC-10AT, an auto injection system SIL – 10A with a 50 µl sample loop, a variable wavelength absorbance UV – VIS detector SPD – 10 set at 365 nm. The analytical component was CLC – ODS. The mobile phase consisted of water: acetonitrile: methanol (59: 29: 12v/v). Analysis was performed at room temperature (25-30°C) and data was recorded in HP DeskJet 670C.

Table1. Various valuable economic products of *Tinospora cordifolia*

Product name	Cure
Guduchi Tablets	General infections, immune disease, Hepatitis, Arthritis and anti- cancerous
Madhu Mehari	Dryness of mouth, Numbness debility, relieves Frequent urination, fatigue, Excessive thirst and maintains the blood sugar
Safe Herbs	Vaginal discharge and also helps in sexual debility
Mussaffen	Blood purifier and skin disease
Rebuild	Anti- stress and anti- oxidant
Septilin	Upper respiratory tract infection
Tonplex	Increases immunity and vitality
Joint & Muscle Excellence Tablets	Eliminate the toxins of joints
Natadadrol	Potent muscle-building androgen
Shila Pravang	Premature ejaculation, erectile dysfunction, to enhance the sexual stamina

Table 2. Qualitative and quantitative estimation of mycotoxins

Mycotoxin Estimated	Solvent System Used for TLC Analysis	Detection/ Wavelength	Colour	Quantitative Estimation	Detector (Wavelength)
Aflatoxins B_1&B_2	Toluene:Isoamyl alcohol: Methanol (90:32:2 v/v)	Long UV	Blue	HPLC	UV/VIS 365nm
Ochratoxin A	Toluene:Ethyl acetate:90% Formic acid(50:40:10v/v)	Long UV	Bluish Green	HPLC	Fluorescence Ex. 333nm, Emm.470nm
Patulin	Toluene:Ethyl acetate: Chloroform(80:70:50v/v) with 1 ml of 90% formic acid	Visible light	Yellow	HPLC	UV/VIS 276nm
Citrinin	Toluene:Ethyl acetate: Chloroform(80:70:50v/v) with 1 ml of 90% formic acid	Long UV	Sky blue	HPLC	Fluorescence Ex. 325nm, Emm.385nm
Zearalenone	Toluene:Ethyl acetate: Formic acid(6:3:1v/v)	Long UV	Blue	HPLC	Fluorescence Ex. 274nm, Emm.440nm
Zearalenol	Toluene:Ethyl acetate: Formic acid(6:3:1v/v)	Long UV	Blue	HPLC	UV/VIS 236nm
Deoxynivalenol	Toluene:Ethyl acetate: Formic acid(6:3:1v/v)	Long UV	Sky blue	HPLC	UV/VIS 229nm

Results and Discussion

During the period of investigation, 22 market samples of *Tinospora cordifolia* were collected in pre-sterilized polythene bags. These samples were screened for the mycobial load by using surface washing technique and three media (Dichloran Rose Bengal Chloramphenicol agar, Dichloran 18% Glycerol agar and Malt Salt agar) of different chemical composition. While recovering surface mycoflora, Dichloran Rose Bengal Chloramphenicol agar medium could trap maximum number of fungal species, Dichloran 18% Glycerol agar medium helped to recover a wide range of non-fastidious xerophilic fungi including most of the *Penicillium* and *Aspergillus* species, whereas, malt salt agar was more useful in recovering the members of *Aspergillus glaucus* group. This indicates that nutritional requirements of various fungi differ and there is no single medium, which can help in the recovery of all the fungi.

During the present investigation, 39 fungal species representing 18 genera were recovered from the dried stem portions of *T. cordifolia* (Table 3). Assessment of surface mycoflora associated with *T. cordifolia* showed the presence of many such fungal species that are widely acknowledged as the most important mycotoxin producers. In view of this, an investigation was undertaken to verify contamination of 8 major mycotoxins from the market samples. These included aflatoxin B_1

(AFB_1), aflatoxin B_2 (AFB_2), ochratoxin A (OTA), patulin (PAT), citrinin (CIT), zearalenone (ZEN), zearalenol (ZOL) and deoxynivalenol (DON).

Perusal of data presented in table 4 shows that dried samples of *Tinospora cordifolia,* were contaminated with aflatoxin B_1 (0.11-1.27 $mgkg^{-1}$), aflatoxin B_2 (0.27-0.77 $mgkg^{-1}$), ochratoxin A (0.48-0.54 $mgkg^{-1}$), patulin (3.75 $mgkg^{-1}$) and citrinin (0.25-0.28). Aflatoxin B_1 was detected from the samples of 6 districts and aflatoxin B_2 was detected from the samples of 3 districts. Other mycotoxins viz., ochratoxins A and citrinin were detected form samples of only 2 districts, whereas patulin was detected from samples of only 1 district of Jammu and Kashmir. Presence of aflatoxins in maximum number of samples is probably because *Aspergillus flavus* was recovered from all these samples. In addition, no contamination of *Fusarium* species and their toxins viz., zearalenol, zearalenone and deoxynivalenol was detected from the investigated samples.

Table 3. Percentage frequency of mycoflora recovered from dried market samples of *Tinospora cordifolia* stem portions collected from various districts of Jammu and Kashmir state

Fungal species	Percentage frequency	Isolation of fungal species on different media		
		DRBC	DG-18	MSA
Alternaria alternata	12.5	-	+	+
Aspergillus candidus	12.5	+	+	+
A. deflectus	12.5	+	+	+
A. flavus	100.0	+	+	+
A. fumigatus	37.5	+	+	+
A. japonicus	62.5	+	+	+
A. niger	87.5	+	+	+
A. niveus	37.5	+	+	-
A. panamensis	12.5	-	+	-
A. sydowii	87.5	+	+	+
A. tamarii	62.5	+	+	+
A. terreus	12.5	+	+	-
A. terricola var. *americana*	12.5	-	+	+
Chaetomium brasiliense	12.5	-	+	-
C. globosum	12.5	+	-	-
C. indicum	12.5	-	+	-
C. olivaceum	12.5	+	+	+
Circinella simplex	12.5	+	-	-
Cladosporium cladosporioides	12.5	-	+	-
C. sphaerospermum	62.5	-	+	+
Curvularia brachyspora	12.5	+	-	+
C. lunata	37.5	+	+	-
Emericella nidulans	25.0	-	+	+
Eurotium amstelodami	87.5	+	+	+
E. chevalieri	62.5	+	+	+
E. rubrum	25.0	+	+	+
Geotrichum candidum	12.5	+	-	-
Mucor geophilus	12.5	+	+	+
Paecilomyces variotii	12.5	+	-	+
Penicillium chrysogenum	50.0	+	+	+
P. citrinum	62.5	+	+	+
P. oxalicum	12.5	-	+	+
P. paxilli	12.5	+	+	-
Phyllosticta sp.	12.5	+	-	-
R. stolonifer	75.0	+	+	+
Scopulariopsis brevicaulis	50.0	+	-	+
Sepedonium niveum	12.5	-	+	-
Syncephalastrum racemosum	37.5	+	+	+
Trichoderma harzianum	12.5	-	+	-
Total number of fungal species recovered	39			

+, Presence -, Absence

This study clearly indicates that xerophilic aspergilli and penicillia are commonly associated with the dried stem of *Tinospora cordifolia.* Among the various *Aspergillus* species, *Aspergillus flavus* was detected from all the samples. In addition, some other xerophilic species belonging to *Eurotium, Phyllosticta, Geotrichum, Emericella, Scopulariopsis, Trichoderma, Paecilomyces* and *Sepedonium* were also recovered from the investigated stem of the medicinal plant. Realizing the importance of the quality of dehydrated medicinal plants, a large number of workers have recently engaged themselves in the study of surface mycoflora of various herbal drug plants during storage and marketing (Toma and Abdullah, 2013; Stevic *et al.*, 2012). These workers also reported diverse range of fungal species belonging mainly to *Aspergillus, Penicillium, Rhizopus, Chaetomium, Fusarium, Eurotium* and *Cladosporium.*

Table 4. Analysis of mycotoxin contamination in market samples of dried stem portions of *Tinospora cordifolia*

Place of Collection	Samples found positive for mycotoxin contamination (mgkg^{-1})							
	Aflatoxins		Ochratoxin A	Patulin	Citrinin	Zearalenone	Zearalenol	Deoxynivalenol
	Afla B_1	Afla B_2						
Kathua	-	-	-	-	-	-	-	-
Kathua	-	-	-	-	-	-	-	-
Kathua	-	-	-	-	-	-	-	-
Jammu	-	-	-	-	-	-	-	-
Jammu	0.48+0.09	-	-	-	-	-	-	-
Jammu	-	-	-	-	-	-	-	-
Jammu	-	-	-	-	-	-	-	-
Jammu	-	-	-	-	0.25+0.06	-	-	-
dhampur	-	-	-	3.75+0.39	-	-	-	-
dhampur	0.22+0.03	-	-	-	-	-	-	-
Rajouri	-	-	0.54+0.03	-	-	-	-	-
Rajouri	0.30+0.09	0.27+0.05	-	-	-	-	-	-
Poonch	-	-	-	-	0.28+0.20	-	-	-
Poonch	0.73+0.13	0.50+0.08	-	-	-	-	-	-
Poonch	-	-	-	-	-	-	-	-
Doda	0.11+0.06	-	0.48+0.07	-	-	-	-	-
Doda	-	-	-	-	-	-	-	-
Srinagar	-	-	-	-	-	-	-	-
Srinagar	-	-	-	-	-	-	-	-
Srinagar	-	-	-	-	-	-	-	-
Leh	1.27+0.37	0.77+0.19	-	-	-	-	-	-
Leh	-	-	-	-	-	-	-	-
Number of positive samples	6	3	2	1	2	0	0	0
Percent of positive samples	27.3	13.6	9.1	4.5	9.1	0.0	0.0	0.0

-, Not detected

The presence of mycoflora, which are known to produce toxic metabolites (mycotoxins) were detected from the dried samples of *T. cordifolia* and many of them have hazardous effects (IARC, 1993). A survey of literature indicated occurrence of mycotoxin contaminants from dried medicinal plants of various Indian states viz., Bihar (Roy and Chourasia, 2001), Uttaranchal (Singh, 2003), Jammu and Kashmir (Sharma *et al.*, 2013).

In the present investigation, dried stem samples of *T. cordifolia* were detected to be contaminated with very high concentration of aflatoxins B_1 and B_2 in comparison to the permissible tolerance limits. However, the magnitude of aflatoxin contamination varies with the type of dehydrated medicinal plant, storage practices, geographical factors, seasonal changes and varying aflatoxigenic potential of associated *A. flavus* strains (Bilgrami, 1984) Few other Indian researchers have also reported aflatoxin contamination in amounts exceeding permissible limits from varied types of dehydrated medicinal plants and their formulations (Roy and Chourasia, 2001; Singh, 2003). Similarly, there are reports of dried medicinal plants with aflatoxin contamination from other countries also (Elshafie *et al.*, 2003; Wongwiwat *et al.*, 2004).

Very few samples of dried stem portions of *T. cordifolia* were detected to be OTA contaminated (0.48 to 0.54 mgkg^{-1}) and they belonged to Rajouri and Doda districts. Reports of OTA contamination from other medicinal plants and their crude products have been provided by other ethanomycologists also (Aziz *et al.*, 1998; Efuntoye, 1999; Santos *et al.*, 2013; Singh, 2003). IARC (1993) has classified OTA in group 2B because of its toxicity on humans. Committee on Toxicity of Chemicals in Food, Consumer Products and Environment (COT) considered OTA as a genotoxic carcinogen and proposed that its level in food be reduced to the lowest (COT, 1997). The joint expert committee of Food Additives of the WHO and FAO set a provisional maximum intake of 100 µgkg^{-1} body weight. However, few countries have legislative limits ranging from 5 to 50 µgkg^{-1} (Webley *et al.*, 1997).

Another mycotoxin detected from the dried sample of *T. cordifolia* consisted of patulin. There is no earlier report of patulin contamination from dried medicinal plants. However, very high patulin contamination has been reported from rotted rosaceous fruits and their products (Beretta *et al.*, 2000).

In the present investigation, detection of aflatoxins B1 and B2, ochratoxin, patulin and citrinin from market samples of dried stem of *T. cordifolia* clearly indicates that its powdered formulations are not completely safe for direct human consumption. In view of the detected mycotoxin contamination, an urgent need for proper drying and storage of crude herbals is urgently required.

Acknowledgement

The authors are thankful to Head, Department of Botany, University of Jammu for providing laboratory facilities.

References

IARC. 1993. Some naturally occurring substances: Food items and constituents, heterocyclic aromatic amines and mycotoxins. Monograph by International Agency for Research on Cancer, Lyon, France. 56: 244-395.

Adhvaryu, M.R., Reddy, M.N. and Vakharia, B.C. 2008. Prevention of hepatotoxicity due to anti-tuberculosis treatment: A novel integrative approach. *World J. Gastroenterol.* 14: 4753-4762.

Aziz, N.H., Youssef, Y.A., El-Fouly, M.Z. and Moussa, L.A. 1998. Contamination of some common medicinal plant samples and spices by fungi and their mycotoxins. *Botanical Bull. Aca. Sinica.* 39: 279-285.

Akerele, O. 1993. Nature's Medicinal Bounty: don't throw it away. *World Health Forum.* 14: 390-395.

Beretta, B., Gaiaschi, A., Galli, C.L. and Restani, P. 2000. Patulin in apple-based foods: occurrence and safety evaluation. *Food Addit. Contam.* 17: 399-406.

Bilgrami, K.S. 1984. Mycotoxins in dry fruits and spices. Technical report of UGC Project, New Delhi. pp. 1-97.

COT. 1997. (Committee on Toxicity of Chemicals in Food, Consumer Products and the Environment). Statement on ochratoxin A in driedwine fruits. Department of Health, London: pp. 1-4.

Efuntoye, M.O. 1999. Mycotoxins of fungal strains from stored herbal plants and mycotoxin contents of Nigerian Crude herbal drugs. *Mycopathologia.* 147: 43-48.

Elshafie, A.E., Al-Siyabi, F.M., Salih, F.M., Omar, T.B., Al-Bahry, S.N. and Al-Kindi, S. 2003. The mycobiota of herbal drug plants in Oman and possible decontamination by gamma radiation. *Phytopathologia Mediterranea.* 42: 149-154.

Gijtenbeek, J.M.M., Vanden Bent, M.J. and Vecht, C.J. 1999. Cyclosporine neurotoxicity. *J. Neurol.* 246: 339-346.

Hooker, J.D. 1897. Flora of British India, The University of California: 1-149.

Kaul, M.K. 1997. Medicinal plants of Kashmir and Ladakh (Temperate and cold arid Himalaya). Indus Publishing Company, New Delhi. pp. 1-173.

Kaul, M.K. 2010. High altitude botanicals in integrative medicine – case studied fron Northwest Himalayas. *Indian J. Trad. Kno.* 9: 18-25.

Nagaprashanthi, C., Khan, P.R., Chand, K.G., Aleemuddin, M.A. and Begum, G.R. 2012. *In vitro* Antimicrobial activity of *Tinospora cordifolia* and its phytochemical screening. *Int. J. Pharm. Tech. Res.* 4: 1004-1008.

Roy, A.K. and Chourasia, H.K. 2001. Mycotoxin contamination in herbal seed samples under storage and their prevention. pp. 393-407. *In*: Seed technology and seed pathology. (Eds. Singh, T and Agrawal, K.), Pointer Publishers, Jaipur, India.

Santos, L., Marín, S., Sanchis, V. and Ramos, A.J. 2013. Mycotoxin in medicinal/aromatic Herbs-a Review. *Boletín Latinoamericano y del Caribe de Plantas Medicinales y Aromáticas* 12: 119 - 142.

Sharma, S., Sumbali, G. and Sharma, V. 2013. Investigations on the mycoflora and mycotoxin contamination of dried medicinal leaves of *Azadirachta indica* A. Juss. and *Justicia adhatoda* Linn. from Jammu and Kashmir state (India). *Indian J. Adv. Res.* 1: 131-138.

Singh, J.P. and Kainsa, R.L. 1983. Microbial flora of grapes in relation of storage and spoilage. *Indian Phytopath.* 36: 72-76.

Singh, P.K. 2003. Mycotoxin elaboration in Triphala and its constituents. *Indian Phytopath.* 56: 380-383.

Stevic, T., Pavlovic, S., Stankovic, S. and Savikin, K. 2012. Pathogenic microorganisms of medicinal herbal drugs. *Arch. Biol. Sci.* Belgrade. 6: 49-58.

Stoloff, L., Nesheim, S,. Yin, L., Rodricks, J.V., Stack, M. and Campbell, A.D. 1971. A multimycotoxin detection method for aflatoxins, ochratoxins, zearalenone, sterigmatocystin and patulin. *J. Assoc. Off. Anal. Chem.* 54: 91-97.

Toma, F.M and Abdullah, N.Q.F. 2013. Isolation and identification of fungi from spices and medicinal plants. *Res. J. Env. & Earth Sci.* 5: 131-138.

Webley, D.J., Jackson, K.L. and Mullins, J.D. 1997. Mycotoxins in food : A review of recent analyses. *Food Australia.* 49: 375-379.

Wongwiwat, T., Fazeli, E.R., Supatra, P. and Josef, B. 2004. Contamination of aflatoxin in herbal medicinal products in Thailand. *Mycopathologia.* 158: 239-244.

IMPACT OF CROP VARIETIES ON HOUSEHOLD FOOD SECURITY IN *HAOR* AREAS OF BANGLADESH

M.A. Kashem*, M.G. Farouque and P.C. Roy

Abstract

The purpose of the study was to identify the crop varieties in the Haor area of Netrokona district to minimize the food vulnerability situation during food crisis. The study was conducted in four selected villages of Madan and Khaliajuri Upazilas under the Netrokona district. Data were collected from randomly selected 400 farmers from the study area in the year 2011. The findings of the study revealed that farmers usually cultivate only three HYV *boro* rice varieties (BR 19, BRRI dhan 28 and BRRI dhan 29) in the Haor area under study. In *aman* season farmers mostly cultivate local varieties and they get minimum yield. Only few vegetables found to cultivate by the farmers in Haor area. The vegetables are: onion, garlic, chilli, bottle gourd, sweet gourd, lady's finger and tomato. Among these onion, garlic and chilli are mostly cultivated. Many farmers reported that during the flood they had to face serious problems in respect of communication, diseases, increased price of different commodities, and lack of adequate government safety net programs/relief. Farmers' lack of flood preparedness has been observed while some farmers remained uncared about the occurrence of flood and consequently they did not take any preparedness for flood.

Keywords: Crop Variety, Food Security, Sustainable, Haor, Flood

Department of Agricultural Extension Education, Bangladesh Agricultural University, Mymensingh, Bangladesh
Corresponding author's email address: kashem@bau.edu.bd (M.A. Kashem)

Introduction

The Haor region is the most vulnerable area of Bangladesh. It is situated at the north-eastern part of the country. There are as many as 423 small or large Haors in Bangladesh. Every two or three years, this region has to face the curse of either flash flood or over flood. For example, in 2010 the Haor region has experienced flash flood that broke the backbone of the farming community. Flood causes severe economic loss and social disruption as well as food insecurity to many people, particularly living in the Haor area. To address this situation of farmers, the primary innovator and the researchers invented so many technologies to escape the affects of flash flood or over flood. Farmers might have particular cropping scheme to combat the consequences of flood that need to be identified. Exploring and organizing of the flood resistant or short duration varieties of crop technologies may have impact in minimizing the effects of flood if they are displayed at the farmer's fields through demonstration and training. Crop production practices, economic activities and over all livelihoods of the farmers of Haor areas are quite different from those of the other parts of the country. The cropping practices particularly Boro rice crop mainly depends on nature (Alam *et al.*, 2011).

The available statistics indicate that, the total cultivated area in those Haor districts is about 1.26 million hectares of which 0.68 million ha (nearly 66%) is under Haor. Almost 80 percent of this area (i.e. 0.68 million ha) is covered by Boro rice, while only about 10% area is covered by T. Aman production (Huda, 2004). In the Haor areas, hybrid rice is also grown (Das, 2004). This area is gradually increasing in different locations of Haor districts (Husain *et al.*, 2001). So, there is a great possibility of growing MV rice as well as other rice and non-rice crops in the Haor areas (Alam *et al.*, 2011).

Haor area is most vulnerable area of our country. It has indeed become imperative to exploit the crop production potentiality of the large Haor areas; it is because those areas usually remain under-utilized with quite low cropping intensity (Jabber and Alam, 1996). In every year, crop is damaged by flash flood or over flood. As a result, food security towards sustainable livelihood of Haor people is threatened. The technologies for that situation are very needed to identify and organize for ensuring household food security towards sustainable livelihood. Keeping these facts in mind, the present study has been undertaken to fulfill the following specific objectives:

1) Identify the flood tolerant crop varieties in the Haor areas of Netrokona district;
2) Ascertain the problems faced by the rural poor households in food security during floods in Haor area;
3) Identify measures normally taken by the farmers to minimize the food vulnerability situation in Haor area;
4) Gather farmers' opinion to overcome the sudden food crisis caused by flood and other natural disasters in Haor areas.

Methodology

The study was conducted in Madan and Khaliajuri Upazilas under the Netrokona district in the year 2011. The physical, social and cultural heritages of the people of this area are similar in many cases with other Haor areas of the country. All the household in Madan and Khaliajuri Upazilas under Netrokona district were considered as the population of the study. The total number of households in Madan and Khaliajuri Upazilas was almost 4,000. Among them 10 per cent i.e. 400 households were randomly selected as the respondents.

The empirical data for the study were collected through pre-tested structured interview schedule from the farmers of the selected villages during 10 March to April 2011. Data were collected by the field level officers of the Women Development Organization (WDO), Netrokona under the closed monitoring and supervision of the Project Director and Principal Investigator. Quantitative data were collected through direct interviewing with farmers through interview schedules while the qualitative data were collected through FGDs (Focus Group Discussions).

Measurement of food vulnerability index (FVI)

Food vulnerability index (FVI) of the flood affected people of Haor areas were measured by the following formula:

FVI = HVFS X 2 + MVFS X 1 + NAVFS X 0

Where,
FVI = Food Vulnerability Index
HVFS = High vulnerability to food security by the respondents
MVFS = Moderate vulnerability to food security by the respondents
NAVFS= Not at all vulnerable to food security by the respondents

Thus, the computed value of Food Vulnerability Index (FVI) could range 0 to 800, 0 indicating no vulnerable situation and 800 indicated the highest vulnerable condition among the households.

Findings and Discussion

Flood tolerant crop varieties, which were cultivated by the farmers' in Haor areas

Farmers were asked to indicate flood tolerant crop varieties based on their long experience in these areas. They have identified as flood tolerant crop varieties in the Haor areas, which were quite usual to occur floods in the Haor areas. The flood tolerant crop varieties as identified by the farmers in Haor area are presented in Table 1.

Table 1. Flood tolerant crop varieties as identified by the respondents in *Haor* areas

Crops	Name of the crop variety	Respondents (n=400)	
		Number	Percentage
A. Boro Rice	BR 19	193	48.25
	BRRI dhan 28	176	44.00
	BRRI dhan 29	314	78.5
	Hira Dhan (Hybrid)	240	60.00
	Sonar Bangla (Hybrid)	157	39.25
B. Aman Rice	Local Variety	254	63.50
	Nazirshail	73	18.25
	Hashi	62	15.50
	Shahjalal	45	11.25
	Mangal	23	5.75
	BRRI dhan 52	18	4.50
	BRRI dhan 51	12	3.00
C. Jute	Deshi	129	32.25
	Tosha	124	31.00
D. Vegetables	Onion	260	65.00
	Garlic	248	62.00
	Chilli	157	57.75
	Bottle gourd	192	48.00
	Sweet gourd	168	42.00
	Lady's finger (Okra)	231	39.25
	Tomato	122	30.50

Various crop varieties in different seasons are effective in flood affected Haor areas. In case of boro rice, it was observed that about more than three-quarters (78.50%) of the respondents of the study area agreed for cultivation of BRRI dhan 29 as flood tolerant variety, followed by 48.25 percent for cultivation of BRRI dhan 19 and 44.0 percent for BRRI dhan 28. However, in case of hybrid rice varieties, 60.0 percent of the respondents agreed for Hari dhan and 39.25 percent for Sonar Bangla as flood tolerant rice varieties in the Haor area.

In respect of aman rice cultivation, it was found that most of the farmers (63.5%) cultivate only various local rice varieties as flood tolerant. However, 18.15 percent of the farmers opined for Nazirshail, and 15.5 percent for Hashi as flood tolerant aman rice varieties. Among the high yielding aman rice varieties, the BRRI dhan 51 got the lowest preference as flood tolerant variety. Some farmers also opinioned that they prefer to cultivate deshi and tosha jute to save the crops from flood.

In terms of vegetables, the findings reveal that about two-thirds (65.0%) of the farmers gave their opinion in favour of onion as flood tolerant crop followed by 62 percent of the farmers who were in favour of cultivating garlic. The next flood tolerant crop as identified by the respondents was chilli; about 58 percent farmers were in favour of the crop. The other flood tolerant crop varieties identified by the respondents was bottle gourd (identified by 48 percent farmers) followed by sweet gourd (42%), lady's finger (39.25%), and tomato (identified by (30.5 percent).

Problems faced by the respondents in securing food and other necessities during severe flood in Haor areas

An attempt was made to identify the constraints faced by the respondents in securing food during severe flood in the Haor areas. Respondents have indicated quite a large number of constraints. However, the major constraints have been presented in Table 2.

Table 2. Problems faced by the respondents in securing food and necessities during severe flood in *Haor* areas

Sl. No.	Problems as identified by the responents	Respondents (n=400)	
		Number	Percentage
1	Communication problem	400	100.00
2	Prevalence of diseases	400	100.00
3	Prices of all necessary commodities goes up very high	400	100.00
4	Lack of Government safety net programs/relief immediately after flood	400	100.00
5	Uneven distribution of Government relief	380	95.00
6	Miserable living conditions	370	92.5
7	Unavailability of day labour	360	90.00
8	Sheltering and feeding domestic animals and birds	310	77.50
9	Spreading of various crimes	150	37.50
10	Droping out of school children	125	31.25

During and immediately after the flood the farmers in the Haor areas have to face enormous problems in relation to food security and meet up other necessities of life. The findings revealed that the severity of problems in respect of communication, prevalence of diseases, increasing the price of different commodities, and lack of adequate government safety net programs/relief immediately after flood were the highest and the intensity of problems as cited by the respondents were also equal (100%). In order of importance, the other problems were: uneven distribution of government fund (opined by 95 percent respondents), miserable living conditions (identified by 92.5 percent), unavailability of day labour (identified by 90 percent), and sheltering and feeding domestic animals and birds (identified 75.5 percent farmers).

Measures taken by respondents in Haor areas to minimize the food vulnerability

Respondents were asked how they tackled the food vulnerability, especially when the area is affected by devastating flood. Respondents indicated various mitigation strategies depending on the nature and extent of damage due to flood. However, FVI (Food Vulnerability Index) was calculated in order to ascertain the importance of the measures taken by the respondents to minimize the food vulnerability. The findings have been furnished in Table 3.

Table 3. Measures proposed by the respondents in *Haor* areas to minimize the food vulnerability

Sl. No.	Measures taken by the respondents	High	Medium	Not at all	FVI
1	Government safety net programme	400	0	0	800
2	Meet up the family needs by storing sweet potato	260	118	22	638
3	Storing the dry fish to meet up the family needs	260	114	26	634
4	Meet up the family needs by leasing own household assets	250	128	22	628
5	Rearing calves as a wealth to overcome the problem of food deficit	250	126	24	626
6	Raising ducks for earning money to use it during food crisis	236	148	16	620
7	Selling goats to overcome the problem at the time of food shortage	248	120	32	616
8	Rearing of poultry birds as a wealth to overcome the food problem	242	130	28	614
9	Meet up the family needs by leasing own ornaments	240	128	32	608
10	By using small savings of the women to buy foods at the time of food vulnerability	243	120	37	606
11	Meet up the family needs by day labour	234	120	46	588
12	Going outside own district for searching works	230	120	50	580
13	Meet up the family needs by leasing own land	210	138	52	558
14	Taking credits from the NGO	135	180	85	450
15	Getting aids from the GOs and NGOs	87	104	209	278

The findings demonstrated that government's safety net programmes topped the list (having FVI of 800) followed by storing sweet potato (FVI=638), and storing the dry fish (FVI=634), leasing own household assets to others (FVI=628), rearing calves (FVI=626), raising ducks (FVI=620), selling goats (FVI=616), rearing poultry birds (FVI=614), leasing own ornaments (FVI=608), and using own small savings of the women (FVI=606). The FVIs, however, ranged from 278 to 800; the lowest FVI was in respect of getting aids from the GOs and NGOs.

Suggestions to overcome the sudden food crisis caused by flood and other natural disasters in Haor areas

Respondents under the study were asked to highlight the suggestions to overcome the sudden food crisis caused by flood and other natural disasters in the Haor area. They were specifically requested to indicate what prior measures the respondents would have to take and what measures they expect from the government to tackle the situation. The findings have been presented in Table 4.

Table 4. Suggestions of the respondents to overcome the sudden food crisis caused by flood and other natural disasters in *Haor* areas

Sl. No.	Suggestions to overcome the sudden food crisis due to flood	Respondents (n=400)	
		Number	Percentage
1	Strengthening government safety net programmes	360	90.00
2	Working as a labourers	345	86.25
3	Increasing the amount of relief	340	85.00
4	Storing dry foods	310	77.5
5	By leasing or selling own wealth/ ornaments	285	71.25
6	By selling poultry birds	280	70.00
7	Dependents on NGO aids and credits	260	65.00
8	By selling goats	260	65.00
9	Meet up by preserved foods	250	62.50
10	By selling cows and calves	150	37.50

The findings in Table 4 dictates that the overwhelming majority of the farmers opined for the strengthening government safety net programmes as a measure to overcome the unforeseeable food in future due to flood and other natural disasters. This was followed by working as a labourers (86.0%), increasing the amount of relief (85.0%), storing dry foods (77.5%), leasing or selling own wealth/ ornaments (71.0%), and by selling poultry birds (70%). Slightly more than one-third farmers opined in favour of selling cows and calves as a measure to overcome food crisis in future.

Conclusion

In the Haor areas, the most important crop is boro rice. Farmers often cannot harvest boro rice due to early flash floods, hailstorms or other kind of disaster. However, there is shortage of high yielding short duration boro rice varieties. Only three HYV boro rice varieties have been reported as suitable for cultivation in the Haor. Hence, it is concluded that more short duration boro rice varieties need to be developed by the concerned research institutes.

In aman season farmers mostly cultivate local varieties and they get minimum yield. Consequently, they become food insecured. This fact leads to the conclusion that so long, the farmers would continue to cultivate local aman varieties, the chances of becoming food secured would be less. Therefore, the concerned agencies should give due cognizance of this fact and take proper steps to generate high yield potential aman rice varieties appropriate for cultivation in aman season.

Only few vegetables were found to cultivate by the farmers in Haor area. The vegetables are: onion, garlic, chilli, bottle gourd, sweet gourd, lady's finger, and tomato. Among these onion, garlic and chilli are mostly cultivated. Therefore, the concerned research institute may take necessary steps for generating more vegetables suitable for cultivation in the Haor areas.

Acknowledgements

The study was funded by the by Ministry of Science and Information & Communication Technology, Government of the Peoples' Republic of Bangladesh. The authors gratefully acknowledge the concerned Ministry for providing the financial support for conducting the study.

References

Alam, M.S., Quayum, M.A. and Islam, M.A. 2011. Crop Production in the Haor Areas of Bangladesh: Insights from Farm Level Survey. *The Agriculturist*. 8 (2): 88-97.

Das, D.C. 2004. Experiences with Modern and Hybrid Rice Varieties in Plain Land Ecosystem in the Mymensingh Region. Proc. of the 20th National Workshop on Rice Research and Extension in Bangladesh. Bangladesh Rice Research Institute, Gazipur-1701. pp. 63-65.

Huda, M.K. 2004. Experience with Modern and Hybrid Rice Varieties in Haor Ecosystem: Emerging Technologies for Sustainable Rice Production. Proc. of the 20th National Workshop on Rice Research and Extension in Bangladesh. Bangladesh Rice Research Institute, Gazipur-1701. pp. 94-97.

Husain, A.M., Hossain, M. and Janaiah, A. 2001. Hybrid Rice Adoption in Bangladesh: A Socio-economic Assessment of Farmers' Experiences, Research Report, Social Sciences Division, International Rice Research Institute (IRRI), LosBanos, Philippines. pp. 55-59.

Jabber, M.A. and Alam, M.S. 1996. Adoption of Modern Rice Varieties in Bangladesh. *J. Agri. Econ.* 16 (2): 77-95.

CONSTRAINTS AND OPPORTUNITIES FOR CULTIVATION OF *Moringa oleifera* IN THE ZIMBABWEAN SMALLHOLDER GROWERS

S.M. Mudyiwa[1], C.T. Gadzirayi[1], J.F. Mupangwa[2], J. Gotosa[1] and T. Nyamugure[1]

Abstract

The study sought to establish the constraints to Moringa agronomic practices and marketing in the smallholder farming sector of Zimbabwe. Purposive sampling selected 96 respondents across four study districts in Zimbabwe who were interviewed using an open-ended questionnaire. Eight focus group discussions disaggregated by gender were carried out to supplement primary data. Growers asserted *Moringa oleifera* production problems of livestock damage, theft, seed supply, inefficient cultivation practices resulting in inherent low productivity and poor marketing. The growers coped through intensive monoculture of *M. oleifera*, enhanced local seed supply from establishing their own seed orchards, product value addition and organized associations. Moringa can be used to promote climate change resilient agricultural systems and will be a solution to environmental problems.

Keywords: *Moringa oleifera*, Multipurpose Tree, Constraints, Growers

[1]Faculty of Agriculture and Environmental Science, Bindura University of Science Education, P. Bag 1020, Bindura, Zimbabwe

[2]Department of Animal Science, Faculty of Agriculture, University of Swaziland, P.O. Luyengo, Luyengo, Swaziland

*Corresponding author's email: silasmudyiwa@yahoo.com (S.M. Mudyiwa)

Introduction

Smallholder growers grow subsistence and cash crops in their rainfed, complex and resource deprived fields often combining the cultivation of the crops with scattered multipurpose trees and realising a wide range of benefits (Radovich, 2009). *Moringa oleifera* is one such multipurpose tree of global interest and is grown in combination with agricultural and horticultural crops by smallholder growers and this give growers a wide range of benefits (Palada and Chang, 2003; Radovich, 2009). It is a suitable tree for traditional agroforestry in the home because of its versatility (Odee *et al.*, 2001; Palada and Chang, 2003; Nduwayezu *et al.*, 2007).

M. oleifera has widespread use in agricultural industry and medicine. The tree benefits from mycorrhizal nitrogen–fixing association but there are no known residual nitrogen benefits for the next crop. Pod shucks and seed kernel press cake can be used as mulch and enhances soil fertility when they decompose (Prat *et al.*, 2002). It can be used in livestock as a biocide. All parts of the Moringa tree are used in natural medicine. The tender leaves and pods contain a wealth of essential disease-preventing substances, for example, it is said to have known immune boosting ability (FAO, 1988; Ncube, 2006; Smith and Eyzaguirre, 2007). Moringa can be used as a source of oil and dye for water purification and wood fuel (Folkard and Sutherland, 1994).

In Zimbabwe, *M. oleifera* has the potential to be adopted as food for humans as well as a ruminant fodder. It can therefore fill in both human and animal nutrition gaps for the resource poor smallholder growers in need of food and livestock feed. The animals are usually a source of sustainable livelihoods. Livestock production in addition to crop production is a strategy employed to raise farm income and reduce the food insecurity experienced by rural households due to the decline in agricultural sector (Jera and Ajayi, 2008). Considering the difficult agro-ecological conditions, which prevail in most parts of Zimbabwe where smallholder growers live, challenges encountered by the growers seem to be important in successful Moringa production.

With *M. oleifera* having such myriad of benefits this study sought for analysis and documentation of the 'invisible information' of its production constraints in agro-forestry management systems under the smallholder farming sector in various agro-ecological regions of Zimbabwe and highlight opportunities. This information will act as basis for supporting the small scale farming communities in their quest to secure better livelihoods.

Materials and Methods

The study area

The study was conducted in Binga, Bindura, Mutoko and Shamva districts of Zimbabwe (Fig 1). These represent the main areas in the country where Moringa trees are grown and promoted for livelihoods improvement by World Agroforestry (ICRAF) and Environment Africa (E- Africa). Table 1 shows the physical characteristics of the study districts.

The interview sites were chosen after liaison with project coordinators of the developmental agents and district extension staff of Agritex and Forestry Commission (FC). At least three wards were chosen in each district for the study (Table 1). In Binga district, Binga Trees Trust staff was involved in identifying study sites as it works in collaboration with Environment Africa.

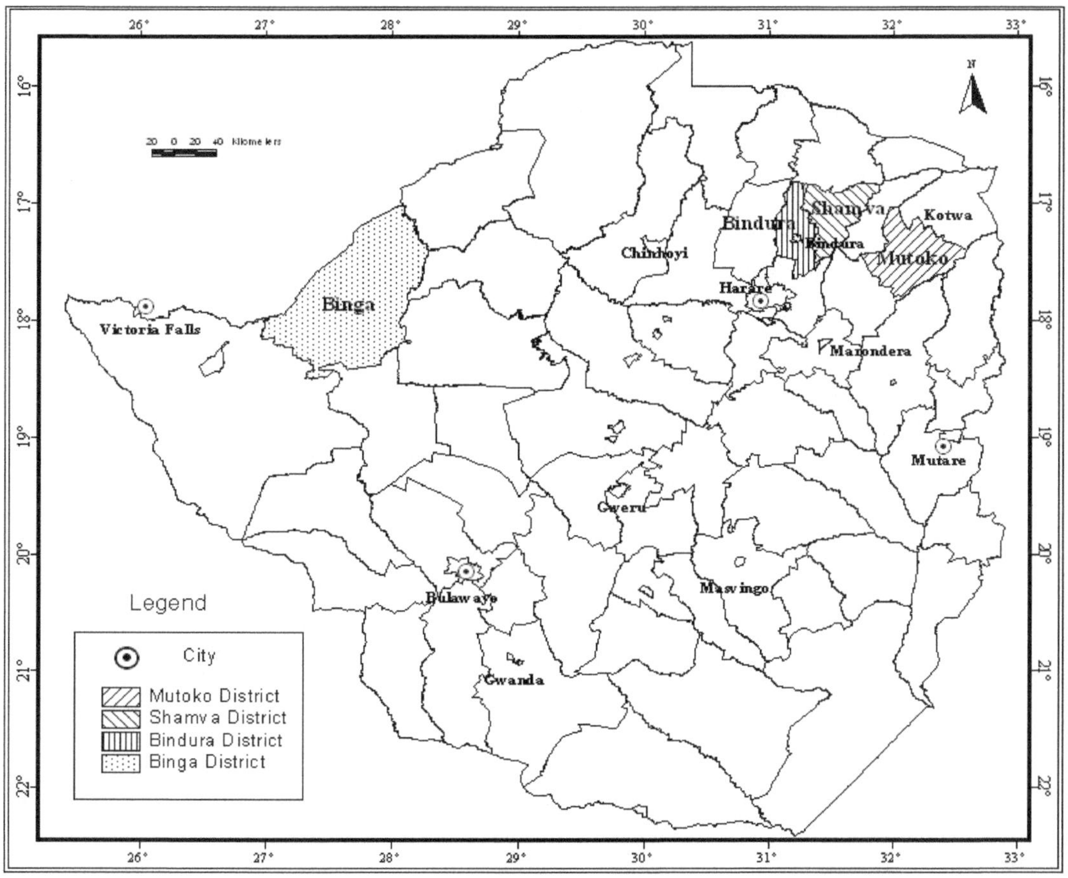

Fig. 1. Map of Zimbabwe showing location of study districts

Table 1. Physical characteristics of the study districts

Variable	District			
	Bindura	Binga	Mutoko	Shamva
Natural Region	2a	5	3	2b
Rainfall (mm/yr)	750-1000	350-500	600-700	650-1000
Maximum Temperature Range	25-28°C	27-30°C	25-30°C	25-27°C
Location	17° 08′ S and 31° 21′ E	17° 45′ S and 27° 30′E	17° 10′ S and 32° 30' E	17° 10′ S and 31° 40′ E
Dominant Soils	Moderately shallow to deep coarse grained sands to sandy clay loams derived from granitic rocks	Nutrient deficient deep Kalahari sands	Moderately shallow to deep coarse grained sands to sandy loams derived from granitic rocks	Moderately deep to deep reddish brown clays derived from mafic rocks
Wards visited	Ward 9, 10, 13 and 15	Ward 7, 8, 24 and 10	Ward 9, 13 and 14	Ward 17, 18, 20 and 28

Data collection

The study employed focus group discussion, questionnaires and field observations to collect data. The snowball technique was used to identify 96 growers to be interviewed in the targeted wards. Purposive sampling was used to select participants basing on the presence of Moringa trees on their properties. Eight focus group discussions disaggregated by sex were conducted, two per district. Each focus group consisted of at least 8 members. This was to triangulate and capture gender differences. Questionnaires were administered to growers who were growing *M. oleifera* trees, to capture data on challenges of Moringa production. Direct field observations were made to supplement and ground truth information collected through questionnaires and interviews.

Data analysis

Focus group discussion data were analysed qualitatively through content analysis description and this with observations complemented questionnaire primary data collected. Data from questionnaires were analysed using Statistical Analysis Systems (SAS) version 9.13 to come up with descriptive statistics of the responses to the questionnaires.

Results and Discussion

Gender and age of Moringa growers

A total of 96 growers in the four districts were interviewed. The age and gender characteristics of the growers are as shown in Table 2. The modal age of the interviewees was the 30-39 years, whilst more (+2) male informants participated in the study.

Table 2. Age and gender characteristics of the Moringa growers

Attribute	Categories	Number of respondents
Gender	Male	49
	Female	47
Age	20 – 29	19
	30 – 39	36
	40 – 49	24
	50 – 59	10
	60 and above	7
Location	Bindura district	23
	Binga district	23
	Shamva district	25
	Mutoko district	25

Moringa production constraints

Lack of knowledge on efficient production practices

Field observations revealed that Moringa was mostly grown scattered around homesteads (52%), in association with fruit trees in family orchards (14%) without irrigation and in home gardens (34%) where trees were grown on the edges. Binga district had one thriving home garden where Moringa was intercropped with different horticultural crops. The trees would benefit from irrigation water while horticultural crops would also benefit from the shade and this saved on irrigation water requirements due to reduced evapotranspiration (Prat *et al.*, 2002). Soil fertility would be replenished through addition of organic manure from leaf litter reducing the need to apply inorganic fertilizers.

The growers lack efficient production systems, which make use of improved models. They still used their indigenous cultivation practises to grow Binga, Mutoko, Mozambique, Malawi, and Tanzania provenances. The cultivation of Moringa in Zimbabwe needs to evolve to such an extent that growers can focus on commercial production of Moringa products like in other regions, for example, West Africa especially Niger and northern Nigeria where farmers have developed efficient farming systems that are adapted to their capacities and constraints. There, the growers have taken Moringa to the fields and grow it intensively as a crop for leaf and pod. They have abandoned subsistence production to venture into monoculture of Moringa at very high tree densities on small pieces of land or they do it in various agroforestry patterns that give them better yields (Gamatie, 2001; Saint Sauveur, 2001; Rajangam *et al.*, 2002).

Competition for land with other food crops

Growers had small portions of land under Moringa production. The highest proportion (55%) of growers had less than 5 trees per homestead under Moringa production. Binga district had the highest proportion (4%) of growers with 20 or more trees per household. Bindura district had also the highest proportion (19%) of farmers with less than 5 trees per homestead.

Growers in Bindura, Mutoko and Shamva indicated that they felt a lot of social pressure deciding growing Moringa at their homesteads because many people stigmatise HIV and AIDs in Zimbabwe especially in the communal areas. Most people became interested in the tree when they were given the hope that it can reverse HIV status (Ncube, 2006) leading to the planting of one or two trees at the backyards without allocating it larger pieces of land as they do to their traditional crops. In Binga Moringa has been incorporated in their traditional system of primary health care and everyday nutrition hence a substantial number of trees per household.

Moringa's practical uses are uncommon to most parts of Zimbabwe except Binga because it is the region where the tree had been grown for a long time and naturalised in Zimbabwe (Maroyi, 2006). Furthermore, it is of concern that cultural knowledge concerning Moringa is absent in most districts leading to its small allocation of production areas and stigmatisation. With the knowledge at hand from developmental agents rural communities should demystify Moringa so that promotion can be fully made and Moringa can be widely grown and used as in India and other parts of Africa and realise its socio-economic values.

Source of seed

Growers had difficulties in the procurement of germplasm. They sourced their planting material from Non Governmental Organizations (NGO) such as ICRAF, E- Africa, Development Aid from People to People (DAPP) and Binga Trees Trust, local growers and across Zimbabwe, Forestry Commission, and private nurseries. The highest proportion (51%) of the growers got their planting material from NGOs. In Bindura and Binga districts, the highest proportion of growers got planting material from other local growers whilst in Shamva and Mutoko districts it was from NGOs. Private nurseries and Forestry Commission contributed very little as sources of germplasm as shown in Fig. 2.

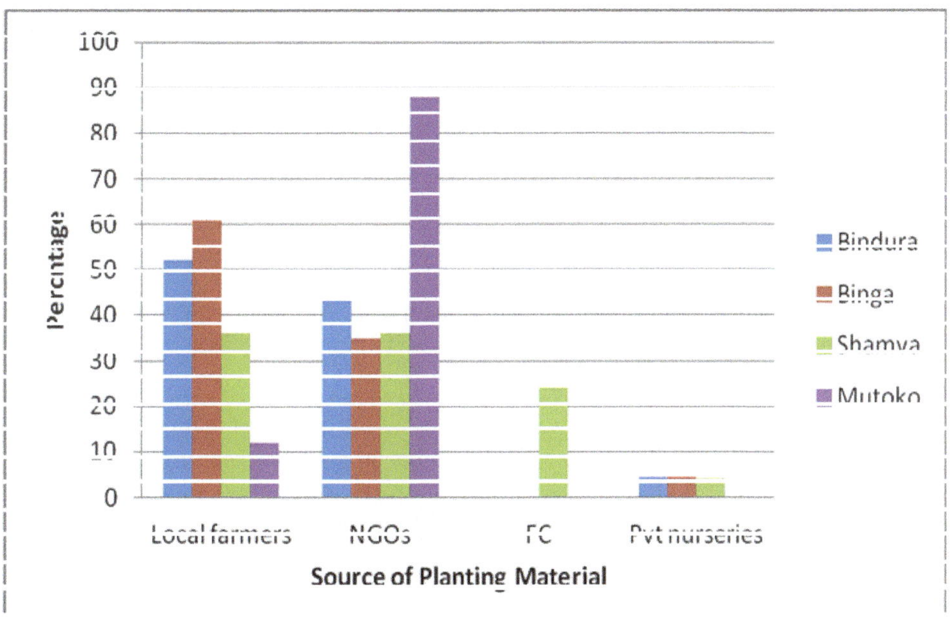

Fig. 2. Source of planting material by district

Where there is lack of information on sources of germplasm and planting material is limited, what is availed to growers is usually what they grow (Brandi *et al.*, 2007) irrespective of the quality of the planting material. Donor agencies especially ICRAF, E- Africa (Mutoko district), Binga Trees Trust (Binga district) and DAPP (Bindura and Shamva districts) distributed the seed themselves or through Agritex and FC extension officers working with growers in the respective communities.

Growers in Binga and Bindura districts also played a major role in the supply of planting material though none of the growers was actively involved in seed production. Their trees were seldom pruned or pollarded and none of them was intentionally managed for seed production resulting in production of poor quality seed with low viabilities. Field observations have shown that most growers had no nurseries of their own or in groups except for a grower in Binga who had one which had different species of trees for her use in the family orchard shown in Plate 1 below:

Plate 1. Grower owned nursery

Forestry Commission contributed very little as sources of planting material due to reduced mobility of its staff in offering extension services and operational constraints. This reduced their ability to carry out their institutional mandate. Little was sourced from private nurseries due to financial constraints. According to respondents, seedlings from private nurseries were expensive so growers resorted to free seed or seedlings handouts from either government or donor agencies.

Shortage of water for irrigation during dry months

Most growers (30%) faced irrigation water shortage during dry periods. Female growers were mostly affected by problem of lack of irrigation water during dry months because they had no permanent access water points like males who are able to dig deep wells that could supply irrigation water throughout the year. In Mutoko farmers had shallow wells that dry up during the dry season or their gardens were near seasonal streams.

Animal damage

Growers (20.8%) had problems of browsing by livestock during the dry season. The animals would destroy the trees after planting either by browsing the leaves and removing biomass or by physically trampling over the plants. Most orchards were not protected making them prone during off season when livestock is allowed to roam around looking for feed. The livestock could also break into gardens during winter season on attraction to the green Moringa foliage.

Low yield levels

The majority of the growers (70%) had no knowledge of their yields (fresh leaf, bark and seed) from their trees. This was a result of sporadic usage of the plant parts by the growers. Moringa production is done at subsistence level so growers' fresh leaf yield levels were too low as compared to other regions though there are no specific average yields in Moringa production (Makkar and Becker, 1999; Manh *et al.*, 2005). In Binga growers produced leaf powder due to the influence of the programme of reducing malnutrition in children (Ncube, 2006) although the production of powder was insufficient in quantity and quality and also there was no control in the preparation and storage.

Theft

Growers experienced theft of their Moringa crop. Growers in Binga revealed that thefts were rampant during periods when Moringa was viewed as a cure for HIV and AIDS prompting thieves to steal the tree parts for sell in towns to unsuspecting patients. Though thefts were experienced, there are no known well defined rural and urban markets or wholesalers in Zimbabwe for the sale of fresh leaves, pods and powder as in countries of West Africa and India.

Pests and diseases affecting the tree

Growers experienced problems with sap sucking pests (red spider mites and aphids), caterpillars, termites and blight though a high proportion of growers encountered no pest and disease problems. Most growers in Shamva district had no problems except a few (8%) with problems of caterpillars and in Mutoko district, caterpillars and termites were the most troublesome (Fig. 3). Bindura growers had problems with sap sucking and leaf eating pests and in Binga, red spider mite and aphids were a problem as these pests are attracted to the green foliage.

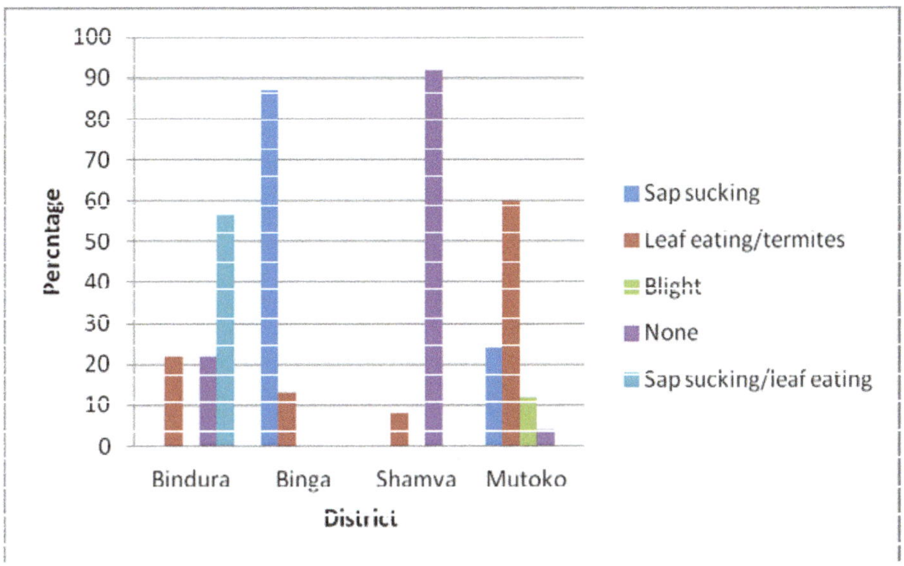

Fig.3. Pests and diseases prevalence by districts

Marketing constraints

Lack of markets

Growers have no assured markets for Moringa tree products and information regarding available markets is scarce. There is need to develop quality standards and efficient packaging and storage of Moringa products so that they meet both local and global markets if opened up. Improved marketing can also be done by creation of producer associations.

Opportunities in Moringa production

There are many opportunities that smallholder growers can grab considering the difficult agro-ecological conditions that prevail to secure a better livelihood from Moringa farming.

Transformation of barren land

Most smallholder growers live in marginal areas with lots of barren land near their fields or homesteads. Moringa can utilise marginal land as it is resilient to harsh growing environments, including drought and poor soil quality. Competition for land with other food crops will be reduced and land will be utilized effectively for both livestock and human food requirements enhancing food security. This will promote climate change resilient agricultural systems and will be a solution to environmental problems.

Increased livelihood security and diversification

The tree is very fast growing, with normal growth ranging from 3-5 meters per year if left uncut. Growers will quickly realize benefits from the tree in a very short period for both him and livestock. With Moringa tree parts, retaining high percentages of vital nutrients throughout the year (Melesse *et al.*, 2012) it is an advantage for livestock production, which experiences great seasonal variations in quality and quantity of forage. Moringa, thus can be used as source of inexpensive protein to supplement poor quality forage in the dry season. This will help sustain and enhance livestock production systems and boost livelihoods.

The land reform programme that started in year 2000 has afforded some marginalized growers to bigger, more productive pieces of land which they can also use in Moringa production for both human and livestock (ruminant and non ruminant) uses. The animals are usually growers' source of sustainable livelihoods together with crop production, a strategy employed to raise farm income and reduce the food insecurity experienced by rural households due to the decline in agricultural sector (Jera and Ajayi, 2008). However, some people may fail to take up Moringa for household use or primary health care options it will still boost productivity and livelihood through improved livestock and crop production. Intercrops of Moringa have led to a realization of high yields in horticultural crops and its organic fertilizer has been found to be more efficient in plant growth enhancement than organic matter from other plant compost (Prat *et al.*, 2002; Emmanuel *et al.*, 2011) and it is a biocide.

Creation of Moringa producer associations

Women mostly participate in donor funded livelihood projects especially nutritional gardens though smaller number of men is found as members. The growers should utilize these nutritional gardens for Moringa production because the gardens have reliable water sources (boreholes) installed, security (wire fences) and management. The developmental agents teach growers better cultivation practices for better yields and post harvest management. The damages caused by livestock and humans will be minimal in the communal nutritional gardens which are cared for by the groups as each of the group members will put an eye on the project. Women also cater for the nutritional needs of the family and their acceptance of Moringa will guarantee its acceptance by the whole family.

Linkage to markets and value addition

Developmental agents are initiating, supporting and promoting projects that are aimed at value added Moringa products. The government and development agents can help open up markets and also explore export opportunities for Moringa products as growers need an assured market for their products for them to produce Moringa as a cash crop as they did for *Jatropha curcas*. Communication of market information should be done clearly to growers so that they can plan and implement tree growing strategies that increase productivity.

Demystifying Moringa

Cultural sensitivities in some regions should be checked as they render Moringa unacceptable to growers and Moringa tree will remain under-cultivated. The belief that it is utilized by people with HIV/AIDS should be rooted out in the communities and Moringa introduced as part of daily diet. Some communities like Binga traditionally grow and use Moringa so developmental agents and extension officers can popularize Moringa with few challenges after incorporation of the different socio–cultural beliefs that may exist from one community to another. Whether moringa is used for nutritional purposes or as a medicinal health remedy, it can boost food and health security of Zimbabwe. People can use this plant regularly in part of their daily diet to fight illnesses and avoid consuming it during illness. Modern primary healthcare facilities may not be easily accessible to the communal growers and services are expensive. This can be the best option for nutritional and medicinal requirements.

Water purification

Moringa seed can also be used as a water purifier both for the urban and rural inhabitants (Ali *et al.*, 2010). The common water purifier alum salt is an expensive option for Zimbabwe urban councils. Thus, the use of the seed can improve the availability of clean water to both rural and urban areas in a country that has a down turn of the economy since early 2000 and battling to procure enough water treatment chemicals. This will reduce exposure to waterborne diseases after using untreated water especially at household level.

Conclusion

Despite the potential of Moringa, its production has remained limited due to lack of efficient cultivation practices, livestock damage, theft, seed supply, and marketing problems. It is of no doubt that Moringa may constitute great economic and strategic value to the smallholder growers of Zimbabwe if they implement systematic tree management and marketing programmes. The current practices being used by the growers need to be altered in order to promote their production. Moringa has potential of becoming the lifeline of the small holder growers in Zimbabwe given their history of failed crops.

Recommendations

There is need to build local capacity of growers to collect and use seed. This is cheap for the growers and with time, growers can set up their own seed distribution systems with quality seed produced by them. There is need to build capacity of the growers on processing of products to add value by building appropriate technologies that suit the local requirements and situations basing on scientific knowledge to increase productivity.

Acknowledgements

The authors are greatly indebted to World Agroforestry Centre (ICRAF), Environment Africa and Binga Trees Trust for supporting this project. Many thanks go to the farmers who provided valuable information about the Moringa tree.

References

Ali, E., Muyibi, S., Salleh, H., Alam, M. and Salleh, M. 2010. Production of natural coagulant from *Moringa oleifera* seed for application in treatment of low turbidity water. *J. Water Resource & Protection.* 2 (3): 259-266.

Brandi, E., Lilleso, J.P.B., Moestrup, S. and Kisera, H.K. 2007. Do organizations provide quality seed to smallholders? A study on tree planting in Uganda, by NGOs and CBOs. Development and Environment No.8. Forest and Landscape, Denmark. p. 32.

Emmanuel, S.A., Emmanuel, B.S., Zaku, S.G. and Thomas, S.A. 2011. Biodiversity and agricultural productivity enhancement in Nigeria: application of processed *Moringa oleifera seeds* for improved organic farming *Agric. Biol. J. North America.* 2 (5): 867-871.

FAO. 1988. Non timber uses of selected arid zone trees and shrubs. FAO Conservation Guide 19. FAO. Rome. pp. 92-101.

Folkard, G.K. and Sutherland, J.P. 1994. *Moringa oleifera*: A multipurpose tree. *Footsteps.* 20: 14-15.

Gamatie, M. 2001. *Moringa oleifera* management systems in the Niger valley: The case of Sarandoarea.www.moringanews.org/actes/gamatie_en.doc

Jera, R. and Ajayi, O.C. 2008. Logistic modeling of smallholder livestock farmers' adoption of tree based fodder technology in Zimbabwe. *Agrekon.* 47 (3): 379-392.

Makkar, H.P.S. and Becker, K. 1999. Plant toxins and detoxification methods to improve feed quality of tropical seeds. *Asian-Aus. J. Anim. Sci.* 12 (3): 467-480.

Manh, L.H., Dung, N.N.X. and Ngoi, T.P. 2005. Introduction and evaluation of *Moringa oleifera* for biomass production and as feed for goats in the Mekong Delta. *Livestock Res. Rural Dev.* 17 (9), Article No. 104. Retrieved May 31, 2013, from http://www.lrrd.org/lrrd17/9/manh17104.htm.

Maroyi, A. 2006. The utilization of *Moringa oleifera* in Zimbabwe, *J. Sustainable Dev.* 8: 172-185.

Melesse, A., Steingass, H., Boguhn, J., Schollenberger, M. and Rodehutscord, M. 2012. Effects of elevation and season on nutrient composition of leaves and green pods of *Moringa stenopetala* and *Moringa oleifera. Agroforestry Systems.* 86: 505-518.

Ncube, D. 2006. Moringa programmes in Binga District: 10 years of experience. Binga Trees Trust. Binga. pp. 1-4.

Nduwayezu, J.B., Chamshama, S.A.O., Mugasha, A.G., Ngaga, Y.N., Khonga, E.B. and Chabo R.G. 2007. Comparison in seed kennel sizes and early growth performance of different *Moringa oleifera* provenances in Southeast Botswana. *Discov. Innov.* 19: 52-58.

Odee, D.W., Muluvi, G.M., Machua, J., Olson, M.E. and Changwony, M. 2001. Domestication of Moringa Species in Kenya, Development potential for Moringa products. Workshop proceedings October 29th – November 2nd. 2001. Dar es Salaam. Tanzania.www.moringanews.org/actes/odee_en.doc

Palada, M.C. and Chang, L.C. 2003. Suggested cultural practices for Moringa. AVRDC. 545: 1-5.

Prat, J.H., Henry, E.M.T., Mbeza, H.F., Mlaka, E. and Satali, L.B. 2002. Malawi Agroforestry Extension Project Marketing and Enterprise Project Report. Agroforestry Publication No. 47. pp. 53-62.

Radovich, T. 2009. Farm and forestry production and marketing profile for Moringa (*Moringa oleifera*). *In:* Elevitch, C.R. (ed.) Specialty Crops for Pacific Island Agroforestry. Permanent Agriculture Resources (PAR), Holualoa, Hawaii. http://agroforestry.net/scps.

Rajangam, J., Azahakia Manavalan, R. S., Thangaraj, T., Vijayakumar, A. and Muthukrishan, N. 2002. Status of production and utilization of Moringa in southern India. Development potential for moringa product, Workshop proceedings held on 29 October–2 November 2002. Dar es Salaam. Tanzania. http://www.moringanews.org/actes/rajangam_en.doc.

Saint Sauveur, A. 2001. Moringa exploitation in the world: State of knowledge and challenges. Dar es Salaam. Tanzania. www.moringanews.org/actes/saintsauveur_en.doc

Smith, F.I. and Eyzaguirre, P. 2007. African leafy vegetables: Their role in the world health organisation's global fruit and vegetables initiative. *African J. Food Agric. Nutri. & Dev.* 7 (3): 1-17.

39

ADOPTION OF SELECTED WHEAT PRODUCTION TECHNOLOGIES IN TWO NORTHERN DISTRICTS OF BANGLADESH

M.S. Rahman[1] and Zerina Haque[2]

Abstract

The study was conducted in two major wheat growing areas of Bangladesh to determine the adoption level and factors affecting the adoption of wheat production practices in the study areas during 2011. Descriptive statistics along with multiple regression technique was used to achieve the objectives. The results revealed that most of the farmers (60.91%) in the study areas were cultivating Shatabdi variety of wheat. Adoption level of seed rate, TSP and MoP application were found to be very low. On the other hand, production practices like time of wheat sowing and number of irrigation were highly adopted by the farmers. Most of the farmers (69.09%) applied TSP below the recommended dose while 81.82% of the farmers applied MoP over the recommended dose. The study also revealed a positive and significant relationship between adoption and the variables like education, experience and extension contact. Lack of proper information and technical knowledge were the major problems that hinder the adoption of wheat production technologies in the study areas. Adoption gaps are needed to be eliminated to enhance the productivity as well as net return of wheat cultivation.

Keywords: Adoption, Production Technology, Northern Bangladesh

[1]Assistant Professor, Department of Management and Finance, Faculty of Agribusiness Management, Sher-e Bangla Agricultural University, Dhaka, Bangladesh
[2]Ex-MS Student, Bangabandhu Sheikh Mujibur Rahman Agricultural University, Salna, Gazipur, Bangladesh
*Corresponding author's email: saadrhmn@yahoo.com (M.S. Rahman)

Introduction

Agriculture is the backbone of the economy of Bangladesh. Climate of Bangladesh is strongly favorable for different crop production. However, most of the farmers of the country still follow the traditional crop production technologies (Pandit *et al.*, 2007). Therefore, they receive lower yield and cannot meet the food demand of the country. Thus, the technology adoption by the farmers is an essential pre-requisite for economic prosperity in developing country likes Bangladesh. Adoption may be defined as the integration of an innovation into farmers' normal farming activities over an extended period of time (Feder *et al.*, 1985). Increasing population pressures, traditional farming systems and small farm holding are not enough to support growing number of people in household. Whereas, the one who adopts improved crop production packages can produce ample food for as many as thousands of people. Successful adoption of improved agricultural technologies could stimulate overall economic growth through inter sectoral linkages while conserving natural resources (Sanchez *et al.*, 2009). Past studies (Dixon *et al.*, 2006; Wanyama *et al.*, 2010) showed that adoption of improved production practices may help the farmers to get higher amount of yield and had impact on household food security and income.

An important objective of rural development is to increase agricultural productivity with a view to enhancing farmer's income and standard of living. Improved practices provide the main venue for increasing productivity in the country's agriculture (Edna *et al.*, 2009). Rural development can be promoted through stimulating the adoption of improved production practices. Bangladesh Agricultural Research Institute has developed improved production practices for various crops. However, this study only considered the wheat crop as it is one of the most important cereal crops in Bangladesh. It is grown on about 3,764,24 hectares of land with an average production of 9,01,490 MT (BBS, 2010). At present, high-yielding varieties occupy 98% of the area (Hossain, 1984). Still every year Bangladesh needs to import large amount of wheat grains to meat up local demand of about 3.5 million ton (Baksh *et al.*, 2009). Several studies at home and abroad (Hossain and Bruce, 1992; Kamruzzaman *et al.*, 2001; Iqbal Muhammad *et al.*, 2002; Walford, 2002; Lee David, 2005; Gardebroek, 2006; Serra *et al.*, 2008; Baksh *et al.*, 2009; Singh and Chahal, 2009) have been conducted on the adoption of different agricultural technology and crop variety. However, very few of them considered the adoption level of production practices of wheat in Bangladesh. Keeping all these factors in consideration the study was undertaken with the following specific objectives.

Specific objectives

i. To determine the level of adoption of improved wheat production practices
ii. To find out the factors affecting the adoption decision and
iii. To identify the problems that limits the acceptability of improved wheat production practices.

Materials and Methods

Two major wheat growing districts of Bangladesh namely Dinajpur and Thakurgaon were purposively selected for the study. A total of 110 sample taking 55 from each district were selected for the study by using random sampling technique. The study was mainly based on primary data collected through face to face interview during the month of February to April, 2011. Collected data were then summarized, tabulated and analyzed to fulfill the objectives of the study.

Analytical technique

Mostly descriptive statistics like average percentage were used to achieve the objectives of the study. Adoption level of wheat production practices were evaluated against the recommended doses given in Krishi Projukti Hatboi (2006) published by Bangladesh Agricultural Research Institute.

Calculation of adoption level

The level of adoption was measured by computing adoption scores for recommended technologies. Scores given to each technology that varied from 1 to 0 according to the adoption of the suggested technology. A respondent farmer could get a score of "1" for adopting the technology. On the other hand, farmer could receive a score of "0" for not adopting the technology. The mean score became the index of level adoption of the recommended technologies. On the basis of the score that earned by the farmers were categorized as high, medium, low and very low level of adoption respectively.

Multiple regression analysis

To find out the factors responsible for adoption of wheat production practices a multiple regression analysis was conducted considering pooled adoption scores of a farmer as dependent variable. Following type of multiple regression model was used for this study;

$$Y = \beta_0 + \beta_1 x_1 + \beta_2 x_2 + \beta_3 x_3 + \beta_4 x_4 + \beta_5 x_5 + \beta_6 x_6 + u_i$$

Where,

Y = Pooled adoption score
X_1 = Education (Year of schooling)
X_2 = Age of the respondent (years)
X_3 = Experience is farming (years)
X_4 = Farm size (ha)
X_5 = Household size (Actual number of members of the household)
X_6 = Extension contact (score)
u_i = Error term

Results and Discussion

Socio-economic profile of the respondents

It is revealed from Table 1 that average age of respondents was more or less similar in both the study areas. About 83% of the respondents were literate in the study areas. Most of them (39.09%) had secondary level of education. Average family size was found to be 5.78 which was higher than the national average. Average farm size of Thakurgaon (1.59 ha) was found higher compared to Dinajpur district (1.47 ha). Farmers of Dinajpur were more experienced in wheat farming than that of Thakurgaon. Average experience in wheat farming was found to be 13.74 years in all areas.

Table 1. Profile of the respondents

Items	Dinajpur (*n = 55*)	Thakurgaon (*n = 55*)	All areas (*n = 110*)
Demographic Characteristics			
Average age of respondents (yrs)	41 (12.7)	40 (10.4)	40 (11.5)
Level of education (%)	-	-	-
Illiterate	10.91	21.82	16.36
Primary	34.55	21.82	28.18
Up to SSC	34.55	43.64	39.09
HSC	16.36	9.09	12.73
Above HSC	3.64	3.64	3.64
Average Family size (No/family)	5.80 (2)	5.76 (2.5)	5.78 (2.2)
Land use (ha)	-	-	-
Own cultivated land	1.18 (1.1)	1.32 (1.9)	1.25 (1.6)
Average farm size	1.47 (1.1)	1.59 (1.9)	1.53 (1.5)
Wheat cultivated area	0.39 (0.3)	0.44 (0.6)	0.42 (0.5)
Experience in wheat cultivation (yrs)	14.65 (6.3)	12.96 (5.7)	13.81 (6)

Note: Figures in the parentheses indicates standard deviation

Farmers growing different varieties

Majority 60.91% of the farmers in both of the study areas were grown Shatabdi variety followed by Prodip (34.55%). More than 67% of the farmers in Thakurgaon grown Shatabdi variety while it was about 54.55% in Dinajpur district (Table 2).

Table 2. Distribution of farmers according to prevailing wheat varieties in the study area

Variety name	Dinajpur *(n = 55)*	Thakurgaon *(n = 55)*	Both areas *(n = 110)*
Prodip	45.45	23.63	34.55
Shatabdi	54.55	67.27	60.91
Showrab	-	9.09	4.55

Seed rate

It was found that majority (87%) of the respondents' exceded the recommended amount of seed (120 kg ha^{-1}). None of the farmers applied the seed lower than the recommended dose. Only 12.73% of the respondents were using recommended quantity of seed. The mean score of adoption was found only 0.13 which indicates very low level of adoption in both the areas (Table 3).

Table 3. Seed rate of wheat used by the sample farmers in the study areas

Location	Recommended rate	Above recommended rate	Below recommended rate	Mean	Adoption level*
Dinajpur *(n = 55)*	7 (12.73)	48 (87.27)	-	0.13	Very low
Thakurgaon *(n = 55)*	7 (12.73)	48 (87.27)	-	0.13	Very low
All areas *(n = 110)*	14 (12.73)	96 (87.27)	-	0.13	Very low

Note: Figures in parentheses indicates percentage of total
*Adoption level was categorized for mean score ≤ 0.25 as very low, between 0.26 to 0.50 as low, between 0.51 to 0.75 as medium and between 0.76 to 1.00 as high

Time of sowing

It is evident from Table 4 that most of the farmers were sowing wheat crop during the recommended period (2nd week of Nov. to 1st week of Dec.). More than 85% of the farmers sowing wheat crop during the recommended period while about 14% of the farmers sowing wheat after the recommended period. Mean score of adoption was found 0.85 which indicated high level of adoption.

Table 4. Time of wheat crop sowing in the study areas

Location	During recommended period	After recommended period	Mean	Adoption level*
Dinajpur *(n = 55)*	47 (85.45)	8 (14.55)	0.85	High
Thakurgaon *(n = 55)*	47 (85.45)	8 (14.55)	0.85	High
All areas *(n = 110)*	94 (85.45)	16 (14.55)	0.85	High

Note: Figures in parentheses indicates percentage of total
*Adoption level was categorized for mean score ≤ 0.25 as very low, Between 0.26 to 0.50 as low, between 0.51 to 0.75 as medium and between 0.76 to 1.00 as high

Seed treatment

It is revealed from the Table 5 that almost all the farmers did not treat the seed before sowing wheat. More than 96% of the farmers did not treat wheat seed before sowing in Thakurgaon while it was 91% in Dinajpur district. Therefore, there exists a scope for filling the gap in the adoption of this technology. It will not only protect the crop from primary diseases infestation but will also augment the productivity of wheat in the study areas.

Table 5. Percentage of farmers adopting seed treatment

Items	Dinajpur (*n* = 55)	Thakurgaon (*n* = 55)	All areas (*n* = 110)
Yes	9.09	3.64	6.36
No	90.91	96.36	93.64

Fertilizer application

Urea

It is evident from Table 6 that 39.09% of the farmers used recommended dose of urea (180-220 kg ha^{-1}). About 32% of the farmers used urea over the recommended level and 29% farmers used urea lower than the recommended dose. Overall adoption score of urea was found 0.39, which indicated low level of adoption.

Table 6. Application of urea by the sample farmers

Locations	Recommended dose	Above recommended dose	Below recommended dose	Mean	Adoption level*
Dinajpur (*n* = 55)	21 (38.18)	20 (36.36)	14 (25.45)	0.38	Low
Thakurgaon (*n* = 55)	22 (40.00)	15 (27.27)	18 (32.73)	0.40	Low
All areas (*n* = 110)	43 (39.09)	35 (31.82)	32 (29.09)	0.39	Low

Note: Figures in parentheses indicates percentage of total
*Adoption level was categorized for mean score ≤ 0.25 as very low, between 0.26 to 0.50 as low, between 0.51 to 0.75 as medium and between 0.76 to 1.00 as high

TSP

Most of the farmers (69.09%) used TSP lower than the recommended dose (140-180 kg ha^{-1}). Only 20.91% of the farmers used recommended dose of TSP. Overall adoption score of TSP application was found 0.21. It indicated very low level of adoption.

Table 7. Application of TSP by the sample farmers

Location	Recommended dose	Above recommended dose	Below recommended dose	Mean	Adoption level*
Dinajpur (*n* = 55)	11 (20.00)	4 (7.27)	40 (72.73)	0.20	Very low
Thakurgaon (*n* = 55)	12 (21.82)	7 (12.73)	36 (65.45)	0.22	Very low
All areas (*n* = 110)	23 (20.91)	11 (10.00)	76 (69.09)	0.21	Very low

Note: Figures in parentheses indicates percentage of total
*Adoption level was categorized for mean score ≤ 0.25 as very low, between 0.26 to 0.50 as low, between 0.51 to 0.75 as medium and between 0.76 to 1.00 as high

MoP

About 86% of the farmers of Thakurgaon applied MoP more than the recommended dose (40-50 kg ha^{-1}) whereas more than 78% of the farmers of Dinajpur used MoP more than the recommended dose. On an average, about 82% of the farmers applied MoP more than the recommended dose. Only 13.64% of the farmers applied recommended dose of MoP. As a result adoption level of MoP application found to be very low in the study areas with an adoption score of only 0.14 (Table 8).

Table 8. Application of MoP by the sample farmers

Location	Recommended dose	Above recommended dose	Below recommended dose	Mean	Adoption level*
Dinajpur (*n* = 55)	8 (14.55)	43 (78.18)	4 (7.27)	0.15	Very low
Thakurgaon (*n* = 55)	7 (12.73)	47 (85.45)	1 (1.82)	0.13	Very low
All areas (*n* = 110)	15 (13.64)	90 (81.82)	5 (4.55)	0.14	Very low

Note: Figures in parentheses indicates percentage of total
*Adoption level was categorized for mean score ≤ 0.25 as very low, between 0.26 to 0.50 as low, between 0.51 to 0.75 as medium and between 0.76 to 1.00 as high

Top dressing of urea

Almost 85% of the farmers applied top dressing of urea within the recommended period (at the time of first irrigation). Only 15% of the farmers applied urea after the recommended period. Overall adoption score was found to be 0.85, which indicated high level of adoption (Table 9).

Table 9. Time of top dressing of Urea in the study areas

Location	During recommended period	After recommended period	Mean	Adoption level*
Dinajpur *(n = 55)*	48 (87.27)	7 (12.73)	0.87	High
Thakurgaon *(n = 55)*	45 (81.82)	10 (18.18)	0.82	High
All areas *(n = 110)*	93 (84.55)	17 (15.45)	0.85	High

Note: Figures in parentheses indicates percentage of total
*Adoption level was categorized for mean score ≤ 0.25 as very low, between 0.26 to 0.50 as low, between 0.51 to 0.75 as medium and between 0.76 to 1.00 as high

Number of irrigation

The recommended number of irrigation to wheat crop is 2-3 times depending on the type of soil (Krishi Projukti Hatboi, 2006). The present study had found that 83.64% of the farmers applied recommended number of irrigation while about 16% of the farmers applied less than the recommended number of irrigation. Overall adoption level of irrigation number was found to be high in the study areas.

Table 10. Number of irrigations applied by the sample farmers

Location	Recommended	Above recommended	Below recommended	Mean	Adoption level*
Dinajpur *(n = 55)*	44 (80.00)	1 (1.82)	10 (18.18)	0.80	High
Thakurgaon *(n = 55)*	48 (87.27)	-	7 (12.73)	0.87	High
All areas *(n = 110)*	92 (83.64)	1 (0.91)	17 (15.45)	0.84	High

Note: Figures in parentheses indicates percentage of total
*Adoption level was categorized for mean score ≤ 0.25 as very low, between 0.26 to 0.50 as low, between 0.51 to 0.75 as medium and between 0.76 to 1.00 as high

Factors affecting the adoptions

Results of the regression analysis showed that 51% of variation in adoption could be explained by the independent variables included in the model. The co-efficient of education, experience and extension contact were significant and positively related with the adoption of wheat production practices. Coefficient of education implies that respondent with higher education adopted more. This agrees with the findings of Mittal Surabhi and Praduman Kumar (2000). Similarly, Coefficient of extension contact suggests that the more the extension contact with the farmers, the more information on the improved wheat production practices reach to the farmers. Hence, farmers are likely to adopt more. This echoes with the findings of Nnadi and Akwiwu (2005) and Singh *et al.* (2000). Coefficient of farmer's age was negative and significant at 10% level of significance. If age of farmers' increases by 1 year, keeping other factors constant, adoption would be decreased by 0.19 units. It implies that younger farmers were more interested to adopt new production technologies.

Table 11. Factors affecting the adoption of wheat production practices

Variable	Co-efficient	Standard error	t-value
Constant	1.96***	0.280	6.99
Education	0.166**	0.069	2.38
Age	-0.193*	0.127	-1.82
Experience	0.024*	0.015	1.58
Farm size	-0.021	0.048	-0.431
Household size	-0.010	0.032	-0.317
Extension contact	0.521***	0.049	10.51
Adjusted R^2	0.51		
F-value	21.73***		
Observations	110		

Note: *, ** and *** indicates significant at 10%, 5% and 1% level of significance

Problems of adoption

Table 12 shows the seriousness of problems that hinder adoption of wheat production practices. Almost 90% of the farmers indicated inadequate information media as a problem of adoption in the study areas. Mean score of this problem was 2.07, which indicated that it is a serious problem. In this context, extension communication can play a vital role. Lack of technical knowledge was also a serious problem that hinders the adoption of wheat production practices. Low level of income, lack of access to credit facilities and fear & suspicion were identified as less serious problem since the mean score were 1.97, 1.86 and 0.43 respectively.

Table 12. Problems of adoption in the study areas

Problems	VS	S	MS	LS	NS	Mean	Remarks*
Inadequate medium as a source of information	27 (24.55)	18 (16.36)	12 (10.91)	42 (38.18)	12 (10.91)	2.07	Serious
Lack of technical knowledge	14 (12.73)	37 (33.64)	16 (14.55)	33 (30.00)	9 (8.18)	2.11	Serious
Low level of income	11 (10.00)	29 (26.36)	34 (30.90)	18 (16.36)	17 (15.45)	1.97	Less Serious
Lack of access to credit facilities	9 (8.18)	36 (32.73)	18 (16.36)	25 (22.73)	20 (18.18)	1.86	Less serious
Fear & suspicion	1 (0.91)	5 (4.55)	5 (4.55)	18 (16.36)	79 (71.82)	0.43	Less serious

Note: Figures in parentheses indicates percentage of total
VS = Very serious, S= Serious, MS = Moderate serious, LS = Less serious, NS = Not serious
*The midpoint of this frequency table is 2. As a decision rule, any mean score that is ≤ 2 = less serious and that >2 = serious

Conclusion and Recommendations

Findings of the study suggested that remarkable gaps existed in the application of various technologies in wheat cultivation. Adoption level of seed rate, Urea, TSP, MoP was found to be very low whereas the adoption of sowing time and number of irrigation was found to be high. Different factors like education, age and extension contact had positive and significant effect on the adoption of wheat production practices. Inadequate medium of information, lack of technical knowledge impede the adoption of wheat production practices in the study areas. If the adoption gaps of some of the recommended technologies are properly eliminated, farmers could receive maximum return. Regular visit by extension agencies to disseminate modern production technologies, training to improve technical knowledge, credit with low interest can play a vital role in this regards. Besides, more research must be done to define appropriate wheat research priorities in the country.

References

Baksh, M.E., Kabir, M.J. and Kispatta, K. 2009. Socio-economic assessment of wheat variety Shatabdi in some selected areas of Bangladesh. *Bangladesh J. Agril. Res.* 34 (2): 215-226.

BBS. 2010. Year book of Agricultural Statistics of Bangladesh, Bangladesh Bureau of Statistics, Ministry of Planning, Dhaka, Bangladesh. 129 p.

Dixon, J., Nalley, L., Kosina, P., La Rovere, R., Hellin, J. and Aquino, P. 2006. Adoption and economic impact of improved wheat varieties in the developing world. *J. Agril. Sci.* 144 (6): 489–502.

Edna, C. Matthews-Njoku, Adesope, O.M. and Iruba, C. 2009. Acceptibility of improved crop production practices among rural women in aguata agricultural zone of Anambra State, Nigeria. *African J. Biotech.* 8 (3): 405-411.

Feder, G., Just, R.E. and Zilberman, D. 1985. Adoption of agricultural innovations in developing countries. *Econ. Dev. & Cultural Change.* 33 (2): 255-298.

Gardebroek, C. 2006. Comparing risk attitudes of organic and non-organic farmers with a bayesian random coefficient model. *European Review of Agril. Econ.* 33: 485–510.

Hossain, A.B.S. 1984. Wheat production in Bangladesh: its constraints and research priorities, *In:* Proc. of the International symposium on wheat for more tropical environments, Mexico D.F. pp. 59-62.

Hossain, S.M.A. and Bruce, R.C. 1992. Patterns and determinants of adoption of farm practices: some evidence from Bangladesh. *Agril. System.* 38 (1): 1-15.

Iqbal Muhammad, M., Azeem Khan and Munir Ahmed. 2002. Adoption of recommended varieties: a farm level analysis of wheat growers in irrigated Punjab. *The Pakistan Dev. Review.* 41 (1): 29-48.

Kamruzzaman, M., Fakhrul Islam, S.M., Begum, M.A.A., Shiblee, S.M.A., Kibria, M.G. and Ray, S.K. 2001. Adoption level of wheat technology and the grower's knowledge gap in Bangladesh. *Pakistan J. Biol. Sci.* 4(1): 1-6.

Krishi Projokti Hatboi. 2006. An Hand book of Agro-technology, 4th edition, Bangladesh Agricultural Research Institute, Gazipur, Bangladesh. pp. 13-14.

Lee David, R. 2005. Agricultural sustainability and technology adoption: Issues and policy for developing countries. *American J. Agril. Econ.* 87 (5): 1325-1334.

Mittal Surabhi and Praduman Kumar. 2000. Literacy, technology adoption, factor demand and productivity: An econometric analysis. *Indian J. Agril. Econ.* 55 (3): 490-499.

Nnadi, F.N. and Akwiwu, C.D. 2005. Rural women's response to selected crop production technologies in Imo State Nigeria. *Global Approaches Ext. Pract.* 1 (1): 47-52.

Pandit, D.B., Baksh, M.E., Sufian, M.A., Harun-ur-Rashid, M. and Islam, M.M. 2007. Impacts of participatory variety selection in wheat on agro-economic changes of wheat farmers in Bangladesh. *Bangladesh J. Agril. Res.* 32 (3): 335-347.

Sanchez, P.A., Denning, G.L. and Nziguheba, G. 2009. The African green revolution moves forward. *Food Security.* 1: 37-44.

Serra, T., Zilberman, D. and Gil, J.M. 2008. Differential uncertainties and risk attitudes between conventional and organic producers, the case of Spanish COP farmers. *Agril. Econ.* 39 (2): 219-229.

Singh, K., Singh, J.P. and Singh, P. 2000. Adoption behaviour of small farmers in Bharatpur district of Rajasthan. *J. Agril. Sci. Res.* 36: 51-56.

Singh, M. and Chahal, S.S. 2009. A study on the extent of adoption of various recommended technologies in wheat cultivation in Punjab. *Agril. Econ. Res. Review.* 22: 349-354.

Walford, N. 2002. Agricultural adjustment: adoption of and adaptation to policy reform measures by large-scale commercial farms. *Land Use Policy.* 19: 243-257.

Wanyama, J.M., Nyambati, E.M., Mose, L.O, Mutoko, C.M., Wanyonyi, W.M. Wanjekeche, E. and Rono, S.C. 2010. Assessing impact of soil management technologies on smallholder farmers' livelihoods in North Western Kenya. *African J. Agril. Res.* 5 (21): 2899-2908.

Permissions

All chapters in this book were first published in IJARIT, by IJARIT Research Foundation; hereby published with permission under the Creative Commons Attribution License or equivalent. Every chapter published in this book has been scrutinized by our experts. Their significance has been extensively debated. The topics covered herein carry significant findings which will fuel the growth of the discipline. They may even be implemented as practical applications or may be referred to as a beginning point for another development.

The contributors of this book come from diverse backgrounds, making this book a truly international effort. This book will bring forth new frontiers with its revolutionizing research information and detailed analysis of the nascent developments around the world.

We would like to thank all the contributing authors for lending their expertise to make the book truly unique. They have played a crucial role in the development of this book. Without their invaluable contributions this book wouldn't have been possible. They have made vital efforts to compile up to date information on the varied aspects of this subject to make this book a valuable addition to the collection of many professionals and students.

This book was conceptualized with the vision of imparting up-to-date information and advanced data in this field. To ensure the same, a matchless editorial board was set up. Every individual on the board went through rigorous rounds of assessment to prove their worth. After which they invested a large part of their time researching and compiling the most relevant data for our readers.

The editorial board has been involved in producing this book since its inception. They have spent rigorous hours researching and exploring the diverse topics which have resulted in the successful publishing of this book. They have passed on their knowledge of decades through this book. To expedite this challenging task, the publisher supported the team at every step. A small team of assistant editors was also appointed to further simplify the editing procedure and attain best results for the readers.

Apart from the editorial board, the designing team has also invested a significant amount of their time in understanding the subject and creating the most relevant covers. They scrutinized every image to scout for the most suitable representation of the subject and create an appropriate cover for the book.

The publishing team has been an ardent support to the editorial, designing and production team. Their endless efforts to recruit the best for this project, has resulted in the accomplishment of this book. They are a veteran in the field of academics and their pool of knowledge is as vast as their experience in printing. Their expertise and guidance has proved useful at every step. Their uncompromising quality standards have made this book an exceptional effort. Their encouragement from time to time has been an inspiration for everyone.

The publisher and the editorial board hope that this book will prove to be a valuable piece of knowledge for researchers, students, practitioners and scholars across the globe.

List of Contributors

Iftekhar Ahmad
Department of Food Engineering and Tea Technology, Shahjalal University of Science and Technology, Sylhet, Bangladesh

Mizanur Rahman
Department of Food Engineering and Tea Technology, Shahjalal University of Science and Technology, Sylhet, Bangladesh

Md. Mahfuzur Rahman
Department of Food Engineering and Tea Technology, Shahjalal University of Science and Technology, Sylhet, Bangladesh

Md. Mehbub Mustain Alam
Department of Food Engineering and Tea Technology, Shahjalal University of Science and Technology, Sylhet, Bangladesh

Md. Shakawat Hussain
Food Technology Division, Institute of Food and Radiation Biology, Atomic Energy Research Establishment, Savar, Dhaka, Bangladesh

Hari Pada Seal
Department of Agricultural Chemistry, Faculty of Agriculture, Bangladesh Agricultural University, Mymensingh-2202, Bangladesh

Mohammad Amdad Ali
Department of Agricultural Chemistry, Faculty of Agriculture, Bangladesh Agricultural University, Mymensingh-2202, Bangladesh

Md. Usuf Ali
Department of Agricultural Chemistry, Faculty of Agriculture, Bangladesh Agricultural University, Mymensingh-2202, Bangladesh

Mosammat Hasina Akhter
Department of Agricultural Chemistry, Faculty of Agriculture, Bangladesh Agricultural University, Mymensingh-2202, Bangladesh

Fowzia Sultana
Department of Poultry Science, Faculty of Animal Husbandry, Bangladesh Agricultural University, Mymensingh-2202, Bangladesh

A. H. M. Kohinoor
Freshwater Station, Bangladesh Fisheries Research Institute, Mymensingh, Bangladesh

M. M. Khan
Freshwater Station, Bangladesh Fisheries Research Institute, Mymensingh, Bangladesh

S. Yeasmine
Freshwater Station, Bangladesh Fisheries Research Institute, Mymensingh, Bangladesh

P. Mandol
Senior Assistant Chief, Ministry of Fisheries & Livestock, Bangladesh

M. S. Islam
Freshwater Station, Bangladesh Fisheries Research Institute, Mymensingh, Bangladesh

D. A. Jahan
Freshwater Station, Bangladesh Fisheries Research Institute, Mymensingh, Bangladesh

M. G. Hussain
Freshwater Station, Bangladesh Fisheries Research Institute, Mymensingh, Bangladesh

M. J. Islam
Department of Agricultural Chemistry, Hajee Mohammad Danesh Science and Technology University, Dinajpur-5200, Bangladesh

A. Akter
Department of Agricultural Chemistry, Hajee Mohammad Danesh Science and Technology University, Dinajpur-5200, Bangladesh

M. Kamrujjaman
Department of Mathematic and Physics, Hajee Mohammad Danesh Science and Technology University, Dinajpur-5200, Bangladesh

A. Siddiqa
Department of Agricultural Chemistry, Hajee Mohammad Danesh Science and Technology University, Dinajpur-5200, Bangladesh

S. Aktar
Department of Agricultural Chemistry, Hajee Mohammad Danesh Science and Technology University, Dinajpur-5200, Bangladesh

Liu Aiju
School of Recourse and Environmental Engineering, Shandong University of Technology, 12 Zhangzhou Road, Zibo 255049, China

Wang Honghai
Zibo entry-exit inspection and quarantine bureau, 63 Liuquan Road, Zibo 255033, China

Gao Peiling
School of Recourse and Environmental Engineering, Shandong University of Technology, 12 Zhangzhou Road, Zibo 255049, China

Xu Hong-zhi
Analysis & Testing Center, Shandong University of Technology, 12 Zhangzhou Road, Zibo 255049, China

Sumon M. Hossain
Department of Biotechnology and Genetic Engineering, Mawlana Bhashani Science and Technology University Tangail-1902, Bangladesh

A. K. M. Mohiuddin
Department of Biotechnology and Genetic Engineering, Mawlana Bhashani Science and Technology University Tangail-1902, Bangladesh

B. K. Chakraborty
Mathsya Bhabon, Department of Fisheries, Dhaka-1000, Bangladesh

N. N. Nur
Programme Manager, NETZ, Dhaka, Bangladesh

M. Ibrahim Khalil
Plant Pathology Division, Bangladesh Institute of Nuclear Agriculture, Mymensingh-2202, Bangladesh

M. Bahadur Meah
Department of Plant Pathology, Bangladesh Agricultural University, Mymensingh-2202, Bangladesh

Mirza Mofazzal Islam
Plant Breeding Division, Bangladesh Institute of Nuclear Agriculture, Mymensingh-2202, Bangladesh

M. A. Hoque
Senior Scientific Officer, F M P Engineering Division, BARI, Gazipur, Bangladesh

M. A. Wohab
Principal Scientific Officer, F M P Engineering Division, BARI, Gazipur, Bangladesh

M. Moniruzzaman
Biotechnologist (Field Research), Bangladesh Jute Research Institute, Manikgonj, Dhaka, Bangladesh

M. S. Alam
Scientific Officer (Plant Breeding), Regional Agricultural Research Station, BARI, Akbarpur, Moulvibazar

J. A. Rashid
Executive, R&D (Biotechnology/ Agriculture/ Microbiology/ Enzyme), MATEX Bangladesh Ltd. Dhaka

S. N. Begum
Senior Scientific Officer, Plant Breeding Division, Bangladesh Institute of Nuclear Agriculture, Mymensingh

M. M. Islam
Principal Scientific Officer & Head, Biotechnology Division, Bangladesh Institute of Nuclear Agriculture, BAU campus, Mymensingh, Bangladesh

M. H. Hossain
Senior Scientific Officer, Plant Pathology Division, Regional Agricultural Research Station, BARI, Comilla, Bangladesh

I. Hossain
Professor, Department of Plant Pathology, Bangladesh Agricultural University, Mymensingh, Bangladesh

M. K. Alam
Scientific Officer, Plant Breeding Division, ORC, Regional Agricultural Research Station, BARI, Ishurdi, Pabna, Bangladesh

U. K. Nath
Professor, Department of Genetics and Plant Breeding, Bangladesh Agricultural University, Mymensingh, Bangladesh

M. A. K. Azad
Principal Scientific Officer, Genetics and Plant Breeding Division, Bangladesh Institute of Nuclear Agriculture, Mymensingh, Bangladesh

M. A. Alam
Scientific Officer, Plant Breeding Division, Regional Agricultural Research Station, BARI, Ishurdi, Pabna, Bangladesh

A. A. Khan
Scientific Officer, Plant Breeding Division, WRC, Regional Agricultural Research Station, BARI, Ishurdi, Pabna, Bangladesh

A. Oloruntoba
Department of Agricultural Extension, Rural Development & Gender Studies, University for Development Studies, Tamale, Ghana

J. O. Oladeji
Department of Agricultural Extension and Rural Development, University of Ibadan, Ibadan, Nigeria

E. F. Odedele
Department of Agricultural Extension and Rural Development, University of Agriculture, Abeokuta, Nigeria

S. A. A. Nahid
Faculty of Fisheries, Chittagong Veterinary and Animal Sciences University, Bangladesh
Faculty of Fisheries, Bangladesh Agricultural University, Mymensingh 2202, Bangladesh

P. J. G. Henriksson
CML, Institute of Environmental Sciences, Leiden University, The Netherlands

M. A. Wahab
Faculty of Fisheries, Bangladesh Agricultural University, Mymensingh 2202, Bangladesh

M. A. Siddiky
Department of Environmental Sciences, Jahangirnagar University, Dhaka, Bangladesh
Horticulture Research Centre, BARI, Gazipur-1701, Bangladesh

P. K. Sardar
On Farm Research Division, BARI, Khulna, Bangladesh

M. M. Hossain
On Farm Research Division, BARI, Khulna, Bangladesh

M. S. Khan
CSO (Ret'd), Soil Science Division, BARI, Gazipur-1701, Bangladesh

M. Khabir Uddin
Department of Environmental Sciences, Jahangirnagar University, Dhaka, Bangladesh

A. H. M. Kohinoor
Freshwater Station, Bangladesh Fisheries Research Institute, Mymensingh, Bangladesh

D. A. Jahan
Freshwater Station, Bangladesh Fisheries Research Institute, Mymensingh, Bangladesh

M. M. Khan
Freshwater Station, Bangladesh Fisheries Research Institute, Mymensingh, Bangladesh

M. S. Islam
Freshwater Station, Bangladesh Fisheries Research Institute, Mymensingh, Bangladesh

M. G. Hussain
Freshwater Station, Bangladesh Fisheries Research Institute, Mymensingh, Bangladesh

M. Z. H. Talukdar
Department of Fisheries Management, Bangladesh Agricultural University, Mymensingh-2202, Bangladesh

M. Shahjahan
Department of Fisheries Management, Bangladesh Agricultural University, Mymensingh-2202, Bangladesh

M. S. Rahman
Department of Fisheries Management, Bangladesh Agricultural University, Mymensingh-2202, Bangladesh

S. P. Baliyan
Department of Agriculture, Livingstone Kolobeng College, Gaborone, Botswana

P. S. Baliyan
Department of Accounting and Finance, University of Botswana, Gaborone, Botswana

K. S. Madhava Rao
Department of Statistics, University of Botswana, Gaborone, Botswana

J. O. Akintonde
Department of Agricultural Extension and Rural Development, Ladoke Akintola University of Technology, P.M.B.4000, Ogbomoso, Oyo State, Nigeria

O. A. Akinboye
Department of Agricultural Extension and Rural Development, Ladoke Akintola University of Technology, P.M.B.4000, Ogbomoso, Oyo State, Nigeria

C. O. Farayola
Agricultural Development Management, Department of Agricultural and Rural Management Training Institute (ARMTI), P.M.B. 1343, Ilorin Kwara State, Nigeria

O. S. Akintola
Department of Agricultural Extension and Rural Development, Ladoke Akintola University of Technology, P.M.B.4000, Ogbomoso, Oyo State, Nigeria

M. N. Uddin
Agriculture Extension Officer, Department of Agricultural Extension, MOA, Bangladesh

M. A. Bakr
Ex-CSO, Plant Pathology Division, Bangladesh Agricultural Research Institute, Gazipur, Bangladesh

M. R. Islam
Professor, Department of Plant Pathology, Sher-e-Bangla Agricultural University, Dhaka, Bangladesh

M. I. Hossain
Training, Implementation and Monitoring Officer (TIMO), IFDC, Bangladesh

A. Hossain
Assistant Professor, Department of Plant Pathology, Sythet Agricultural University, Bangladesh

H. K. Barman
Department of Horticulture, Bangladesh Agricultural University, Mymensingh, Bangladesh

M. N. Siddiqui
Department of Biochemistry and Molecular Biology, Bangabandhu Sheikh Mujibur Rahman Agricultural University, Gazipur-1706, Bangladesh

M. A. Siddique
Department of Horticulture, Bangladesh Agricultural University, Mymensingh, Bangladesh

M. S. Roni
Department of Horticulture, Bangabandhu Sheikh Mujibur Rahman Agricultural University, Gazipur-1706, Bangladesh

M. Nuruzzaman
Department of Horticulture, Bangladesh Agricultural University, Mymensingh, Bangladesh

E. N. Ajani
Department of Agricultural Extension, Faculty of Agriculture, University of Nigeria, Nsukka, Nigeria

E. A. Onwubuya
Department of Agricultural Extension, Faculty of Agriculture, University of Nigeria, Nsukka, Nigeria

E. N. Ajani
Department of Agricultural Extension, Faculty of Agriculture, University of Nigeria, Nsukka, Nigeria

E. A. Onwubuya
Department of Agricultural Extension, Faculty of Agriculture, University of Nigeria, Nsukka, Nigeria

N. Nupur
Department of Fisheries Management, Bangladesh Agricultural University, Mymensingh-2202, Bangladesh

M. Shahjahan
Department of Fisheries Management, Bangladesh Agricultural University, Mymensingh-2202, Bangladesh

M. S. Rahman
Department of Fisheries Management, Bangladesh Agricultural University, Mymensingh-2202, Bangladesh

M. K. Fatema
Department of Fisheries Management, Bangladesh Agricultural University, Mymensingh-2202, Bangladesh

M. Al-Amin
Department of Genetics and Plant Breeding, Bangladesh Agricultural University, Mymensingh 2202, Bangladesh

M. M. Islam
Biotechnology Division, Bangladesh Institute of Nuclear Agriculture (BINA), BAU Campus, Mymensingh 2202

S. N. Begum
Plant Breeding Division, Bangladesh Institute of Nuclear Agriculture (BINA), BAU Campus, Mymensingh- 2202

M. S. Alam
Department of Genetics and Plant Breeding, Bangladesh Agricultural University, Mymensingh 2202, Bangladesh

M. Moniruzzaman
Department of Genetics and Plant Breeding, Bangladesh Agricultural University, Mymensingh 2202, Bangladesh

M. A. K. Patwary
Department of Genetics and Plant Breeding, Bangladesh Agricultural University, Mymensingh 2202, Bangladesh

C. O. Farayola
Agricultural Development Management Department, General Management Division, ARMTI, Ilorin, Km 18, Ilorin Ajase-Ipo Highway, P.M.B. 1343, Ilorin, Kwara State, Nigeria

V. Okpodu
Rural Development and Gender Department, Entrepreneurial Development and Management Division, ARMTI, Ilorin, Km 18, Ilorin Ajase-Ipo Highway, P.M.B. 1343, Ilorin, Kwara State, Nigeria

O. O. Oni
Agricultural Development Management Department, Agricultural Finance and Rural Credit Division, ARMTI, Ilorin, Km 18, Ilorin Ajase-Ipo Highway, P.M.B. 1343, Ilorin, Kwara State, Nigeria

M. A. Hoque
Senior Scientific Officer, F M P Engineering Division, Bangladesh Agricultural Research Institute, Gazipur-1701, Bangladesh

M. A. Hannan
Senior Scientific Officer, Seed Technology Division, Bangladesh Agricultural Research Institute, Gazipur-1701, Bangladesh

M. Hasanuzzaman
Department of Animal Science and Nutrition, Chittagong Veterinary and Animal Sciences University, Chittagong, Bangladesh

M. Ahaduzzaman
Department of Medicine and Surgery, Chittagong Veterinary and Animal Sciences University, Chittagong, Bangladesh

M. R. Begum
Department of Agricultural Economics and Social Science, Chittagong Veterinary and Animal Sciences University, Chittagong, Bangladesh

M. M. Rahman
Department of Food Engineering and Tea Technology, Shahjalal University of Science and Technology, Sylhet-3114, Bangladesh

M. A. Kalam
Department of Food Engineering and Tea Technology, Shahjalal University of Science and Technology, Sylhet-3114, Bangladesh

M. A. Salam
Department of Chemical Engineering and Polymer Science, Shahjalal University of Science and Technology, Sylhet-3114, Bangladesh

M. R. Rana
Department of Food Engineering and Tea Technology, Shahjalal University of Science and Technology, Sylhet-3114, Bangladesh

V. Anburaja
PG and Research Department of Botany, National College, Tiruchirappalli - 620 001, Tamil Nadu, India

V. Nandagopalan
PG and Research Department of Botany, National College, Tiruchirappalli - 620 001, Tamil Nadu, India

Md. Azharul Islam
Assistant Professor, Graduate Training Institute, Bangladesh Agricultural University, Mymenshingh, Bangladesh

Shamima Sultana
College of Development Alternative (CODA), 14 A Dhanmondi, Dhaka 1205, Bangladesh

Md. Asaduzzaman
Environmental and Analytical Laboratories, Faculty of Science, Charles Sturt University, Wagga Wagga, NSW2678, Australia
School of Agricultural and Wine Sciences, Faculty of Science, Charles Sturt University, Wagga Wagga, NSW 2678, Australia

J. J. Musa
Department of Agricultural and Bioresources Engineering, Federal University of Technology, P M B 65 Minna, Nigeria

G. A. Fumen
College of Agriculture, Ahmadu Bello University, Zaria, Nigeria

F. Siddika
Department of Fisheries Management, Bangladesh Agricultural University, Mymensingh-2202, Bangladesh

M. Shahjahan
Department of Fisheries Management, Bangladesh Agricultural University, Mymensingh-2202, Bangladesh

M. S. Rahman
Department of Fisheries Management, Bangladesh Agricultural University, Mymensingh-2202, Bangladesh

S. Sharma
Department of Botany, University of Jammu, B.R. Ambedkar Road, Jammu-180006, India

G. Sumbali
Department of Botany, University of Jammu, B.R. Ambedkar Road, Jammu-180006, India

V. Sharma
Department of Botany, University of Jammu, B.R. Ambedkar Road, Jammu-180006, India

M. A. Kashem
Department of Agricultural Extension Education, Bangladesh Agricultural University, Mymensingh, Bangladesh

M. G. Farouque
Department of Agricultural Extension Education, Bangladesh Agricultural University, Mymensingh, Bangladesh

P. C. Roy
Department of Agricultural Extension Education, Bangladesh Agricultural University, Mymensingh, Bangladesh

S. M. Mudyiwa
Faculty of Agriculture and Environmental Science, Bindura University of Science Education, P. Bag 1020, Bindura, Zimbabwe

C. T. Gadzirayi
Faculty of Agriculture and Environmental Science, Bindura University of Science Education, P. Bag 1020, Bindura, Zimbabwe

J. F. Mupangwa
2Department of Animal Science, Faculty of Agriculture, University of Swaziland, P.O. Luyengo, Luyengo, Swaziland

J. Gotosa
Faculty of Agriculture and Environmental Science, Bindura University of Science Education, P. Bag 1020, Bindura, Zimbabwe

T. Nyamugure
Faculty of Agriculture and Environmental Science, Bindura University of Science Education, P. Bag 1020, Bindura, Zimbabwe

M. S. Rahman
Assistant Professor, Department of Management and Finance, Faculty of Agribusiness Management, Sher-e Bangla Agricultural University, Dhaka, Bangladesh

Zerina Haque
Ex-MS Student, Bangabandhu Sheikh Mujibur Rahman Agricultural University, Salna, Gazipur, Bangladesh